Explorations into
The Nature of the Living Cell

Robert Chambers.

ROBERT CHAMBERS
EDWARD L. CHAMBERS

Explorations into The Nature of the Living Cell

PUBLISHED FOR THE COMMONWEALTH FUND
By Harvard University Press
Cambridge · Massachusetts · 1961

LIBRARY OF CONGRESS CATALOG CARD NO. 61-8845
MANUFACTURED IN THE UNITED STATES OF AMERICA

For nearly twenty-five years the COMMONWEALTH FUND, through its Division of Publications, independently sponsored, edited, produced, and distributed books and pamphlets germane to its purposes as a philanthropic foundation. Since July 1, 1951, HARVARD UNIVERSITY PRESS has been the Fund's publisher.

PUBLISHED FOR THE COMMONWEALTH FUND
By Harvard University Press, Cambridge, Massachusetts
Distributed in Great Britain by Oxford University Press, London

Preface

Robert Chambers, my father, was a true experimentalist in that he relied in his work not on the formulation of hypotheses and theories to be tested for their validity, but upon an intuitive grasp of the complexities of the living cell. Many of his important contributions were the result of "hunches" which came to him as he made observations through his microscope. These "hunches" derived from an acute observer's broad experience in the handling of living cells.

Robert Chambers pursued his investigations on living cells with a sense of adventure. This adventurous spirit derived from the harsh, exciting life he spent during his formative years in eastern Turkey, among the wild mountain tribes, the Koords, and among the Armenians, persecuted and massacred by the Turks for their Christian beliefs. Born in Erzroom (or Erzerum), Turkey, in 1881, he was the son of a Presbyterian missionary and educator. Later, my father lived in nearby Bardizag until he attended Robert College in Constantinople. Although it was here that his interest in nature first crystallized, nonetheless he returned to eastern Turkey after graduation, to teach for four years high school subjects to the native youth. Robert Chambers began his career in research at the University of Munich, working under the great zoologist Richard Hertwig, with whom he obtained his Ph.D. degree in 1908.

The major turning point in my father's career occurred while he was working at the Marine Biological Laboratory, Woods Hole, Massachusetts, in 1912, where he was greatly stimulated by George Lester Kite's demonstration that the union of egg and sperm nuclei in starfish ova can be prevented by means of glass microtools. At this time Chambers' close associates at Woods Hole included such remarkable biologists, then youthful men, as T. H. Morgan, E. G. Conklin, A. P. Mathews, E. B. Wilson, F. R. Lillie, and G. N. Calkins. Subsequently, Chambers developed and extensively applied the microsurgical technique, to the point where his name has become synonymous with the micromanipulative method. In 1915 he joined Cornell Medical College, and shortly thereafter was appointed professor of microscopic anatomy. These were highly fruitful years. In 1928 he transferred to the Department of Biology at New York University, where he maintained for twenty years a thriving research center, which attracted students, many of whom are now eminent scientists, from all the countries of Europe, from Asia, and from South America. All who worked with Robert Chambers were infused by his dynamic personality. His warm human qualities transcended even his outstanding scientific achievements.

Woods Hole was the center of his scientific life, and his most important contributions were the result of his summer research activities there. After retirement as Professor Emeritus of Biology at New York University in 1947, he assumed directorship of the Experimental Cell Research Laboratories at Woods Hole. The last several years of his life, until his death in 1957, were devoted to the preparation of this book.

This book brings together the major part of Robert Chambers' many experimental observations. These consisted largely in the study by micromanipulative methods of the properties of living cells. Although most of the results have been described in individual papers, over 240 in number, some of the data have not been previously published. Since the most active phase of Robert Chambers' work started in 1915 and extended through three decades, a real attempt is made to bridge the gap between the time of completion of the original work and the present. This book reveals the wealth of insight which has been achieved by microdissecting and microinjecting living cells.

EDWARD L. CHAMBERS
Department of Physiology and Chairman of the Interdepartmental Program in Cell Physiology, University of Miami

Acknowledgments

The writing of this book was made possible by a grant from The Commonwealth Fund. I am deeply indebted to Mrs. Dorothy S. Obre and Mr. Roger A. Crane of The Commonwealth Fund for the editorial work the manuscript needed to ready it for publication.

I wish to express my gratitude to my colleagues, Dr. W. R. Duryee, Dr. A. R. Moore, and especially Dr. A. H. Whiteley, for reading through the entire manuscript and offering numerous helpful, invaluable suggestions. I am grateful also to Dr. Zolton T. Wirtschafter, who had the photographic portrait taken expressly for the frontispiece of this book.

Finally, it is a pleasure to acknowledge the co-operation of the following editors, publishers, and official bodies similarly concerned, who have granted permission to reproduce copyrighted illustrations: Academic Press, Inc.; *American Naturalist; Annals of Botany;* Company of Biologists Limited (*Journal of Experimental Biology*); *Cytologia;* Dr. Dietrich Steinkopff, Embryologia Society (*Embryologia*); *Experimental Cell Research; Journal of Ultrastructure Research; Kolloid-Zeitschrift;* Marine Biological Laboratory (*Biological Bulletin*); New York Academy of Sciences (*Annals*); Oxford University Press; Paul B. Hoeber, Inc.; Princeton University Press; *Protoplasma;* Rockefeller Institute (*Journal of General Physiology*); Royal Society (*Proceedings*); *Science Reports, Tohôku University;* Springer-Verlag; University of Chicago Press; *Wilhelm Roux' Archiv;* Wistar Institute of Anatomy and Biology (*Journal of Cellular and Comparative Physiology, Journal of Experimental Zoology*).

EDWARD L. CHAMBERS

Contents

List of Figures

List of Tables

Introduction

Scientific endeavor in biology has been engaged, up to the present, largely in the discovery and the application to one another of an ever increasing number of facts. The endeavor is largely cumulative, and the very multiplicity and diversity of the details amassed tend to discourage any generalized thinking. Some have proposed that further fact finding be brought to a halt at least temporarily and that investigators, equipped with the knowledge now on hand, direct their thoughts to an evaluation of the facts and to the gathering of the significant ones into synthetic principles. Many theoretically minded individuals are already attempting this. But have we facts in sufficiency at our disposal? Moreover, should not the evaluation of the data be done by those who not only have vision and philosophy but who are experienced biologists, thoroughly aware of the complexities and limitations of the factual data available?

There are broad gaps in our knowledge of biology. These gaps are so numerous as to raise grave doubt whether the subject can yet be classified as a science in the same sense as are physics and chemistry. It is true that there are things in biology which can be measured, but they are still too few for the logically minded. Indeed, many biologists seem almost ready to admit the impossibility of describing vital phenomena in the quantitative terms of our "accepted" sciences, physics and chemistry.

Physics is so clearly a science based on pure logic that brilliance in this field attends on consecutive reasoning from factual data once established. Within recent years the physicist seems to have filled in at least a first portion of his jigsaw picture; the chemist does not seem to be far behind; but the biologist is still groping! He is as yet too ignorant even to discriminate the significant from the insignificant, the logical from the apparently illogical.

When the modern biologist adapts to his own field the newer developments of physics and chemistry, he must re-examine his biological data. Most important, he must test the significance of these developments in biological as well as in physical and chemical terms. Particularly in problems dealing with cellular physiology and especially the nature of the single living cell, the available data are too scattered and inconclusive. An important reason why little of the data is indicative of real advances is that we are still too prone to draw conclusions without a sufficiency of factual evidence. Many more years of digging for thoroughly substantiated facts are required before the biologist's intuitive powers are sufficiently readied for constructive thought.

Biological investigation, as is true for any scientific pursuit, depends upon the accuracy with which the reasoning mind records what the eye sees. This requires training. It is an interesting thought that during the early period of biological investigation, such training could not have been other than faulty because men's minds were too occupied with the prevailing, non sequitur elements of superstitious dogmatism. Men's minds had not learned to distinguish data based on abstract thought from those based on actual observation. Fact tended to be intermixed with fancy, and it was centuries before the logical sense was developed sufficiently to permit discrimination between the real and the unreal.

The search for objective accuracy was greatly enhanced by such scientific controversies as the famous one which held sway in the seventeenth and eighteenth centuries between the epigenesists and the preformationists. This controversy stimulated considerable research and led to the accumulation of important observational data. The data, dependent for a long time on the limitations imposed by ordinary vision, remained incomplete until after the microscope was invented in the seventeenth century.[1] The revelations of the micro-

[1] The use of a single glass lens for magnifying objects was well appreciated during the sixteenth century, but these lenses were not sufficiently efficient to serve as a stimulus for biological investigation. A bilenticular system of far greater potentialities was devised toward the end of the sixteenth century by Hans and Zacharias Jansen. They constructed a true compound microscope by placing two convex lenses in proper relation to each other with the result that the image formed by the objective was magnified by the ocular. Probably to Robert Hooke belongs the credit for realizing the full importance of magnifying instruments in the study of nature. He used both simple and compound microscopes and in 1665 published the first known account of precise microscopical observations, in his *Micrographia.*

Leeuwenhoek's even more remarkable and comprehensive observations were made using only a single lens, and these are recorded in a series of letters and communications to the Royal Society of London extending over a span of fifty years (1673–1724). His "microscopes" consisted of single glass spherules held between metal plates. The smaller the diameter of the glass sphere, the greater the curvature of its surface and, hence, the scope resulted in a vast array of recorded observations. Much of the data served only to emphasize the intricacy of Nature's patterns and to perplex the early microscopists, who devoted a disproportionate amount of attention to the elucidation of hitherto invisible details of the more minute forms of life, like the flea, the mite, and the louse. Microscopic observation of these animals revealed that the more grossly visible structures were in turn composed of finer tissues of different sorts.

In spite of the steady increase in knowledge of living objects, there was nothing to counter the prevailing conviction that "life" was a metaphysical concept. The general outlook on life was largely theistic, and any discussion regarding it was of the order of the immaterial and metaphysical. A few individuals proposed ideas which could be considered on a more material basis, such as that life-matter is a vital fluid permeating the body (Descartes, 1630) or that it consists of manifoldly divided, living vacuoles (Lorenz Oken, 1810). However, these were hypothetical deductions and hence but little removed from the older metaphysical concepts.

Realism was brought forcibly to the fore by Marie François Xavier Bichat, a brilliant young surgeon of the Hôtel Dieu in Paris, who formulated the thesis that it is essential to have structure upon which life can manifest itself. He declared that the tissues, which he classified into twenty-one sorts, are the true conservators of the life of the body, each possessing its own type of structure (1802). Bichat is recognized as the father of histology. His tissue theory held the attention of the great surgeons of his day, and he built anatomy into a science. Bichat could readily see tissues without magnifying them and he actually objected to the use of the microscope. In common with most microscopists of the day, he assumed that the chief use of the instrument was for revealing details of minute free-living organisms.

Bichat's tissue theory, which solved the problem of human anatomy, proved too narrow to meet the requirements of universality inherent in the developing concepts of organic evolution. Moreover, investigations using improved microscopes indicated that even in what Bichat called elementary tissues there could be found structural units of living matter which were far more fundamental

greater its magnifying power. Leeuwenhoek's ability to see cilia and bacteria, and to determine the size of the human red blood corpuscle at 8 micra depended not only on his ability to make good lenses free of defects, but also on his use of an ingenious contrivance to mount and hold the specimens and a means of illumination (which he never divulged) to provide the necessary contrast between the object and the background (Dobell, 1932). The chief disadvantage of Leeuwenhoek's microscope was that the eye of the observer had to be at an uncomfortably short distance from the spherule which was being used as the magnifying lens.

than any unit presented in Bichat's tissue theory. These living units were, in time, to be recognized as a constant feature of all living organisms, plant and animal.

Living cells, as individuals and in groups, have been continuously observed ever since the invention of the microscope. No particular attention was given to them, and no one regarded them as being significant in any pattern or system of life. This feature may be likened to Tom's experiences in Kingsley's *Water Babies.* After playing with the vulgar trout in the fresh water brooks, Tom, in searching for the water babies, finally arrived at the sea, where, although the water babies were all about him, he mistook them for sea shells, and when he heard their laughter, thought he was listening to the rippling of the waves on the shore.

The cell theory, which attained full development in the first part of the nineteenth century, had its origin in the early studies of plant anatomy. By 1809 Brisseau de Mirbel had shown that all plant structures consist of cells and develop from cellular tissues. At the same time Lamarck[2] extended this generalization to cover all living organisms. Both investigators, although recognizing the universal presence of cells, considered them to be bound together in a continuous membranous tissue. Subsequently, Dutrochet (1824) demonstrated, by utilizing maceration techniques, that the single cell is the fundamental structural and physiological element in all plant and animal tissues. Schleiden's (1838)[3] and Schwann's (1839) principal contribution was to elaborate further this great concept and to attach importance to the cell nucleus (discovered by Robert Brown, 1831).

After establishment of the cell theory, attention was directed to the substance incorporated within cells. By the middle of the nineteenth century biologists had been brought to the realization that the essential component of cells is the "living jelly," or protoplasm. The characteristics of this material were found to be basically similar throughout all living nature.

The scope of the observations on cells was enlarged by the discovery that many dyestuffs penetrate living cells from their environment. This led to vital staining, which made it possible to discriminate structural details only vaguely distinguishable in un-

[2] This was developed principally in Lamarck's *Philosophie zoologique,* published in 1809.

[3] From his studies on the embryo-sac of plants Schleiden (1838) made the erroneous generalization that cells originate by aggregation of "mucus granules" in the cytoplasm. The cell theory became generally accepted only after the universality of cell multiplication by cell division had been established. Both von Mohl (1835) and Meyen (1839) demonstrated the origin of new plant cells by cell division. Remak (1841) concluded from his observations on embryonic blood cells that cell division is the only means of multiplication among animal cells.

stained cells. Later, coagulants were used to fix and harden tissues, and by the latter part of the nineteenth century a variety of complicated techniques had been developed for the observation of killed tissues. These techniques seemed so fruitful of results that the aim of their originators—to visualize live cell structure—was lost sight of. More and more attention was focussed on killed and fixed cytological structures, despite the fact that many of the details, now so clearly visible, might well have been due to the coagulating action of the particular fixative used. The result of this excessive attention to fixed cells was the accumulation of much data of questionable significance. During the past several decades intense preoccupation with fixed tissues has waned. Recently, however, with the advent of electron microscopy and its fascinating revelations, there have been indications that the cycle may repeat itself!

Cyto- and histochemists have applied specialized methods to sections of fixed or frozen-dried tissues. They have ascertained the effects of different chemicals on the ingredients of the tissues and have drawn important conclusions in regard to the chemical nature and localization of various intracellular components. Maceration procedures have also been used for isolating intracellular constituents such as mitochondria. Mitochondria tend to be more resistant to mechanical injury than are other intraprotoplasmic constituents. With maceration procedures, questions remain as to how many of the originally existing internal structures, when removed from the cell, remain unchanged, and to what extent the protoplasmic material isolated from mashed cells is or is not a de novo product of the procedure involved. The investigator's chief concern is to distinguish between the artifact and the real. Protoplasm is highly susceptible to injury. Slight mishandling may induce profound changes in its reaction and may cause the transference of materials from one cell constituent to another. Killing, fixing, and isolating procedures may be useful for an over-all appraisal of the chemical ingredients of the cell, but they may not be reliable guides to the significance of the particulate materials in protoplasm.

The protoplasm of every living cell is functionally a highly organized morphological unit of microscopic dimensions, with its various parts, including its protoplasmic surface film, serving as cooperating partners in harmony with the salt-containing, generally slightly alkaline surrounding aqueous medium. Although the major chemical ingredients of protoplasm are the proteins, very little is known concerning their state in the living cell. To the micrurgist, the term protoplasm has meaning only in so far as it refers to a living entity.

Every effort should be made to adapt methods for use on living cells. We require the perspective of this experimental approach as the newer developments in chemistry and in physics are being applied to biological problems. A wide search needs to be made for cell types best suited for particular experimental purposes and for tissues of which the constituent cells can be isolated and kept alive and active. Pertinent data on the physical nature and behavior of protoplasm as a living entity are at present far too sparse and too sketchy. This is an untried science, full of pitfalls. We lack adequate tools and proper directive forces. The biology of protoplasm must still, for a long time, be a science of probing for factual data.

The micromanipulative technique has made it possible to operate on protoplasm while it is alive and to determine physical and chemical reactions within the cell without destroying its protoplasm. This technique has enabled the biologist, so to speak, to cross a barrier, the protoplasmic surface film, and to learn something more tangible than hitherto of this film and the protoplasmic interior. The unique advantage the micrurgist has is that he can operate on protoplasm and still have protoplasm, with its protoplasmic surface film, in an intact state during his operations.

The material presented in the chapters which follow is concerned principally with what has been learned about protoplasm by micromanipulating living cells: by probing them with microneedles, by cutting away specialized regions or parts of cells, and by microinjecting them with various solutions.

ROBERT CHAMBERS
EDWARD L. CHAMBERS

I

The Protoplasmic Unit

INTRODUCTION TO PART I. Probably the first investigator to recognize living matter in terms of physical structure was Félix Dujardin of Paris; he observed a semifluid, jelly-like substance exuding from the crushed bodies of Infusoria in a drop of water. He recognized that amebae are composed of a similar living jelly (gelée vivante) and he designated this substance "sarcode" in 1835. At that time Dujardin was attempting to correlate the extraordinary variety of structure among the lower forms of life, of which he was making his special study, with the prominently developed organization observed among higher forms. Dujardin's term, sarcode, for the living matter of protozoa and of lower forms became widespread in France. A few years later Johannes Purkinje (1840), an anatomist in Prague, used the term "protoplasm" for the formative substance of which the cells of embryonic animal tissues are composed. Purkinje compared this with the granular material within the cambium cells of plants. Independently, in 1846, Hugo von Mohl, a botanist, used the term protoplasm for the essential living substance, a diaphanous granular material exhibiting streaming motion, which he observed within the cellulose coated plant cell.

To Max Schultze (1861) and de Bary (1859) largely belongs the credit, after exhaustive investigations, for demonstrating that Dujardin's sarcode of the lowest organisms and the protoplasm within the cells of the higher plants and animals are identical. Schultze described the cell as "a small mass of protoplasm endowed with the attributes of life." Thomas Huxley's (1868) eloquent exposition on the physical basis of life brought biologists, in general, to the realization of the identity of protoplasm throughout all living nature. The development of this concept may be considered an outgrowth of the unifying doctrine of organic evolution.

In considering protoplasm as the physical basis of life, it should be realized that protoplasm, as we know it, must have existed ever since life began. As far as we know, living cells and, therefore, their contained life substance, protoplasm, must have been built up through long evolutionary processes. H. S. Jennings, one of our great philosophically minded biologists, presented the case aptly in calling the living cell a history.

Protoplasm implies a substance, which it is not. It is a highly organized unit of living matter, continually in motion whether perceptible or barely perceptible, and with a structure which is delicate and easily destroyed. At the same time, it possesses a persistent, self-perpetuating ability sufficient to out-do all in the inorganic world.

In addition to the self-perpetuating property, protoplasm possesses self-adaptitude. This unique property, which enables protoplasm to change to the extent that it can continue living in a previously lethal environment, presents a problem of practical importance. Examples of such change are the appearance in certain bacteria and insects of strains which are no longer affected by the same lethal agents to which they were previously susceptible.

Every protoplasmic body persists true to form. The plant protoplast, the protoplasm of an ameba or of any metazoan tissue cell, all maintain those peculiar characteristics with which they were originally endowed. Inherent in every type of cell is a high degree of specialization. Therefore, in any experimental study of protoplasm it is necessary to distinguish features which represent a specific adaptation to a given condition from others which are characteristic of the more fundamental properties of protoplasm in general.

Protoplasm exists in intimate relation to the environment. So close is this relation that it is questionable whether protoplasm can be considered without constantly referring to the medium in which it lies. This medium is both aqueous and salt containing and reacts with protoplasm as a fully co-operating partner in the carrying on of the attributes of life. On the other hand, the myriads of multicellular plant and animal forms, including many of their constituent cells, have been equipped through organic evolution with devices of various sorts to enhance their independence of the environment.

To ascertain the intimate structure of protoplasm and the function of its various components much has been learned by analytical methods which involve, at best, destroying protoplasm and examining the disintegrated and isolated remains. The micromanipulative method, however, has a unique advantage. It permits the study of protoplasm while retaining it in a healthy, living condition during operative procedures. The results described in this book have depended largely on micromanipulation.

1

The Protoplasmic Unit

We need to know what part of the cell is the actual protoplasm and we should differentiate it from parts which can be eliminated without depriving the cell of its life-endowed properties. The micrurgist recognizes that every cell consists of two parts, the dispensable and the nondispensable. The dispensable parts which surround the protoplasm externally are the jelly coats, the extraneous membranes, and the intercellular cement. These can be removed either mechanically or chemically without detriment to the life of the cell. Other dispensable parts, existing within the protoplasmic interior, are yolk bodies, oil droplets, vacuoles, and many granular inclusions. These can be segregated by centrifugation and removed from the cell without affecting the life of the protoplasm. All the dispensable parts are by-products of the protoplasmic portion and are continually undergoing replacement. The differentiation between the dispensable and the nondispensable is best shown in a typical plant cell.

In the plant cell the actual living protoplasmic part, the protoplast, is a sharply circumscribed part of the cell, containing the nucleus within its substance, and spread as a thin sheet immediately under the cellulose wall. Where the cells are arranged in many layers, their cellulose walls are held together by means of an intercellular cement consisting primarily of the calcium salt of a pectinate. Both the cellulose wall and the cement are extraneous parts. A third dispensable part is completely incorporated within the protoplast as a large vacuole with its contained liquid sap which, in some cells, may attain macroscopic proportions. For example, in the multinucleated *Valonia* the central vacuole may grow to the size of a pigeon egg or even larger. Because of its central position within the protoplast and its turgescent state, the vacuole keeps the protoplast spread as a microscopically thin layer against the overlying wall of cellulose.

In animal cells the more truly living, or protoplasmic, portion generally constitutes the greater part of the volume of the cell. It is frequently interpenetrated with extraneous inclusions such as yolk bodies, secretion granules, and fluid vacuoles having delicate, film-like envelopes. Because of this feature no part of the animal cell is sufficiently segregated, as is the protoplast of the plant cell, to be recognizable as more purely protoplasmic.

THE PROTOPLASMIC UNIT IN RELATION TO ITS ENVIRONMENT

A major characteristic of protoplasm should be borne in mind. Protoplasm does not exist in bulk. No part of the interior of the protoplasmic unit can survive at more than a microscopic distance from the outer environment. Even the creeping, ameboid plasmodial bodies of the slime molds, which spread over areas as large as a square foot or more, are thin sheets of protoplasm only a few micra in thickness.

An essential feature of living cells is that they derive their nutriment directly from aqueous media. All their gaseous and electrolytic requirements as well as their essential foodstuffs come from aqueous media. It is into their environmental water that the cells dispose of their waste products, either gaseous or otherwise.

The usual microscopic size of the animal cell serves the same purpose as the microscopic thinness of the relatively larger plant protoplast or of the plasmodial sheet by ensuring an adequate opportunity for metabolic exchange with the environment. Even when we consider the large dimensioned, yolk-laden animal ovum, it is the mass of inert yolk which constitutes the major part of its size. The more truly protoplasmic portion is always relegated to a microscopically thin layer over the periphery of the egg.

Multicellular plants and animals, especially those which are land adapted, are equipped with devices to enhance their independence of the environment. On the other hand, a consideration of the protoplasmic unit reveals a situation of a different order. Protoplasm requires a special environment of its own. This consists of an aqueous solution containing electrolytes in definite proportional concentrations, at a hydrogen-ion concentration which can be varied only within narrow limits. Living cells generally exist in aqueous media which are on the alkaline side of neutrality. Fresh water ponds and lakes tend to be alkaline. The pH of the body fluids of metazoa is about 7.4; that of sea water is about 8.2. The buffer which plays the major role in maintaining the pH of the environmental fluids is the carbonic acid-bicarbonate system. The salts of sodium, potassium,

TABLE 1.1. Concentrations of ions in natural waters, sap from central vacuoles of plant cells, and muscle fluid

Ion	Marine		Mammalian		Fresh water	
	(mM/L) sea water bathing *Valonia*[a]	(mM/L) sap of *Valonia*[a]	(mM/L) plasma H_2O of cat[b]	(mM/L) muscle H_2O of cat[b]	(mM/L) pond water bathing *Hydrodictyon*[c]	(mM/L) sap of *Hydrodictyon*[c]
Na	498	90	178	28	1.3	4
K	12	500	5	151	0.019	76
Ca	12	2	3	1	1.08	2
Mg	57	trace	1	15	0.9	—
Cl	580	597	128	18	1.08	55
SO_4	36	trace	—	—	0.77	8
HCO_3	—	—	22	15	—	5

[a] Osterhout (1936); [b] Fenn, Cobb, Manery, and Bloor (1938); [c] Blinks and Nielsen (1940).

and calcium occur in the aqueous fluids which surround living cells in multicellular organisms in approximately the same proportional concentrations as they exist in sea water. The similarity is shown in Table 1.1. The total concentration of salts in sea water is over three times that in vertebrate body fluids. The proportional concentrations of magnesium and sulfate in sea water, however, are considerably higher.

This fact of the similarity between the proportional, although not the total, concentration of the salts in sea water and in the body fluids of the vertebrates gave rise to speculations one of which was presented by A. B. Macallum in 1926. This speculation was that the separation between the body fluids of the organisms and their environment had occurred at a period in organic evolution when the salt concentration of the sea was about one third of what it is today. This idea, although intriguing, has been disputed.[1]

The existence of living cells in naturally occurring fresh waters depends upon the presence of the same salts as in sea water, but in high dilution (Table 1.1). In pond waters, as in sea water and body fluids,[2] the cation sodium occurs in concentrations many times that of potassium.

Although the concentration of sodium in the environment of living cells generally exceeds that of potassium, the reverse holds true within the protoplasmic interior. Nonetheless, cells are gen-

[1] See, for example, Smith (1932); Baldwin (1937).

[2] An exception is the body fluids of certain species of insects, in which the potassium may far exceed the sodium concentration (Boné, 1944; Tobias, 1948). Potassium is also the predominant cation of the corresponding intracellular fluids.

erally permeable to both the sodium and potassium ions. The trapping mechanism for keeping potassium inside may be related to the smaller size of the hydrated potassium ion as compared with the sodium ion with its larger water shell. Since the proteins in protoplasm exist predominantly as anions, the many fixed negative groups may preferentially hold the potassium ions, with their thinner water shells, in the interstices between the protein molecules. The maintenance of a high concentration of potassium in the protoplasmic interior of many cell types has been shown to depend on the generation of metabolic energy. Agents which block the cytochrome system or interfere with oxidative phosphorylation cause cells to lose potassium. Since sodium is gained in approximately equal amounts by the inhibited cells, the concentrations of monovalent cations in the protoplasm and the environment tend toward equality.

Although the intracellular content of potassium is high (Table 1.1) in most cells, there is little information as to its precise location. In this regard it is of interest to note that in many plant cells potassium chloride is accumulated in the vacuolar sap, as in *Valonia,* a marine coenocytic plant cell, and *Hydrodictyon,* a fresh water coenocytic plant cell (Table 1.1). It is possible that in animal cells also potassium is accumulated in the vacuoles, which are of small size and distributed throughout the protoplasmic interior. Recently, mitochondria isolated in an optimal state of preservation have been shown to retain potassium against unfavorable concentration gradients.[3] Maintenance of the higher concentration within the mitochondria occurs only under conditions which permit the continuation of oxidative phosphorylation.

An important feature is that the electrolytes in protoplasm exist in a state of "dynamic equilibrium" with the electrolytes of the environment. For example, although potassium is held preferentially in the protoplasm, when its concentration is decreased in the external medium the result is a leaking of potassium from the cell. Thus, in spite of the fact that the intracellular concentration of salts is dependent upon cellular metabolism, alteration of the proportional concentration of salts in the environment results in changes in concentration of the salts in the protoplasmic interior.

THE EXTERNAL COATINGS

Externally applied noncellular material is a feature of both animal

[3] Macfarlane and Spencer (1953); Spector (1953); Bartley and Davies (1954). The latter authors have drawn attention to the fact that a small fraction of the mitochondrial potassium is retained even in the absence of oxidative phosphorylation and resists being washed out.

and plant cells. In ova the material serves to form enveloping membranes. In multicellular forms where the cells exist in cohering groups so as to form glands, sheets, or tubular structures such as ducts and blood capillaries, there is a cement-like material between the constituent cells. This layer, in animal cells, is the salt of a proteinate, and in plant cells, of a pectinate, combined with sodium and calcium. The degree of stiffness of this cement material varies with the proportional concentrations of sodium or of calcium in the combination. The more calcium, the more adhesive and stiff is the cement, and the more firmly do the contiguous cells adhere together. The monovalent cation promotes the water solubility of the cements. Just as animal cells survive when the intercellular cement is dissolved, so also individual plant cells can survive the absence of their pectate coats.

A common error is to confuse the extraneous coat which lies closest to the surface of the protoplasmic body with the actual protoplasmic surface film. With microneedles it is possible to remove the adhering external coats without causing detriment to the viability of the underlying protoplasm. Moreover, such coatings tend to have an appreciable thickness and can be detected also by the fact that they stiffen in the presence of calcium but soften and, in many instances, dissolve away in the presence of excess sodium or potassium.

THE PROTOPLASMIC SURFACE FILM

At the protoplasmic surface is a water-immiscible surface film having no appreciable thickness at the highest magnification of the light microscope and, as long as the protoplasm is alive, continuous over the protoplasmic body. It separates the protoplasmic aqueous phase from the external aqueous environment and possesses properties of restricted permeability. The integrity of this layer conditions the life property of the protoplasm. The presence of this film may be demonstrated by the fact that when noncoagulating, aqueous solutions of dyes which do not penetrate from outside the cell are microinjected, they diffuse readily through the interior of the protoplasm, but, on coming into contact with the surface film, do not pass out into the external medium. When the surface film is torn by rapid and repeated thrusts of the microneedle, a wave of disintegration sweeps around the protoplasm, and its complete destruction ensues. The disruption occasioned by small tears tends to be localized. A surface film quickly re-forms, closing the gap and walling off the healthy part of the protoplasm from the cytolyzed region. Repair of a localized tear of the protoplasmic surface requires the presence in the external medium of the appropriate proportional concentrations of monovalent (sodium or potassium) and divalent (calcium) cations. The low surface tension at the outer surface of the protoplasmic surface film indicates that hydrophilic groups in the film project into the external medium. Micromanipulative experiments reveal that the protoplasmic surface film of a cell which has been divested of its extraneous coats and immersed in an isosmotic solution of salts behaves as if it were a liquid, irrespective of whether the salts are sodium, potassium, or calcium. The film should not be confused with any of the extraneous coatings, since these differ from the film by stiffening in the presence of calcium salts.

THE CYTOPLASM

Except for the cell nucleus, which may lie anywhere within the protoplasmic body, many or all the microscopically visible granules and vacuoles can be removed without detriment to the protoplasm. The cytoplasmic matrix is then made evident and shows no structure when viewed in the field of the light microscope.

Micrurgical experiments reveal that the continuous phase of the protoplasmic interior is aqueous. Oils, when injected into the fluid regions of cytoplasm, immediately round up as spherules. Aqueous solutions of the monovalent salts and of water-soluble dyes such as the sulfonphthaleins, which are noncoagulating, when microinjected, diffuse readily through the interior of the protoplasmic body. Echinoderm eggs and amebae, torn in aqueous solutions of monovalent salts, disperse. This is because the cytoplasmic matrix, being aqueous, merges with the aqueous solution of the environment.

Microinjection experiments have shown that the pH of the continuous aqueous phase of the cytoplasm is slightly to the acid side of neutrality. This was determined to be at a pH of 6.8 ± 0.2 by injecting aqueous solutions of Clark and Lubs' hydrogen-ion colorimetric indicators. The buffering capacity of the cytoplasmic matrix is considerable, and more than a momentary change of the hydrogen-ion concentration, as measured by colorimetric indicators, results in disintegration of the protoplasm. As long as a cell is living the pH of its matrix is maintained constant, irrespective of changes in the alkalinity or acidity of the external medium. On the other hand, the aqueous contents of the cytoplasmic vacuoles show variable pH values and have little or no buffering power. They readily become alkalinized or acidified by penetrating bases and acids, respectively; but the more highly buffered cytoplasmic matrix does not. When any part of the cytoplasm is mechanically injured, an acid of injury occurs, resulting in a fall of the observed pH to a value of 5.2 to 5.4. This fall in pH is either localized or general, in accordance with the extent of the tear.

The remarkable constancy of the intraprotoplasmic pH among the great variety of cells tested indicates a fundamental feature of living matter in regard to the physico-chemical state of the protoplasmic proteins. A main characteristic of these proteins, which are amphoteric, is that, due to the buffers in the intraprotoplasmic continuous aqueous phase, they are maintained on the alkaline side of their isoelectric points as proteinates. The anionic state of at least the majority of the intraprotoplasmic proteins has been indicated by the fact that neutral solutions of the chlorides of divalent cations (for example, calcium, magnesium) and of basic dyes (for example, neutral red), when microinjected, induce coagulation of the injected zone. Furthermore, as long as the microinjection does not alter the pH of the protoplasmic interior, the sodium and potassium salts of such anions as sulfate and picrate, and of acid dyes, fail to induce coagulation. When cells are injected with a weak acid such as picric acid, no reaction occurs as long as the buffers of the protoplasmic matrix are not upset, and the normal intraprotoplasmic pH persists. This is because both picrate and proteinate are anions. If, however, the amount of acid introduced is enough to overcome the buffering capacity of the continuous aqueous phase, coagulation occurs because of the lowering of the pH below the isoelectric points of the cytoplasmic proteins and their conversion to cationic salts. The proteins then precipitate as protein picrates (Pollack, 1927).

The toxicity of heavy metallic salts, in general, is to be expected from the readiness with which the metallic cations react chemically with the dispersed proteinates as they exist normally in the protoplasm (Reznikoff, 1926).

Although the cytoplasmic proteins of the living cell are chemically reactive insofar as their interactions with salts and metals are concerned, they are chemically inert at oil–cytoplasm interfaces. Surface denaturation (that is, unfolding of protein molecules) does not occur at oil–cytoplasm interfaces as long as the protoplasm is uninjured. This is shown by the absence of adsorption films when drops of paraffin oil are injected into living cells such as amebae or echinoderm eggs (Figure 1.1a).[4] However, if deterioration of the protoplasm occurs, for example, because cytolysis is induced, the proteins are altered so that they accumulate at oil–water interfaces, causing the oil drops to wrinkle (Figure 1.1b).[4]

[4] Kopac (1938a, b; 1950).

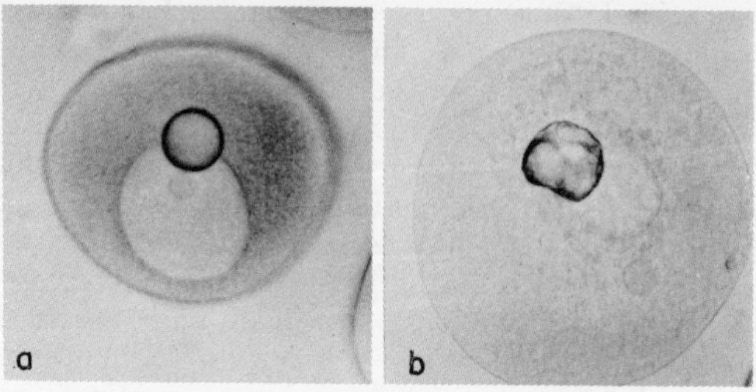

FIGURE 1.1. (a) Immature, living starfish egg microinjected with drop of paraffin oil. Oil drop spherical, showing no reaction between oil and healthy cytoplasm. (b) Same, after cytolysis of egg has been induced by stabbing germinal vesicle. Oil drop becomes wrinkled during cytolysis because of interfacial adsorption of denaturing proteins.

Another indication of the inertness of the protoplasmic proteins is the absence of detectable protein error in the colors assumed by pH indicators microinjected into the cytoplasm of living cells.

The reducing intensity of cytoplasm as determined by microinjecting a series of reversible oxidation–reduction indicators is considerable, even under aerobic conditions.[5] At a pH value of 6.8 to 7.0, the aerobically determined oxidation–reduction potential of the cytoplasm of the ameba and various marine ova was measured at −0.072 volts. This aerobic value expresses an average between the relative velocities of the reducing and oxidizing reactions within the cell. Under anaerobic conditions the oxidation–reduction potential is shifted far to the negative side. The lower limit could not be determined, but this is definitely more negative than −0.125 volts, and, in some cases at least, more negative than −0.167 volts.[5]

Proteins in the protoplasmic interior presumably exist as molecules or as micellae, dispersed in the continuous aqueous phase. If there is a fibrillar network,[6] the network is freely open to the through passage of aqueous solutions. Cell inclusions move about

[5] Cohen, Chambers, and Reznikoff (1928); Chambers, Pollack, and Cohen (1929); Chambers, Cohen, and Pollack (1932); Chambers (1933b).

[6] Moore (1935, 1945) postulated that fibrillar elements essential to life are present in the protoplasm of the plasmodium, or slime mold. He observed that the plasmodium passes of its own accord through filter pores 1 micron in diameter, but cannot live if forced through pores less than 200 micra in diameter. Moore supposed that pressing the plasmodium through pores caused the fracture of fibrillar elements at least 200 micra in length, resulting in death of the protoplasm. Another possibility is that the protoplasmic surface film is unable to form itself with sufficient rapidity over protoplasmic strands forced through too fine a mesh.

anywhere through the continuous aqueous phase of the protoplasmic interior. Upon application of centrifugal acceleration, there is no evidence, except in gelled structures, that the granules meet obstacles in their movement through the cytoplasm.[7]

Particles and granules suspended in protoplasm show continual and incessant movement. This movement is of two sorts: a physical, Brownian motion which is exhibited by all particles suspended in any liquid, and a translational, or directional, streaming motion which is peculiar to protoplasm and occurs in all types of cells. Brownian movement becomes prominent only after deterioration and cytolysis.

Structural orientation can be demonstrated in the protoplasmic interior of certain cells by observing the intracellular growth of ice crystals. A cell can be subcooled to a very low temperature without occasioning freezing in its protoplasmic interior. However, if a micropipette with an ice crystal projecting from its tip is inserted into the subcooled cell, internal freezing is induced, and crystal formation spreads through the cell from the point of insertion (Chambers and Hale, 1932). If there is any structural orientation of the protoplasmic components, the ice-crystal formation follows a visible pattern. This has been observed in isolated segments of the frog muscle fiber, the interior of which, in the healthy living state, appears homogeneous except for cross-striations. The effect of "seeding" the subcooled fiber with a micropipette tipped with an ice crystal was the formation of longitudinally arranged, slender columns of ice which started from the site where the fiber had been punctured and spread in both directions to the two opposite ends of the fiber (Figure 1.2). In another case the subcooled muscle fiber was twisted before inserting the micropipette. The induced internal freezing caused the appearance of twisted columns of ice. Another case was the fresh water ameba, in which there was no evidence of structural orientation. When the ameba was "seeded," ice crystals formed and spread through the subcooled ameba in all directions in the form of fine, feathery crystals.

The physical state of protoplasm varies in different regions of the same cell. The term "gel" has been ascribed to the solid state, and the term "sol" to the fluid state. Protoplasm in both states is

[7] The movement of crystals in "jerks" through the cytoplasmic interior of amebae subjected to centrifugal acceleration, observed by Harvey and Marsland (1932), was probably due either to a bumping of the inclusions along the irregular inner margin of the plasmagel or to the presence of gelated regions in the endoplasm. The possibility that the endoplasmic reticulum, in cells in which this system of membranes is well developed, might affect the movement of granules through the cytoplasm under the influence of centrifugal acceleration, or even affect the rate of diffusion of aqueous solutions through the cytoplasm, has not as yet been sufficiently investigated.

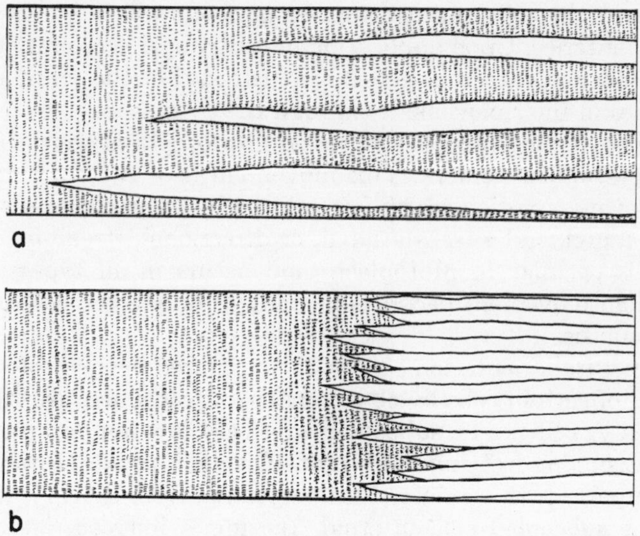

FIGURE 1.2. Subcooled living frog muscle fibers at $-6°C$. Internal freezing induced by "seeding" has resulted in longitudinally arranged ice columns indicating internal structure. (a) When fibers were "seeded" with ice-tipped micropipette, three ice columns formed and advanced in sarcoplasm. Cross-striations visible in unfrozen part. (b) Another fiber treated as in (a), showing numerous internal ice columns with tapering tips.

freely permeable to aqueous solutions, including water-soluble dyes. Spontaneous changes in the physical state of the cytoplasm from a gel to a sol and vice versa occur in accordance with events in the normal life history of the cell. These changes are associated with streaming movements, as in ameboid movement and gel formation connected with growth of the cytoplasmic aster and cell division.

The gelated regions can be detected by probing with the tip of a microneedle[8] or by microinjecting aqueous suspensions of particulate matter such as carbon. Another method is to test the distortion of shape of injected, innocuous oil drops. Mechanical agitation with a microneedle or the application of hydrostatic pressure[9] causes a reversal of the gel state to that of a sol.

Insertion of the microneedle reveals the presence,[8] in the mature echinoderm egg as in cells in general, of a narrow gelled zone

[8] Chambers (1917a); Hiramoto (1957).
[9] Brown and Marsland (1936); Marsland (1956).

immediately underlying the protoplasmic surface film. This superficially located part of the protoplasmic body is aptly termed the cortex.

A significant finding regarding the physiologic role of the gelled cortex concerns its essentiality, in the echinoderm egg, for fertilization and subsequent development.[10] A mature, unfertilized starfish egg can be decorticated by seizing the egg with a microneedle and compressing it at the edge of a hanging drop of sea water. The compression causes a rupture and an outflow of the more fluid interior, which eventually pinches off (Figure 1.3a). By means of this operation the integrity of the protoplasmic surface film is maintained, and the egg is separated into a gelled cortical remnant (Figure 1.3b, c, to the left) and an endoplasmic sphere, which rounds up because of its fluid nature (Figure 1.3b, c, to the right). The cortical remnant can be sperm fertilized and will undergo development (Figure 1.3b, c, to the left). On the other hand, the endoplasmic sphere, which may or may not contain the egg nucleus, is unresponsive to entering spermatozoa (Figure 1.3b, c, to the right). In another experiment the endoplasmic sphere, before it had pinched off, was cut away in such a way that a portion of the cortical remnant was incorporated in the sphere (Figure 1.4a). The presence of a small part of the cortex now allowed sperm fertilization; subsequent development, however, was abnormal (Figure 1.4b–e). In additional experiments the amount of cortical material incorporated with the endoplasmic sphere was varied. The degree of normality of devel-

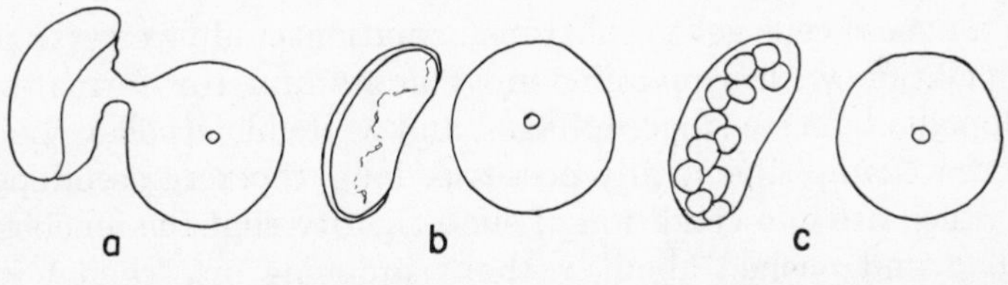

FIGURE 1.3. Separation of mature, unfertilized starfish (*Asterias forbesii*) egg into cortical remnant and endoplasmic sphere. (a) Egg, with its jelly coat shaken off, lying in hanging drop of sea water, is seized by tip of one microneedle and dragged to edge of drop. It is then compressed by other needle to cause fluid endoplasmic sol of egg interior to flow out and form endoplasmic sphere (right) separated by narrow stalk from irregularly shaped gelled cortical remnant (left). (b) Fragments inseminated after endoplasmic sphere containing egg nucleus has been pinched off. Only irregularly shaped cortical remnant forms fertilization membrane. (c) Several hours later. Cortical remnant has undergone cleavage; endoplasmic sphere is inert and nonfertilizable.

[10] Chambers (1921e); Runnström and Kriszat (1952).

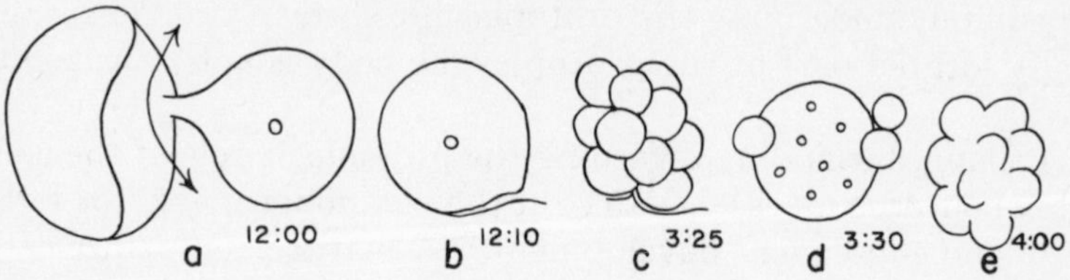

FIGURE 1.4. Same as Figure 1.3, but endoplasmic sphere includes small part of cortical remnant. (a) Arrow shows where cut is made with microneedle. (b) Endoplasmic sphere incorporated with small part of cortical remnant, inseminated. A partial fertilization membrane rises off in region where small part of cortical remnant is incorporated. (c) Same, two hours later. Cleavage furrows form simultaneously over surface. (d) Five minutes later. Sphere has reverted to multinucleated body. (e) Half an hour later. Sphere again attempts to segment.

opment following fertilization was found to be in proportion to the amount of cortical material present.

Changes from the sol to the gel state and vice versa have been observed to occur in accordance with procedures related to cell division. The asters prominent in dividing embryonic cells are gelated bodies which involve centripetal radial streaming of the hyaloplasm. The spindle of the mitotic figure in its relation to cell division undergoes gelation and then solation. Artificial reversal of the gel state of the aster to the sol state can be induced by agitation with the tip of the microneedle. When the agitation ceases, unless this has been too prolonged, the sol state reverts to the gel state, and the astral radiations re-form.

In ameboid cells gel to sol transformations and vice versa occur in association with streaming movements and the formation of pseudopodia (amebae, macrophages, and white blood cells). *Amoeba dubia,* for example, generally possesses long, tapering pseudopodia, all of which are in a condition of such rigidity that the ameba may be rolled and pushed about without inducing any change in its shape. Underlying its thin, tenuous pellicle is a relatively broad granular zone in the gel state. This merges insensibly into the fluid granular interior, the endoplasmic sol. Between the gelled granular ectoplasm and the external pellicle is occasionally a hyaline, fluid peripheral zone. When an aqueous suspension of carbon particles is microinjected into the peripheral hyaline zone, the particles spread throughout this zone, indicating its liquid nature.

The extension of a pseudopodium in an ameba occurs as follows: A localized liquefaction of the gelated ectoplasm takes place at the site where the pseudopod is about to form. If the liquid peripheral hyaline zone is very wide, as occurs occasionally, the granules spread out in it. Usually, however, the pellicle bulges locally, and the granules stream forward only into the bulge. The granules flow to the extreme tip of the bulge, and then flow back immediately beneath the extending pellicle, in the manner of a fountain stream. The back flow is, however, quickly arrested by a peripheral cortical solidification of the flowing material as it comes into contact with the gelled granular ectoplasm extending from the base. This setting process builds up a semisolid wall about a central, freely flowing channel.[11] As the pseudopodium extends still farther, the peripheral back flow at its tip continually adds solidifying material to the top of the hollow, gelled, cylindrical wall (Figure 1.5). With a slowing down and cessation of the axial stream, the tip of the pseudopodium solidifies, and we now have an extended arm, with a solid wall, which is elastic.

The gel state of the granulo-ectoplasm and the sol state of the granulo-endoplasm and of the tips of extending pseudopodia is well shown in an experiment in which an actively swimming small ciliate introduced into the interior of the ameba was found to swim about undisturbed in its unusual surroundings. The ciliate continued its progress within the solated granulo-endoplasm of the middle region of an advancing pseudopodium, but met an obstacle when it approached the gelled cortical region along its side. It is of interest to note that the ciliate was able to escape only at the tip of an advancing pseudopodium where the ectoplasmic gel had solated during the advance. The introduction of the ciliate

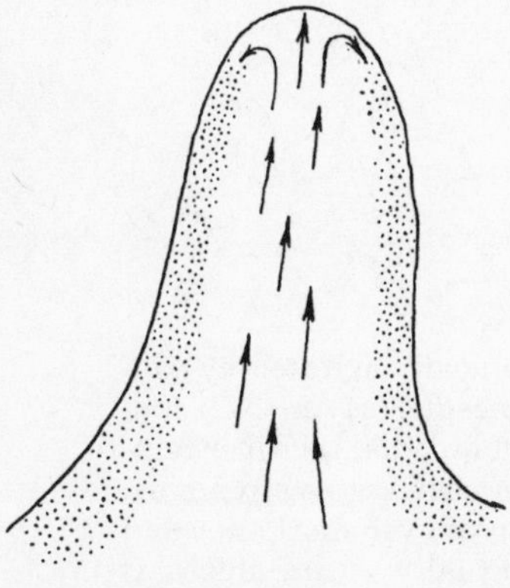

FIGURE 1.5. Sketch of advancing pseudopodium of ameba. Arrows show direction of axial flow to tip of pseudopodium and reversal of flow at tip, with formation of granulo-ectoplasmic gel as advance of pseudopodium progresses. (From Chambers, 1924a.)

[11] The sol-gel changes accompanying pseudopodial formation, retraction, and ameboid movement described by Chambers (1920a, 1924a) are essentially as described by Mast (1926).

into the ameba was done by releasing it from a gastric vacuole in which the ciliate had been trapped. The release was effected by causing a dissolution of the vacuolar membrane through the dispersive action of a solution of sodium chloride microinjected into the vacuole. The innocuousness of the streaming granulo-endoplasm was shown by the rapid revival of the released ciliate from its initially moribund state while inside the intact gastric vacuole.

Mechanical agitation of an ameba by means of an inserted microneedle promotes solation of the gelated regions (Chambers, 1924a). Continued agitation causes retraction of the narrow pseudopodia and solation of the ectoplasmic gel, resulting in the formation of lobate pseudopodia (Figure 1.6b). With continued agitation, these become increasingly broad (Figure 1.6c) until finally the ameba is converted into a sphere with rolling currents which flow under its surface from one pole and turn in at the opposite pole (Figure 1.6d). As fast as the axial currents move forward, the peripheral currents move back, with the result that the ameba as a whole remains stationary. Once an ameba is brought to this condition, the currents may continue for several hours, but death eventually occurs.

THE NUCLEUS

Max Schultze defined the living cell as a body of protoplasm containing a nucleus. The dependence for growth and survival of the cytoplasm on its nucleus has been demonstrated by amputating the nucleated from the non-nucleated part of the cell. Such experi-

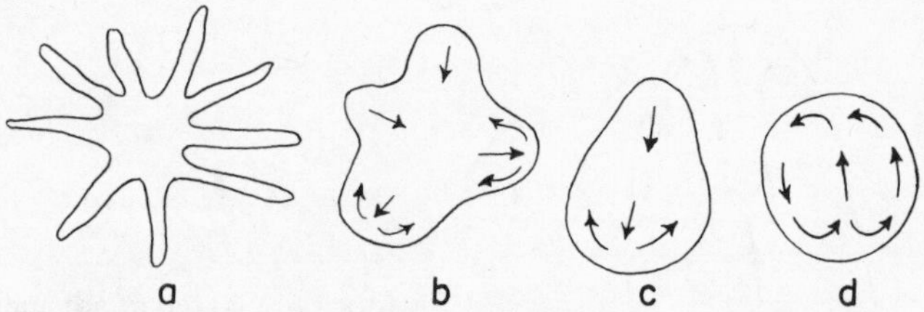

FIGURE 1.6. *Amoeba dubia* mechanically agitated by constant irritation with inserted microneedle. (a) Ameba before agitation, with long, extended pseudopodia, relatively stationary. (b, c) Continued agitation accelerates streaming, with general solation and formation of ever shortening and broadening pseudopodia. (d) Eventually entire ameba is converted into a rolling ball, with axial currents flowing in one direction at same rate that peripheral currents return flow.

ments have been done repeatedly with a variety of cells. Both plant and animal fragments survive for a time, but only the nucleated portion eventually regenerates cytoplasm, which survives indefinitely.

Generally, the non-nucleated portion is relatively short lived, as in the ameba. Anucleate fragments may live as long as two weeks. The lifespan of fasted nucleated fragments is only a little greater (two to three weeks). The anucleate fragments lose the ability to phagocytose living organisms, since ameboid movement becomes sluggish. The fragments tend to round up, and they are unable to adhere to the substrate owing to lack of secretion of a slime-like material over their surfaces.[12] The diminished motility is evidently related to a weakening effect of enucleation on the gelational state of the protoplasmic interior (Hirschfield, 1959). The non-nucleated fragments still retain the ability to incorporate amino acids in proteins, or purines and pyrimidines in nucleic acids, although at diminished rates.[13] All normal activities are regained if a nucleus is retransferred to a non-nucleated fragment, as long as this is not too long delayed (Commandon and de Fonbrune, 1939).

The life span of an enucleated ciliate is generally even less than the ameba's. An enucleated *Stentor* lives 5 days (Tartar, 1956). It is unable to feed and remains alive for about the same period of time as an unfed control animal containing its nucleus. Ciliary activity and contractility of the myonemes are maintained in the enucleated animal. Regenerative capacity, however, is minimal. Although the enucleated *Stentor* is capable of realigning the cortical stripe pattern, it is unable to regenerate the head and mouth. This is in striking contrast to the ability of a nucleated fragment to regenerate a cut-off head. For this purpose, only 1 of the 15 macronuclear nodes suffices. Interestingly, the enucleated *Stentor* utilizes its food reserves at a much slower rate than a starved control with intact nucleus (Tartar, 1956).

In certain green plant forms, notably *Acetabularia,* the non-nucleated fragment is relatively long lived (several months). The anterior anucleate part cut off a nucleated *Acetabularia* cell at a particular period of its life history and exposed to light is capable of considerable growth and differentiation accompanied by net synthesis of protein and ribonucleic acid.[14] The anucleate green plant cell exposed to light is assured a food supply since photosynthetic activity continues. In the absence of light, no growth,

[12] Verworn (1888); Hofer (1890).

[13] Brachet (1955, 1959); Mazia and Prescott (1955).

[14] Hämmerling (1934, 1953); Hämmerling and others (1959); Brachet, Chantrenne, and Vanderhaege (1955).

differentiation, or net synthesis of proteins can occur, although amino acids can still be incorporated in the proteins.[14]

While the presence of a fully constituted nucleus is essential for the continued life of the cell, a recognition of the fact that cellular activities, including synthetic activities and differentiation, can continue in the absence of the nucleus is important. The extent to which this can occur, however, differs widely in different cells and under different conditions. Some of this variability is related to the availability to the organism of foodstuffs after removal of the nucleus. Anucleate fragments of the plant cell *Acetabularia* exposed to light have extraordinary synthetic capacities compared with the starving anucleate fragments of ameba and *Stentor*. Anucleate fragments of the two protozoan forms, however, behave very much like *Acetabularia* in the absence of light: net protein synthesis and growth are stopped in all three.

Another source of variability is the amount, or rate, of utilization of an essential nuclear substance remaining in the cytoplasm after enucleation. We may postulate that this material, required for regeneration, is not retained in sufficient amount in the cytoplasm of a *Stentor* after enucleation, but is stored in considerable amount in the cytoplasm of *Acetabularia* (at least during a certain period of its life history).

Gruber, in 1885, drew attention to the fact that not all nucleated fragments of the ciliate *Stentor* underwent regeneration. Recently, Tartar (1956) corroborated Gruber's old results by finding a certain type of nucleated fragment which failed to regenerate. Control experiments had shown that the apposition of wide and narrow cortical stripes is crucial for formation and differentiation of the oral primordium (Figure 1.7a–c). The nucleated pieces which failed to regenerate (Figure 1.7d), although they survived many days, were cut in such a way as to lack a region of appreciable contrast in the width of cortical stripes. The importance of the experiments on *Stentor* is the demonstration that in addition to the nucleus surrounded by cytoplasm, a certain anisotropy of the cortical cytoplasmic pattern is essential.[15]

The physical state of the interkinetic nucleus, as determined by probing with the tip of a glass microneedle, varies considerably in different types of cells. For example, in the anterior motor nerve cell of *Lophius,* or in the immature egg of a marine echinoderm,

[15] This feature is also indicated in cutting experiments on sea urchin eggs. The smallest nucleated fragment of a sea urchin egg capable of developing to a pluteus is one eighth to one fourth the volume of the whole egg. Below this size, the smaller the fragment the earlier the stage at which development ceases (Loeb, 1906; Tennent, Taylor, and Whitaker, 1929). See also Chapter 5, footnote 1.

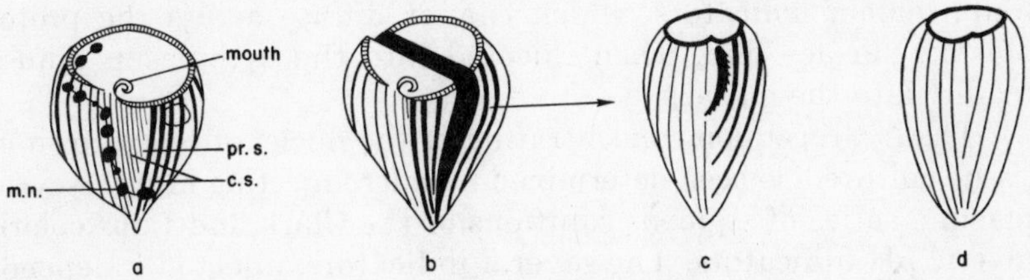

FIGURE 1.7. Regenerative capacity of aboral halves (both nucleated) of *Stentor*. (After Tartar, 1956.) (a) Diagram of intact *Stentor* showing principal structures. Chain of macronuclear nodes (m. n.); cortical stripes (c. s.) of graded widths in cortex; primordium site (pr. s.) at juncture of narrowest and widest stripes. (b) A longitudinal half is cut off to exclude original primordium site and widest and narrowest stripes. (c) A difference in stripe widths nevertheless appears at line of heal because remaining stripes are of graded widths; regeneration primordium appears at region where widest and narrowest stripes are in apposition. Subsequently a normal *Stentor* is formed. (d) Case in which an aboral half failed to regenerate, presumably because of insufficient contrast in width of stripes.

the contents of the spherical, hyaline nucleus are largely liquid; in the ameba and in many protozoa the nucleus is a gelated body. In general, however, in a large variety of metazoan cells the interkinetic nucleus behaves like a liquid body enclosed in a delicate membrane.

The nuclear membrane is a morphologically definitive, relatively inelastic structure, which can be distorted and wrinkled by indentation with the tip of a needle. If the nucleus, especially of an immature marine ovum, is gently compressed, the membrane "gives" at one or more points, and nuclear material protrudes into the cytoplasm enclosed in outpocketings of the nuclear membrane. If, however, the egg is compressed abruptly, the nuclear sap mixes directly with the cytoplasm, and a dramatic cytolytic reaction ensues. Such a reaction also occurs when the nucleus is pricked with the tip of a microneedle (see below and Figure 13.2).[16] The injured nucleus shrinks; the nuclear membrane persists intact in the form of a well-defined sphere.

An important characteristic of the nuclear membrane is its high permeability (Chapter 15). For example, the salts of the sul-

[16] An extraordinary feature is that the resulting injury to the cytoplasm, due to an agent which emanates from the pricked nucleus, does not start where the membrane is first torn, but spreads simultaneously from the entire nuclear surface. Evidently the pricking causes a wave of increased "porosity" to sweep over the surface of the injured nucleus. This resembles puncturing red blood cells with the tip of a microneedle, upon which hemoglobin escapes immediately from the entire surface. The injured red cell membrane, like the injured nuclear membrane, remains intact and may still retain the property of semipermeability (Teorell, 1952).

fonphthalein indicators, which cannot diffuse across the protoplasmic surface film, when injected into the cytoplasm, diffuse readily into the nucleus.

The hydrogen-ion concentration of the nuclei of cells grown in tissue culture has been determined by microinjecting into the cytoplasm a series of aqueous solutions of the Clark and Lubs colorimetric pH indicators. The several indicators, injected independently of one another, diffused from the cytoplasm into the nucleus, in which they all assumed colors which were consistent with a pH of 7.6 to 7.8. A pH value identical with the above has been obtained also for the liquid interior of the germinal vesicle, or nucleoplasm, of the immature ovum of the starfish; in this case, by careful manipulation, the micropipette could be introduced directly into the germinal vesicle without inducing cytolysis. The nuclear contents, therefore, are more alkaline than the cytoplasmic matrix by about 1 pH unit.

The metaphase chromosomes of plants and of insects, isolated by rupturing the cell and liberating the spindle, when seized and stretched between two microneedles behave as if they were elastic gels.[17] In all these cases the physical properties were studied in liquid media containing calcium ions or products of injured cells, which markedly alter the physical properties of the material constituting the chromosomes tested.[18] However, our concern is with the properties of chromosomes in situ, that is, within the nucleus of intact cells in their normal environment. Such studies have been carried out by D'Angelo (1946, 1950) on the giant chromosomes of the salivary gland cells of the *Chironomus* larva immersed in a drop of the larva's own lymph.

The chromosomes and nucleolus in the intact, fresh salivary gland cell are plainly visible, the transverse bands being sharply defined and either beaded or homogeneous in appearance. A hyaline material, the nucleoplasm or nuclear matrix, separates the chromosomes from each other. This hyaline material is differentiated into a central portion of jelly-like consistency in which the chromosomes are imbedded and a more fluid peripheral zone. Fine microneedles can be inserted into the nucleus of an intact cell without causing any visible change.

> As long as the cell is not injured, the chromosomes are not sticky, as revealed by pushing them one against another with the microneedle or by microinjecting carbon particles, which fail to adhere to the chromosomes. When the cell is

[17] Chambers and Sands (1923); Chambers (1924b); Pfeiffer (1940).
[18] Duryee (1941); D'Angelo (1946).

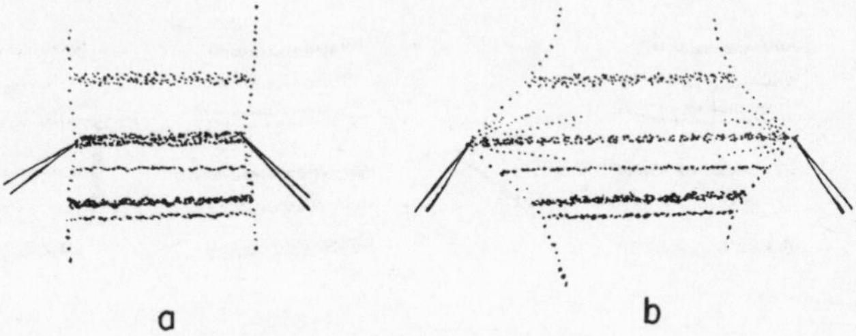

FIGURE 1.8. (a) Sketch of portion of chromosome in intact nucleus of salivary gland cell of *Chironomus* larva, showing two microneedles inserted directly into a band prior to lateral stretching of chromosome within intact cell. (b) Beaded appearance of band when same chromosome is stretched laterally. Note tension lines in interband regions. (From D'Angelo, 1950.)

injured, as by tearing, the chromosomes shrink,[19] stick to each other, and then can be separated only by pulling out thick viscid strands. Carbon particles adhere readily to the now sticky chromosomes.

In the intact, uninjured cell the chromosomes are soft, easily deformable gels. The transverse bands are of stiffer consistency than the interband regions of the chromosomes. When a short region of the chromosome is seized and stretched lengthwise, the hyaline interband regions stretch far more than the band regions. During the stretching, the broad bands often separate into several narrow bands. When the chromosomes are stretched laterally, the bands, which hitherto appeared homogeneous, become beaded (Figure 1.8). When tension is released, in each case the chromosome returns to its original condition. Surrounding the chromosome is a delicate membrane, or surface layer, which can be lifted with a microneedle (Figure 1.9a) or by microinjecting a suspension of carbon particles just under the surface (Figure 1.9b, c).

The multiple nature of the salivary gland chromosome is revealed by the fact that it is made up of numerous longitudinal fibrillae. Nodular swellings are located along the fibrils, which correspond in position to the chromosomal bands. The fibrils can be demonstrated by inserting the tip of the microneedle into the ex-

[19] Shrinkage, or condensation, of the chromosomes upon injury to the nucleus is a general phenomenon. For example, in the living grasshopper spermatocyte the nucleus, typically in the prophase stage, is optically homogeneous. When the nucleus is pricked, chromosome filaments appear. These progressively thicken until they are transformed from characteristic early prophase chromosomes into chromosomes resembling those of the metaphase stage (Chambers, 1924).

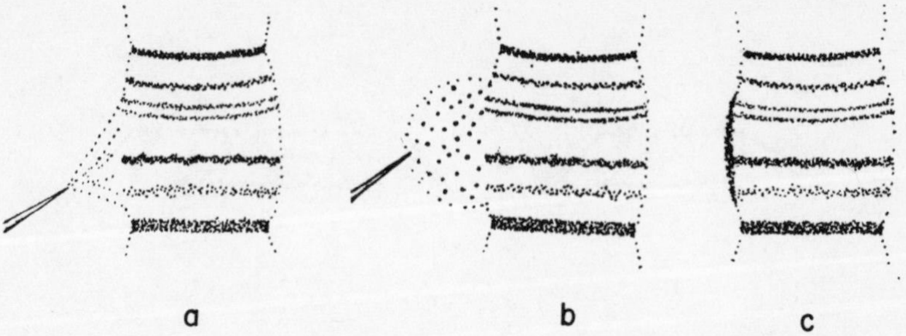

FIGURE 1.9. (a) Membrane-like material lifted by microneedle from portion of chromosome within intact cell. (b) Microinjection of carbon suspension into area subjacent to lifted "membrane" shown in (a). (c) Concentration of carbon particles into narrow zone as result of gradual contraction of "membrane." (From D'Angelo, 1950.)

treme margin of the chromosome; after the surface layer has been removed, fibrils can be stripped off repeatedly from the same exposed region (Figure 1.10).

Although the chromosomes are readily visible in the intact nucleus, they can be rendered invisible by immersing the salivary gland in slightly hypotonic or mildly alkaline media. Since the nuclei are now hyaline, they have the same appearance as the nuclei of the vast majority of interkinetic cells (observed by the light microscope). That the hyalinization is due only to swelling of the chromosomes, their structural integrity being maintained intact, can be shown by first microinjecting carbon particles into the nucleus of a salivary gland cell in its normal medium. When the nucleus is rendered hyaline by hypotonic treatment, the carbon particles show definite alignments, defining the contiguous borders of the adjacent, but now swollen, chromosomes (Figure 1.11).

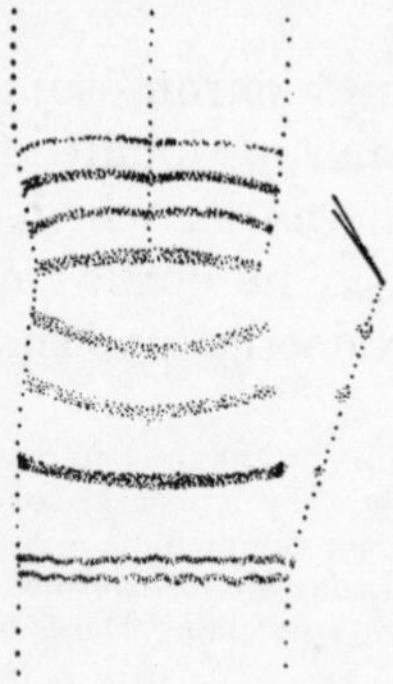

FIGURE 1.10. Single fibril with chromonema-like appearance being removed by microneedle from chromosome within intact cell. (From D'Angelo, 1950.)

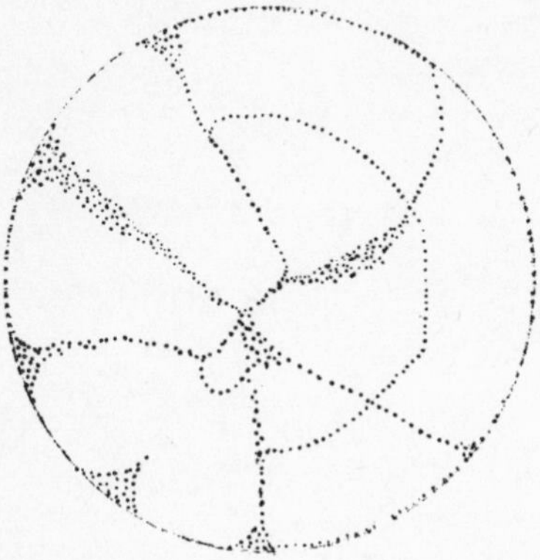

FIGURE 1.11. Alignment of carbon particles outlining contiguous borders of adjacent chromosomes in an optically homogeneous nucleus. Nucleus was first injected with carbon suspension, then hyalinized by immersing salivary gland in hypotonic Ringer's solution. (From D'Angelo, 1950.)

NUCLEOCYTOPLASMIC RELATIONS

At certain periods during the life history of the cell, during maturation of the ovum and when cell division is initiated, the hitherto separated components of the protoplasmic unit merge with one another, the nucleoplasm spontaneously mixing with the cytoplasm. The admixture constitutes the nucleocytoplasm.

The maturating ovum affords an unrivalled example of interaction between nucleus and cytoplasm. In the course of development changes occur from the immature egg with characteristics essentially those of a somatic cell, incapable of responding to the spermatozoon, to the mature ovum, a highly specialized cell, unique in its potentiality of being activated to develop into an organism. The fully grown immature ovum, with its prominent germinal vesicle, contains an unusually large amount of nucleoplasm. As maturation proceeds, erosion of the nuclear membrane occurs, and the nuclear contents, as well as the nucleolar material, slowly intermix with the cytoplasm. The egg chromosomes constitute a very small part of the volume of the germinal vesicle (Figure 1.12).

Included in the contents of the nucleus added to the cytoplasm in the maturating amphibian egg are fragments of the lateral loops of the egg chromosomes.[20] Prior to breakdown of the nuclear membrane, these lateral loops had grown out from chromomere loci to form the "lampbrush" chromosomes; the lateral loops had then sloughed off to be set free in the nucleoplasm. The many nucleoli in the nucleus of the mature amphibian ovum set free in the cytoplasm when the nuclear membrane is eroded had also been previously formed at specific chromosomal sites.

[20] Duryee (1941, 1950); Duryee and Doherty (1954); Guyénot and Danon (1953). The lateral loops, though containing ribonucleic acid, lack desoxyribonucleic acid (Dodson, 1948; Wischnitzer, 1957).

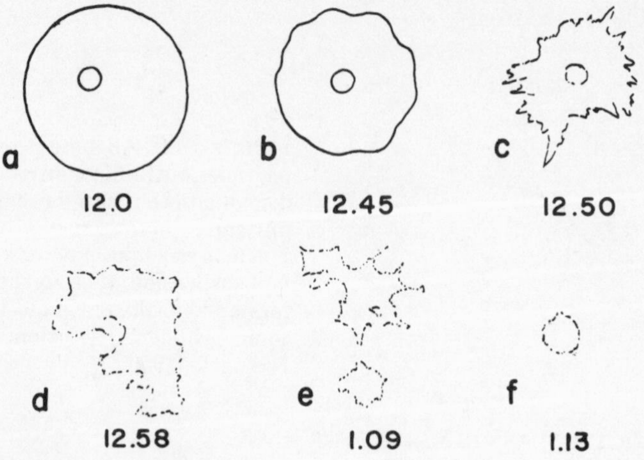

FIGURE 1.12. Successive steps in normal dissolution of germinal vesicle in maturing starfish egg. Only germinal vesicle area is shown. Time given in hours and minutes. (a) Intact germinal vesicle with prominent nucleolus. (b) Nuclear membrane wrinkles, becomes less distinct. (c) Membrane no longer visible, nucleolus fading, nuclear sap diffusing out, cytoplasmic granules invading nuclear area. (d, e) Nucleolus gone, cytoplasmic granules continue to invade nuclear area. Lower segregated region in (e) contains the egg chromosomes. (f) Nuclear sap completely intermixed; only small clear area containing egg chromosomes remains.

The properties of the cytoplasm before and after breakdown of the germinal vesicle are completely different. Anucleate fragments cut away from the starfish ovum immediately prior to dissolution of the germinal vesicle survive up to several days. Although spermatozoa can enter such fragments, they undergo no change after entry, nor does development ensue.[21] Attempts at parthenogenetic activation are equally unsuccessful. On the other hand, anucleate fragments of the starfish ovum (devoid of egg pronucleus) cut away *after* complete admixture of the nuclear contents with the cytoplasm has occurred, disintegrate in several hours if not inseminated, although when fertilized, they cleave and develop into normal (haploid) embryos.[21] The nucleocytoplasm of the mature egg is, so to speak, in an unstable, poised state, yet possessing full capacity for development. However, unless the stimulus to further development is provided (and parthenogenetic stimuli are extraordinarily nonspecific), deterioration rapidly sets in.

Anucleate fragments cut away at intermediate stages, before admixture of nucleoplasm and cytoplasm has been completed,

[21] Delage (1901); Chambers (1921e).

show varying potentialities for development (Chambers, 1921e). These experiments reveal that the developmental capacity of the cytoplasm depends upon the extent to which diffusion of nucleoplasm into the cytoplasm has occurred (Figure 1.13).

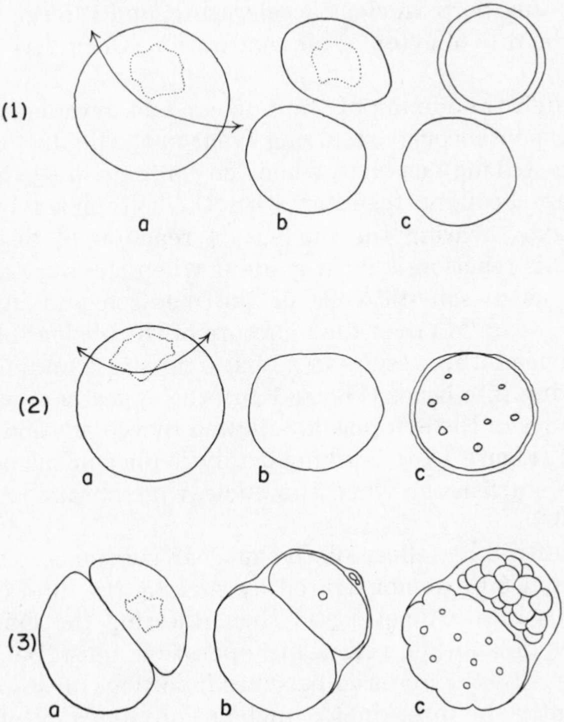

FIGURE 1.13. Starfish eggs at various stages during dissolution of germinal vesicle.

(1) Starfish egg in stage corresponding to (b) in Figure 1.12 cut into two fragments (1a and 1b) and both inseminated. Non-nucleated fragment (lower) contains no material from germinal vesicle and is nonfertilizable (1c), while nucleated fragment (upper) forms fertilization membrane and develops normally.

(2) Same, but egg cut at later stage corresponding to (c) in Figure 1.12. Cut, passing through lower part of nuclear region (2a), results in injury and disintegration of nucleated fragment, leaving non-nucleated fragment (2b). That fragment is fertilizable is shown in (2c) by elevation of fertilization membrane and repeated division of sperm nucleus. Fragment, however, is unable to segment.

(3) Same, but egg cut (3a) at still later stage corresponding to (d) in Figure 1.12. Both fragments fertilized (3b). Nucleated fragment (upper) segments in normal way (3c). In non-nucleated fragment (lower) sperm nucleus divides repeatedly (3c), and periodically furrows appear over surface in attempts at segmentation.

The interaction of cytoplasm and nucleoplasm to achieve a nucleocytoplasm with full developmental capacities requires the lapse of time. While this process of cytoplasmic maturation does not require the presence of either the sperm or egg pronucleus in the starfish egg, nonetheless these bodies exert a rate-determining influence, the sperm nucleus accelerating, and the egg pronucleus slowing, the rate of cytoplasmic maturation (Chambers and Chambers, 1949).

Although the mixing of the nuclear and cytoplasmic components is a spontaneously recurring event in the life history of every cell, it is a striking fact that, when the contents of the interkinetic nucleus are brought together with the cytoplasm by artificial means, as by tearing the nucleus, a reaction of disintegration occurs. This reaction is most evident when the nuclear contents are fluid, as in somatic cells of the metazoa and in immature echinoderm ova. Microneedle puncture of the nucleus of a somatic cell in tissue culture results in a characteristic rounding up of the mitochondria into beads (Figure 1.14), the appearance of degeneration granules in the cytoplasm, followed by coagulation and death of the cell (Figure 1.15). Striking cytolytic phenomena occur in the immature starfish egg when the nuclear membrane is punctured (Figure 13.2).

The destructive effect of the nuclear substance on the cytoplasm can also be demonstrated by sucking the fluid contents of the nucleus into a pipette and then injecting the fluid into the cytoplasm of another cell. This operation must be performed rapidly because the material becomes innocuous on standing.

Evidently the interkinetic nucleus contains a cytolytic agent which cannot readily diffuse through the intact nuclear mem-

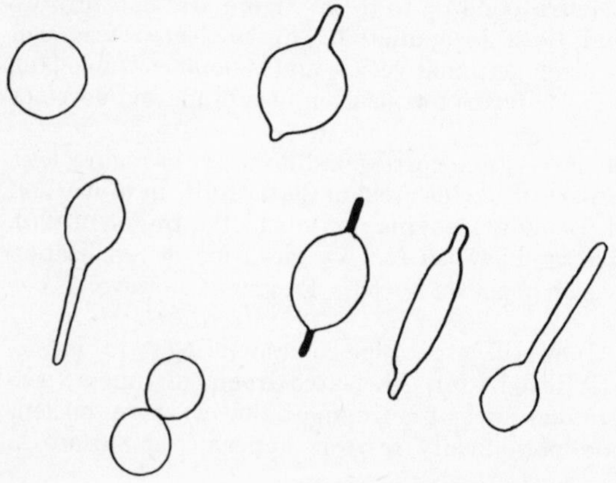

FIGURE 1.14. Swelling of filamentous mitochondria following puncture of nucleus. Swellings shown in several continue to enlarge until mitochondria become converted into spherules. (From Chambers and Fell, 1931.)

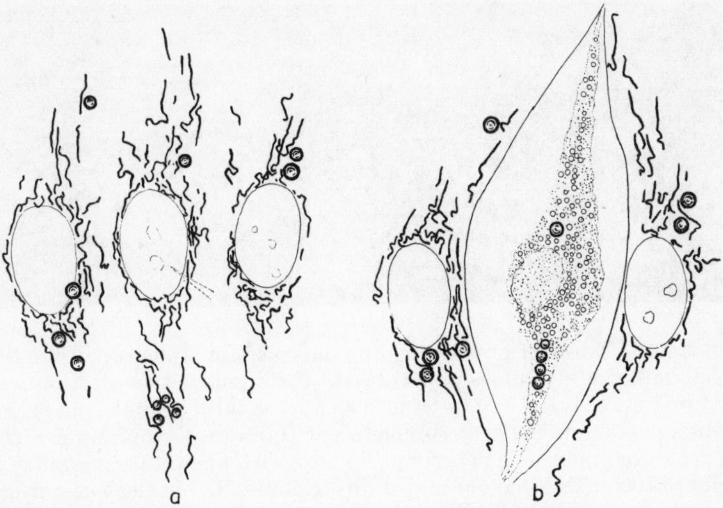

FIGURE 1.15. Effect of puncturing nucleus of one of three intestinal epithelial cells of chick, from sheet in tissue culture. (a) Before nucleus of middle cell is punctured with microneedle. Cell boundaries invisible. Note intact nucleus and filamentous mitochondria of middle cell, also fat droplets in cytoplasm. (b) Same cells 8 minutes after puncture of nucleus of middle cell. Cytoplasm of injured cell has contracted away from its neighbors, revealing cell outlines. Note shrinkage and granulation of punctured nucleus, granulation of cytoplasm, and conversion of mitochondria into degenerating globules which eventually disperse in cell debris. The two neighboring cells remain intact. (From Chambers and Fell, 1931.)

brane.[22] This would be one reason for the existence of the nucleus as a separate morphological entity. The difference in the outcome of spontaneous and artificial nucleocytoplasmic mixing may be related to the slowness of the process when this is a spontaneously occurring event, the cytolytically active agent being inactivated as the mixing occurs.

An extraordinary demonstration of an immediate controlling influence of the nucleus on the cytoplasm becomes evident when one nucleus of a binucleated fibroblast is punctured (Figure 1.16). Contrary to the complete disintegration which results when the nucleus of a uninucleated fibroblast is punctured, when puncturing is done in the presence of a second, intact nucleus, the initial dis-

[22] The nuclear sap can be set free in the cytoplasm by strongly shaking immature starfish eggs *without* causing any reaction of cytolysis. Although shaking causes the germinal vesicle to disappear, nonetheless no rupture has occurred, since the collapsed nuclear membrane can be detected by prodding with the microneedle. Evidently the cytolytic agent must be retained within the confines of the collapsed nuclear membrane, in spite of the fact that most of the nuclear sap passes out into the cytoplasm.

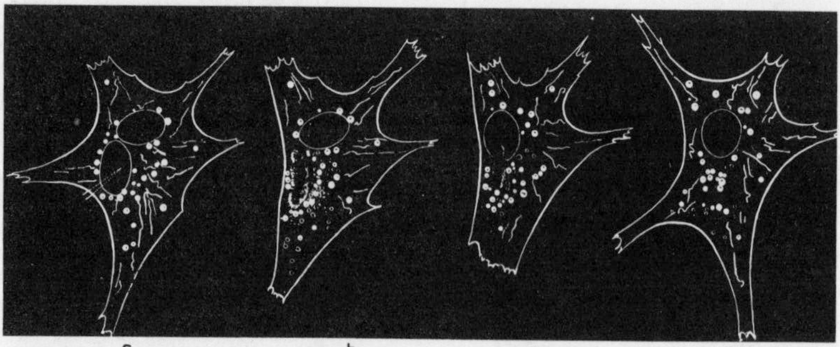

a b c d

FIGURE 1.16. Effect of puncturing one nucleus of a binucleated fibroblast. (a) Cell immediately before puncture. (b) Cell immediately after puncture. Punctured nucleus has collapsed into small, wrinkled granular mass. Adjacent pseudopodium has been completely withdrawn. Cytoplasm in vicinity of degenerated nucleus has become clouded with fine diffuse granules, and slender mitochondria have rounded into globules. Other nucleus normal in appearance. (c) Cell 4 minutes after puncture. Uninjured nucleus is quite unaffected. Granulation of cytoplasm around still more shrunken punctured nucleus is beginning to disappear. (d) Cell 17 minutes after puncture. It is again completely expanded and, except that it is now mononucleated, has regained its former appearance. (From Chambers and Fell, 1931.)

integrative process is soon reversed and full recovery ensues (Figure 1.16b–d). The experiment on the binucleated fibroblast may be explained on the basis that a substance on which the normal healthy state of the cytoplasm depends diffuses out from the interkinetic nucleus. This substance, in its interaction with the cytoplasm, would have the effect of antagonizing the cytolytic action when one nucleus is punctured while the other is left intact.

Another instance of an immediate controlling influence exerted by the nucleus on the cytoplasm is the causal relation between the initiation of streaming in the cytoplasm at or near the surface of an ameba, and the approach of the nucleus to that surface. If the ameba is impaled with the tip of a microneedle inserted into an ingested object, such as the shell of a rotifer, the ameba soon moves away by ejecting the shell with the needle. On the other hand, if the needle is inserted into the nucleus—which, if carried out very carefully, can be done without injuring the nucleus—the ameba may be kept in one position for as long as 24 hours. Each time the random movements of the ameba bring the surface of the ameba close to the stationary nucleus, pseudopodia develop on the surface adjacent to the nucleus. The ameba then reverses its direction. If the impalement injures the nucleus, the ameba rapidly ejects the nucleus and moves away.

The profound controlling influence exerted by the nucleus over cellular processes suggests, in itself, that an interchange of materials must occur between nucleus and cytoplasm. Obviously, transfer of materials synthesized in the nucleus is accomplished when the nuclear membrane undergoes dissolution (during maturation, mitosis). With regard to the transfer of materials during periods when the integrity of the nuclear membrane is maintained, we have already called attention to the fact that the fairly large, highly ionized molecules, the sulfonphthaleins, readily pass across the nuclear membrane (in contrast to their inability to diffuse across the protoplasmic surface). That an upper limit exists in the size of molecules able to diffuse across the nuclear membrane, at least of macromolecular dimensions, is indicated by the retention within the nucleus of a cytolytically active agent and by the fact that the pH of the nuclear sap differs by at least half a pH unit from that of the cytoplasm. Of great interest is the occurrence of an extrusion of intranuclear material into the cytoplasm, during which process the integrity of the nuclear membrane is maintained. This may occur by vesicle formation and discharge of contents of the vesicle into the cytoplasm (Weisz, 1949). Nucleoli have been observed traversing the nuclear membrane of resting somatic cells or everting their contents into the cytoplasm in immature amphibian ova.[23]

[23] Walker and Tozer (1909); Ray (1953); Duryee and Doherty (1954). The latter authors have also described a massive extrusion of ribonucleic acid from the nucleus into the cytoplasm in cancer cells.

II

Extraneous Coats

INTRODUCTION TO PART II. All living cells in nature are surrounded by one or another kind of extraneous membrane or jelly-like coating. These coats may be delicate or tough, thin or thick, soft or hard, elastic or rigid, viscous or plastic. Extraneous coats may be removed mechanically or by suitable chemical solvents so that finally only a limiting surface remains. This surface is the protoplasmic surface film, which cannot be removed without destroying the protoplasmic unit or be torn without damaging the protoplasm. Cells which have been denuded of their extraneous coats remain viable.

Nearly all investigations dealing with the surface of the living cell have been made on cells which possess one or more of their normal complement of extraneous coats. Accordingly, many of the observations which have been made and the physical properties which have been determined as of the cell surface have reference to the extraneous coatings rather than to the actual protoplasmic surface of the cell.

Many of the extraneous coats are stiffened by calcium salts, while they are softened, weakened, or even dispersed in sodium or potassium salts, in the absence of calcium. In contrast, the protoplasmic surface film, per se, is maintained intact in monovalent salt solutions, its fluidity actually being promoted in the presence of calcium. The pronounced differences in the effects of the salts of monovalent and of divalent cations on the intercellular material at the pH of physiological media suggest that some, at least, of the conclusions about the antagonistic action of salts on the preservation of life should be referred to the extraneous intercellular materials rather than to the maintenance of the individual cells.

Polysaccharides have long been known to be the major, and frequently the only, constituent of the extraneous coatings of plant cells. It is hardly surprising, therefore, that polysaccharides are increasingly being found to occupy an important role as components of the extraneous coatings of animal cells. Proteins also enter into the constitution of some of the coatings of plant cells and most of the coatings of animal cells.

2

General Characteristics of Extraneous Coats and Methods of Investigation

GENERAL CHARACTERISTICS

The principal types of coatings which surround living cells are well illustrated in the sea urchin egg. A mature, unfertilized echinoderm egg possesses two well-defined extraneous coats: the jelly layer and the vitelline membrane (Figure 2.1a). The vitelline membrane closely invests the egg and lies directly on the external surface of

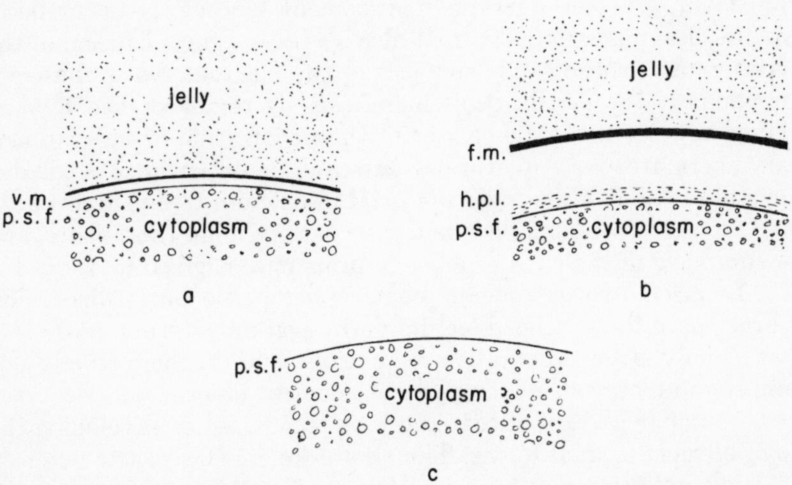

FIGURE 2.1. Diagram showing extraneous coats of sea urchin egg. (a) Unfertilized egg, with broad jelly coat and tenuous vitelline membrane (v. m.) closely applied to protoplasmic surface film (p. s. f.). (b) Fertilized egg. Fertilization membrane (f. m.) has risen; hyaline plasma layer (h. p. l.) has been secreted, covering protoplasmic surface. (c) Denuded egg, unfertilized or fertilized. All coats have been artificially removed, leaving protoplasmic surface film as external bounding layer.

the protoplasmic surface film. It is a thin, delicate differentiated pellicle, which is stable in salt solutions even in the absence of calcium salts. Outside the vitelline membrane is the voluminous, soft, invisible jelly layer of varying dimensions. It can be outlined by placing the cell in a suspension of colloidal carbon. The jelly coat of the echinoderm egg is composed of a polysaccharide with ester sulfate groups. Amino acids, probably in peptide linkage, are also present. In the absence of calcium, the jelly swells and tends to disperse. The jelly substance is claimed to be essential for the fertilization process; this claim has recently been disputed.[1]

On insemination, the vitelline membrane separates from the protoplasmic surface, elevates, and stiffens to become the fertilization membrane (Figure 2.1b; Chapter 4). Although this membrane is stable in the absence of calcium, it will not stiffen in the absence of this salt. Several minutes after insemination a third coating, the hyaline layer, appears, secreted as an exudate of the egg. This material accumulates on the surface previously occupied by the vitelline membrane (Figure 2.1b), possesses no visible structure, and attains a thickness of 2 to 3 micra. At first soft and gelatinous in consistency, it stiffens with continued exposure to sea water and can be pulled out into strands with the microneedle. The hyaline layer serves as an intercellular cement, binding the blastomeres of the cleaving egg together. It also serves to prevent fusion of the contiguous protoplasmic surfaces of the blastomeres as these are formed. If the fertilization membrane is removed, the hyaline layer now serves as the barrier to the penetration of supernumerary spermatozoa.[2] A feature of outstanding importance is the dispersal of the hyaline layer material in calcium-free media. Later, in Chapter 4, we shall discuss these coatings more thoroughly, and a summary of their main physical properties is given in Table 4.1.

In brief, two extraneous coats envelop the unfertilized, and three coats the fertilized, echinoderm egg: an external, loose-textured jelly layer; a definitive membrane; and (in the fertilized egg only) an innermost, relatively dense cement layer.

A similar pattern of coats may be recognized as enveloping the ova of most animal forms. The shell-like chorion of the fish egg, for example, is analogous, in structural relations, to the much thickened and toughened vitelline membrane of the sea urchin egg. In the human egg the jelly layer external to the vitelline membrane, or zona pellucida (Duryee, 1954), corresponds to the jelly

[1] Runnström, Hagström, and Perlman (1959).

[2] Dispersal of the hyaline layer after removal of the fertilization membrane permits the penetration of supernumerary spermatozoa (Sugiyama, 1951; Nakano, 1956).

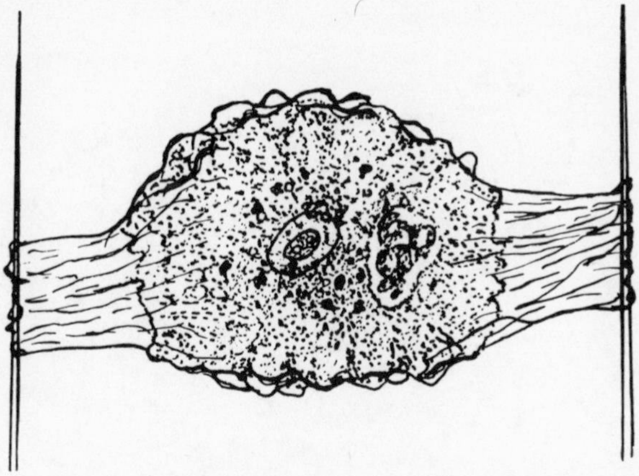

FIGURE 2.2. Living *Amoeba verrucosa* with pellicle lifted on both sides by needles. (From Howland, 1924.)

coat of the echinoderm egg. Material which serves to bind the blastomeres together is present in all embryos. This material, however, may not be sufficiently thick to be visible in the light microscope or sufficiently electron dense to be detected by electron microscopy.

Differentiated membranes, or pellicles, applied to the external true protoplasmic surface of the cell abound. These may be considered analogous, at least in their relationships, to the vitelline membrane of the unfertilized sea urchin egg. Free-living cells generally possess such pellicles. For example, in the protozoan ciliates *Paramoecium* and *Stentor* this pellicle can be lifted off the surface with a microneedle, or it can be seen on blister-like elevations which occur when these organisms are compressed or placed in an abnormal environment. Pricking the blister with a microneedle at once causes it to collapse, without injury to the protoplasmic part of the organism. In some Protozoa, for example *Vorticella* and *Amoeba verrucosa,* the pellicle[3] is tough and difficult to tear with microneedles (Figure 2.2). *Amoeba proteus* possesses a pellicle of intermediate toughness. It can be easily lifted off by microinjecting water beneath it. In *Amoeba dubia* the pellicle is extremely thin and tenuous. Electron photomicrographs of the pellicles of two different protozoa are shown in Figures 2.3 and 2.4.

Among cells of the Metazoa a prominent example of an extraneous differentiated membrane applied at the protoplasmic surface

[3] The plasmalemma, or pellicle, of several different species of amebae contains both protein and polysaccharide (Bairati and Lehmann, 1953; Pappas, 1954).

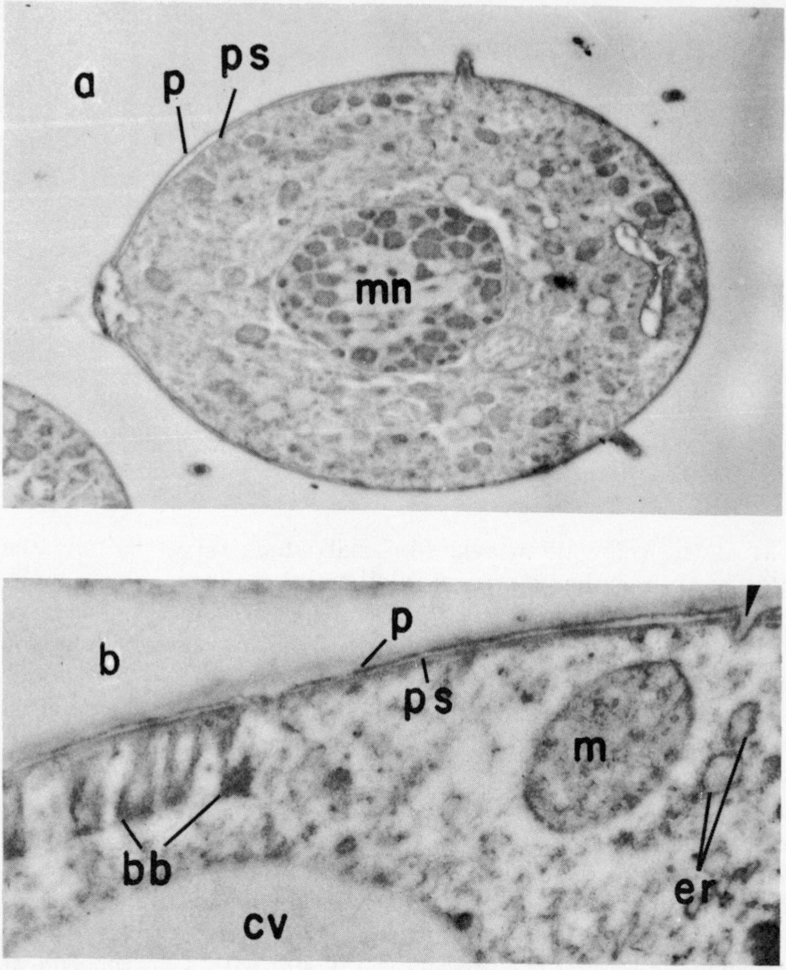

FIGURE 2.3. (a) Low power electron photomicrograph of longitudinal section through *Tokophrya infusionum.* Note that body is covered by two membranes, the pellicle (p) and a membrane at protoplasmic surface (ps). In center of cell a large macronucleus (mn) composed of dense chromatin bodies can be seen. 3,600×. (b) Higher power electron photomicrograph showing relationship between pellicle (p) and membrane at protoplasmic surface (ps). At intervals both membranes come in closer contact by papilla-like invaginations seen at arrow. A mitochondrion is depicted at m, basal bodies at bb, endoplasmic reticulum at er, and part of a contractile vacuole at cv. 22,600×. (Photomicrographs contributed by Maria D. Rudzinska, not previously published.)

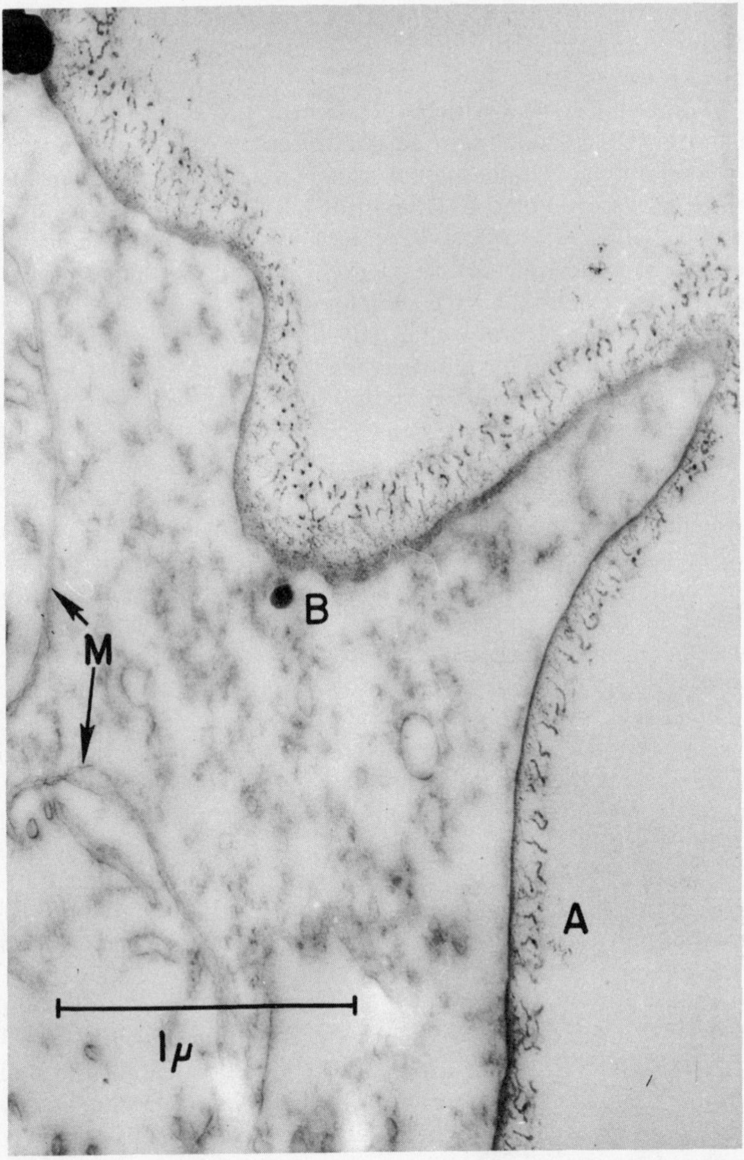

FIGURE 2.4. Electron photomicrograph of plasmalemma of *Amoeba proteus*. Outer portion of plasmalemma has fine fibrous extensions. Cross section of plasmalemma is seen at A; tangential view is seen at B. Thickness of plasmalemmal membrane at A is about 200 Å. Outer filaments extend about 1200 Å. A portion of two mitochondria (M) can be found in the cytoplasm. 45,000×. (Photomicrograph contributed by G. D. Pappas, not previously published.)

is the sarcolemma of the skeletal muscle fiber. This can be raised off the muscle fiber by microinjecting a droplet of oil immediately beneath the surface.

The cellulose wall which encloses the protoplast of plant cells is another prominent example of a differentiated extraneous coat.

If an animal cell does not possess a definitive extraneous membrane, it is surrounded by a layer of jelly-like material, as are the ameboid plasmodia of the Myxomycetes, or by a relatively firm, cement-like substance which serves to hold contiguous cells together, as are epithelial cells in general.

In many of the Protozoa a jelly-like material is secreted external to the pellicle. This material may be adhesive, and it is by this means that the ameba often fastens itself to the substratum. In an ameba which has been stationary for some time, the amount of this slime secreted may be considerable. The extensibility of the material is such that if the ameba is dragged with the microneedle a distance through the water, and is then released, the ameba will be passively drawn back almost to its original position. Interesting cases are certain free-living plant cells, the desmids and diatoms, some species of which glide along a surface by secreting a slime through pores in their rigid cell walls.

If we pack together by centrifugation a number of unfertilized *Arbacia* eggs, we have a condition which resembles the connective tissues of multicellular animals—that is to say, individual cellular elements imbedded in a gelatinous matrix. The ground substance of the connective tissues would represent the aggregated jelly-like coatings of the fibroblasts, produced by their secretory activity. This analogy is more than a merely structural one, in that the egg jelly and the ground substance of connective tissues are both composed of polysaccharides,[4] including at least one which is esterified with sulfate, together with a protein constituent. A similar type of tissue is met with in many marine algae, and the intervening abundant jelly substance consists entirely of polysaccharides esterified with sulfate.[4] The structural role of the jelly substance in multicellular tissues is to bring about the agglomeration of cells in the absence of any specific architectural arrangement. This is in contrast to the intercellular cementing substances, the role of which is to bind cells together in a close-knit, highly organized pattern.

[4] Fibroblasts in tissue culture (Grossfeld, Meyer, Godman, and Linker, 1957) produce hyaluronic acid (polymer of glucuronic acid + acetylglucosamine) and chondroitin sulfate (polymer of glucuronic acid + acetylaminogalactose sulfate).

The polysaccharide component of the egg jelly in several species of sea urchins is polyfucose sulfate, polygalactose sulfate, or a mixture of the two (Vasseur, 1952).

The jelly substances of algal tissues are polygalactose sulfates or polyfucose sulfates (Chapter 3).

In general, the extraneous coat of greatest significance in the constitution of multicellular forms, both plant and animal, is the intercellular cement. Among plants this material is a pectate (a polysaccharide), which binds together contiguous cellulose walls and constitutes the middle lamella of Payen (Figure 3.4). Among animal cells the cement is a protein, or more probably a polysaccharide-protein complex. The intercellular cement material forms very early over the protoplasmic surface of recently divided plant cells or of egg cells shortly after insemination.

The cement substances, in both animals and plants, are stiffened in the presence of calcium salts, while they are weakened and dispersed by sodium or potassium salts. The cement substances occur as organic anions, as proteinates or pectates in their natural aqueous environment, the reaction of which is generally to the alkaline side of neutrality. With calcium present in the environment, the pectates and proteinates combine with calcium to form undissociated salts which are sufficiently stiff to bind cells together. If calcium is removed from the medium,[5] sodium largely replaces the calcium so that the intercellular cement is transformed into the dissociated and soluble sodium proteinates or pectates. The tendency then is for the cells to fall apart, and the multicellular tissues disintegrate into their constituent cells. These continue living, but as isolated cells.

Actually, under natural conditions, with sodium, potassium, magnesium, and calcium in the environment, the cement substance exists in the gel state in combination with all four cations, all of which contribute to the specific physical properties of the cement material. The predominant cations in the complex are the divalent ones, especially calcium, because of their tendency to form undissociated compounds.

The value of calcium as a means for maintaining an intercellular cement substance was indicated by Ringer (1890), who made the statement that bicarbonate of lime offsets the weakening effect of distilled water and monovalent salt solutions on the cement substance which binds together plant and animal cells. His observations were made chiefly on tadpoles and on *Laminaria,* a marine alga; the binding material of the latter he regarded to be bassoria or tragacanth. Wille (1897), also working on *Laminaria,* was con-

[5] Calcium may be very firmly bound in the calcium-cement complex, with the result that the complete absence of calcium from the external medium may be insufficient to bring about dissociation. In these instances, a powerful chelating agent such as versene (which displaces calcium from calcium oxalate) may be necessary, or even treatment with fairly strong acids in the case of calcium pectate.

vinced that the main function of the calcium in the tissue is to maintain the intercellular cement. Herbst (1900) found this to be true for the blastomeres of sea urchin eggs and he was able to disintegrate and then to reconstitute the advanced blastula by varying the calcium content of the sea water in which the embryos were developing. Overton (1904) regarded the intercellular substance as an organic salt which varied in its consistency with variations in the proportions of sodium and of calcium in the medium. A similar conclusion was drawn by Lillie (1906) from his findings that the coherence of the ciliated cells in the gill epithelium of *Mytilus* is lost in isosmotic solutions of sodium salts and that the loss is prevented by the addition of calcium to the solution. Lillie concluded that when the cells were exposed to a solution of sodium chloride alone the organic calcium compound which constitutes the intercellular cement underwent dissociation, with the formation of its soluble sodium salt. Hansteen (1910) made a detailed study of the chemical reactions of salts, especially of calcium, on the pectates which serve as cell-binding material in plants; a discussion of this subject was also made by True (1922).

Gray (1926) drew attention to the stabilizing effect of the magnesium ion on the cement material of the gill epithelium of *Mytilus*. Complete stability, however, required the additional presence of calcium.

The extraneous coatings are generally secretion products of the underlying protoplasm. The coat which lies immediately adjacent to the protoplasm, if removed, is frequently replaced. For example, the blastomeres of marine echinoderm ova which have been denuded of their coatings either mechanically or by exposure to monovalent salt solutions isosmotic with sea water, acquire new hyaline layer, or cement material, when returned to sea water with its full complement of salts. An analogous phenomenon is seen among plant cells. In an hypertonic environment their protoplasts can be made to shrink from their stiff walls of cellulose. The surface of the shrunken protoplast subsequently becomes coated with a new wall of cellulose.

METHODS OF INVESTIGATION IN LIVING CELLS

By successively removing one coat after another, the investigator can ascertain the role each plays by determining in what manner the absence of each coat affects the functional capacity of the cell. He can also investigate the capacity of the denuded cell to replace the removed coatings, since as long as the protoplasmic surface film is intact the cell continues to carry on its vital processes. Furthermore, a great deal can be learned concerning the nature of the coats by ascertaining what kinds of agents cause their dispersal. The investigator who is concerned with the properties of the protoplasmic surface is faced with the problem of differentiating the protoplasmic surface from the extraneous coats, and of removing these coats. We are, therefore, concerned with methods which can be applied to the removal of extraneous coatings from living cells.

Frequently, the extraneous coats are too thin and too closely applied to the protoplasmic surface film to be seen in the living cell. We can, however, establish the presence or absence of such delicate coatings by testing the ease with which cells coalesce with oil drops. This is the method of oil-coalescence. Important information regarding the nature of these extraneous coatings has been obtained by washing cells with various dispersing agents and subsequently determining the ease with which the cells coalesce with oil droplets, as a measure of the extent to which the treatment weakened or removed the tenuous coatings.

Methods for removing coats.

Three separate techniques have been employed to remove extraneous coatings and expose the actual protoplasmic surface layer of sea urchin eggs: the mechanical, to remove coatings which are sufficiently brittle and loosely stretched to be broken off by shaking; the biological, in which, for example, the act of fertilizing an egg induces the lifting of the vitelline membrane, which can then be removed by shaking; and the chemical, involving the use of enzymes or of calcium- and magnesium-free media effective in removing the intercellular cements. A combination of the three techniques is sometimes necessary to ensure removal of the extraneous coatings so as to expose the more truly protoplasmic surface.

A number of proteolytic enzymes, such as papain, the digestive juices of various invertebrates, and especially trypsin (pH optimum slightly to the alkaline side of neutrality), are important agents for weakening or digesting membranes of various kinds.[6] Polysaccharide-splitting enzymes, such as pectinases and pectases, break down the pectic compounds which bind together plant cells. Other polysaccharide-depolymerizing enzymes (for example, hyaluronidase) and polysaccharide sulfatases have not, as yet, been found

[6] The dissociation of cells in tissues by the action of trypsin (Moscona, 1952) involves not only the weakening of intercellular cementing substances but also the digestion of collagenous and reticular fibers and the constituents of basement membranes.

to affect the cementing substances of animal or plant tissues. (We do *not* include here the ground substances of connective tissues and basement membranes.)

Washing unfertilized sea urchin eggs in a solution of isosmotic urea causes the dispersal of their vitelline membranes. This agent also causes the solution of the elevating fertilization membranes of newly fertilized eggs. The elevating membrane, since it has not yet hardened, is susceptible not only to dispersal by urea, but also to mechanical rupture and removal by shaking.

Monovalent salts and weakly acidic solutions (to pH 4)[7] disperse cementing substances, such as the hyaline layer material, by replacing the calcium ion. In cases where calcium is firmly bound in an undissociated calcium cement complex, calcium-binding agents (oxalate; citrate; and especially the powerful chelating agent, versene, see footnote 5) have proved extremely useful in causing solubilization. Even after the extraneous coatings have been fully removed, the surface of the cell will remain in a denuded state only in media which lack divalent cations. Such media ensure the dispersal of the extraneous material which is continually being generated at the protoplasmic surface. Since many of the experimental results discussed in this book depend upon the use of denuded cells, we describe in the following paragraphs the media which are best suited for maintaining the fertilized eggs of *Arbacia punctulata,* as an illustrative example, in a denuded state while at the same time permitting their continued development.

> After the fertilization membranes have been removed by shaking the *Arbacia* eggs strongly (for 30 seconds) during membrane elevation (1 to 2 minutes after insemination), the eggs are washed free of sea water and immersed in a monovalent salt solution. The monovalent salt solution best tolerated by *Arbacia* eggs is an isosmotic mixture of soduim chloride and potassium chloride in proportion of 19 to 1, at pH 6 or 7. In this solution the naked eggs continue to cleave, in percentages closely approaching those of controls in sea water, at the normal rate for many generations; the resulting blastomeres fall apart and continue to exist as isolated living cells. The higher proportion of potassium in this mixture than in sea water is required in order to adequately antagonize the toxic effects of sodium chloride in the absence

[7] High alkalinity of the external medium promotes the dissociation of amphibian blastulae and gastrulae (Holtfreter, 1943a, b). This is due to the weakening or dispersive action of the hydroxyl ion on an external coat to which the embryonic cells normally adhere. Alkaline solutions of monovalent salts cause the dissociation of the gill epithelium of *Mytilus* (Gray, 1926).

of divalent ions. The more acid pH of the monovalent salt mixtures than of sea water is essential, since the alkaline pH (8.2) of sea water is toxic in the absence of the calcium ion.

It is advantageous to add a small amount of isosmotic sodium oxalate to the chloride solutions, since this removes traces of calcium and eliminates the necessity of much washing of the eggs. Actually, this addition is beneficial, since of the two anions, the oxalate ion is less toxic than the chloride alone.

Another useful monovalent salt solution is isosmotic potassium chloride, also at pH 6. When this is used, *A. punctulata* eggs must be transferred from sea water to this solution from 3 to 10 minutes after insemination. Under these conditions 90 to 95 percent of the eggs pass through the early cleavage stages almost synchronously with the control eggs in sea water. The eggs cannot be transferred from sea water to the potassium chloride at any later stages, since this results in the blockage of further development (E. L. Chambers and R. Chambers, 1938, 1949).

Judging from oil-coalescence data (see below), the above described procedure achieves as complete a removal of extraneous materials from the protoplasmic surface as it has yet been possible to achieve.

The oil-coalescence method

The oil-coalescence method[8] is an extremely delicate and objective quantitative method for establishing the presence or absence of delicate, tenuous coatings on the surface of living cells. It is particularly useful in detecting the appearance of an extraneous coating on a previously denuded surface or in detecting the dispersion of a coating in response to various experimental procedures. This method involves examining the ease with which an oil drop, exuding from the tip of a micropipette, coalesces with a cell when the droplet is brought into contact with the surface of the cell.

If an extraneous coat is absent or extremely tenuous, a small droplet of an oil with a relatively low surface tension will be engulfed by the cell. As the coating overlying the protoplasmic surface becomes thicker, the size of the droplet must be increased or the surface tension of the oil increased, or both, to achieve coalescence. Finally, when the extraneous coat reaches a certain limiting thickness or strength, no oil-coalescence can be induced, even with large droplets of an oil with a high surface tension. In

[8] Chambers and Kopac (1937a); Kopac and Chambers (1937); Kopac (1940).

the case of the *Arbacia* egg, it has been found that if coalescence occurs at all, the cell always incorporates the oil. This reaction is rapid, taking less time than the interval between two successive frames of a motion picture film at 72 frames per second (Figure 8.5).

The cell and the oil drop behave similarly to two liquid drops in contact. Such a system may or may not be stable, depending on, among other things, the magnitude of the tension and the diameter of the oil drop. If the system is unstable, the two drops will spontaneously coalesce and produce a stable equilibrium.

The tendencies toward coalescence are proportional to the tension at the interface of the oil–aqueous phase and to the diameter of the freshly applied oil drop. The importance of tension and the size (surface area) of the oil drop indicate reactions involving changes in surface energies. Accordingly, coalescence can occur only when the potential energy of the cell with the oil drop inside is lower than with the oil drop outside. Ideal liquid drops coalesce spontaneously with only infinitesimal differences in the potential surface energies between the two. The cell, however, rarely approximates the behavior of an ideal liquid drop. In order to achieve coalescence, the potential energy difference between the two systems, oil drop in contact with the cell and oil drop inside, must frequently be very great. This is due principally to the presence of an extraneous coating at the surface, which acts as a barrier to oil-coalescence. The potential energy difference required to overcome this barrier to coalescence is suitably referred to as the potential hill. The potential hill can be calculated from the surface tension of the oil and the diameter of the smallest oil droplet which will coalesce with the cell. The thicker or the stronger the extraneous coat, the greater the barrier, or the potential hill, which must be overcome to achieve coalescence. These considerations explain why, if the same oil is used throughout a series of measurements, the more developed the extraneous coat, the greater must be the size of the oil drop to achieve coalescence (Table 2.1). The procedure followed in estimating the ease with which a cell coalesces with oil is as follows.

> The tip of a micropipette previously filled with oil of known tension is placed a short distance from the surface of the cell, which is suspended in an hanging drop in view under the microscope. In the case of the unfertilized, mature *Arbacia* ovum, which averages 70 to 75 micra in diameter, the tip of the micropipette is usually placed 4 to 5 micra from the cell surface. Pressure is then exerted by means of the injection apparatus so that oil exudes to form a small

drop, which enlarges until its surface touches the surface of the cell. The oil drop, now in contact with the surface of the cell, is expanded at a controlled rate until coalescence occurs. At the instant of coalescence, the size of the drop is measured. The contact with the cell, occurring while the oil drop is expanding, ensures the presentation of an uncontaminated surface of the oil to the cell. This is essential, since most oils develop an adsorption film from impurities present in the external medium, and this prevents coalescence.

Application of the oil-coalescence method

Important information regarding the new formation of coats on the developing cell and their physical and chemical properties may be obtained by combining the methods for removal of one or more coats with the oil-coalescence method. The usefulness of these methods is illustrated in the following studies on unfertilized and fertilized *Arbacia* eggs.[8]

Throughout this series of measurements the same oil was used. The control eggs were unfertilized *Arbacia* eggs, washed and cen-

TABLE 2.1. Diameter of smallest drops of cotton seed oil which will coalesce with unfertilized and fertilized *Arbacia* eggs*

	Diameter of drop
Unfertilized, jelly-free eggs, controls, vitelline membranes intact	
1 to 2 hours in sea water	18 to 19
Fertilized eggs without fertilization membranes, in sea water:	
within 2 minutes after insemination	5
10 to 13 minutes after insemination	8
15 minutes after insemination	25
over 15 minutes after insemination	No coalescence
Fertilized eggs without fertilization membranes, immersed in KCl, pH 6.0, at 5 minutes after insemination, left in KCl 30 minutes	1 and less
Unfertilized jelly-free eggs, same as controls, but kept	
12 hours in sea water	9
20 hours in sea water	2

* Adapted from Kopac (1940). The diameters are in arbitrary units, each unit being equal to approximately 5 micra.

trifuged in isosmotic sodium chloride and then returned to sea water. This treatment removed the jelly coats, but not the vitelline membranes. The average diameter of the oil drop at the instant of coalescence was then determined (Table 2.1) as described above for comparison with the treated eggs.[9] Similar measurements were made on eggs in sea water within 3 minutes after fertilization, the fertilization membranes having been removed by shaking immediately after insemination. These newly fertilized eggs coalesced with oil droplets less than one third the diameter of those that would coalesce with the unfertilized eggs (Table 2.1). This indicates that a barrier to oil-coalescence had been eliminated following the fertilization process. Undoubtedly, the vitelline membrane was the barrier which had been removed. Subsequently (more than 3 minutes after fertilization), the ability of the fertilized eggs (without fertilization membranes) to coalesce with oil droplets diminished until, by 15 minutes after insemination, the eggs would no longer coalesce with oil drops (Table 2.1). The decreasing ability to coalesce (as indicated by the increasing size of the droplets needed to achieve coalescence) corresponded closely with the formation and gradual thickening of the hyaline layer on the surface of the egg in sea water. By 6 to 8 minutes after insemination, this layer had already attained a width of about 1 micron.

However, if fertilized eggs without their fertilization membranes were placed in an isosmotic solution of potassium chloride (see above, under "Methods for Removing Coats") instead of in sea water, the ability to coalesce rose greatly, as shown by the very small size of the droplets which would coalesce, in comparison with the control unfertilized eggs possessing vitelline membranes. The great rise in ability to coalesce was undoubtedly due to the fact that the hyaline layer dispersed in the monovalent salt solution.

In another series of experiments aging was shown to have a decided effect on the ability of unfertilized, jelly-free eggs in sea water to coalesce with oil. After standing 8 to 12 hours, the aged eggs would coalesce with oil droplets one half the diameter of those that were engulfed by the unfertilized controls. For eggs remaining in sea water for 20 hours, the values obtained indicated a coalescing ability which approached that of membraneless fertilized eggs in solutions of potassium chloride. This indicates that aging causes the gradual disintegration of the vitelline membrane.

The coalescing ability of fresh, unfertilized eggs also increases

[9] More useful comparisons involve expressing the data in terms of the potential hills required to achieve coalescence as in Kopac's (1940) original work. The magnitude of the calculated potential hill varies directly with the square of the diameter of the oil drop at the instant of coalescence.

markedly when they have been immersed for about 2 minutes in a 1.0 M urea solution and when the eggs are churned with microneedles. Urea disperses, while churning breaks up, the vitelline membrane.

The changes in the oil-coalescing ability of the *Valonia* aplanospore during new formation of an extraneous coat over an initially naked protoplasmic surface is described in the next chapter.

The greatest tendency to coalesce with oil droplets indicates the minimum of extraneous coatings. Fertilized *Arbacia* eggs, with their fertilization membranes removed, immersed in monovalent salt solutions shortly after fertilization, and the newly formed aplanospore of *Valonia* offer the closest approach thus far to exposure of the actual protoplasmic surface layer of cells.[10]

[10] An interesting study would be to correlate the oil-coalescence data on denuded cells with the electron microscopic picture of the surface of these cells.

3

Extraneous Coats of Plant Cells

In contrast to our fragmentary knowledge concerning the structure and composition of the extraneous coatings of animal cells, we know a great deal about the fine structure and chemistry of these coatings in plant cells. We shall first describe, as an illustrative example, the structure of the cell wall of *Valonia*. We shall then proceed with a consideration of the structural features of the walls of plant cells generally. Finally, the properties of the cementing materials found in plant cells will be described.

CELL WALL OF VALONIA

As an illustrative example of the major structural features of plant cell walls, we shall first examine the wall of the green marine alga, *Valonia ventricosa*. When mature, this interesting plant consists largely of an almost spherical vesicle whose size varies from that of a pea to that of a small hen's egg. The vesicle is a single coenocytic cell enveloped by a relatively thick wall composed of cellulose and amorphous, probably pectic, material. Covering the inner surface of the wall is a very thin layer of multinucleated protoplasm, which encloses the enormous central vacuole.

The mature cell wall

Examination of the wall by electron microscopy[1] reveals that it consists of cellulose fibrils of uniform thickness laid down in parallel alignment, in successive lamellae (Figure 3.1). An amorphous material fills the interstices between the cellulose fibrils. The direction of orientation of the fibrils, while consistent in each lamella,

[1] Preston (1952); Steward and Mühlethaler (1953). The untreated wall shows little structure. The cellulose framework can be observed only after removing the amorphous component by boiling the walls in acid and alkali and peeling off one or several lamellae as extremely thin sheets. Cross sections of the whole wall are obtained after embedding in methacrylate.

FIGURE 3.1. Electron photomicrographs of mature wall of *Valonia ventricosa*. (a) Surface view of extremely thin, stripped-off portion of wall, 10,400×, showing three successive lamellae with directions at approximately 120°; mucilaginous material incompletely removed. (b) Cross section through complete wall, 5,200×, showing many lamellae with fibrillar directions tending to be repeated in every fourth lamella; mucilaginous material removed. (From Steward and Mühlethaler, 1953.)

shifts abruptly from one lamella to the next through 120°. The successive shifts in direction are achieved with such accuracy that through the entire thickness of the wall, containing 40 to 50 lamellae, the fibrils in every fourth lamella run parallel to each other. The three orientations of the cellulose fibrils are inherent in the structural symmetry of the cell as a whole. This is shown by the fact that two poles of symmetry exist in each cell, located opposite to each other,[2] with the three fibril systems converging at each pole.

Formation of a new cell wall

When the turgid wall of a mature vesicle is punctured (Kopac, 1937, 1940), the multinucleated protoplast reacts by fragmenting into numerous mono- or polynucleated ameboid bodies, the aplanospores. Each aplanospore subsequently develops into a new adult. The first step in this development is the formation of a new cell wall on the initially naked protoplasmic surface of the aplanospore. The nakedness of the protoplasmic surface of the aplanospores, when first formed in sea water, is revealed by the readiness with which they fuse with oil drops (Kopac, 1937, 1940; Chapter 8). Their oil-coalescing ability at this early stage approximates that

[2] Preston (1952); Wilson (1955).

of truly naked sea urchin eggs in monovalent salt solutions (page 48). The tendency of the aplanospores to coalesce with oil drops, however, rapidly decreases, until by 4 hours after the initiation of their formation, coalescence could not be induced. Evidently during the intermediate period of decreasing ability to coalesce with oil droplets, the aplanospore built up at its surface a soft mucilaginous material or a friable cellulose net, which could still be swept aside by the penetrating oil droplet. Finally, a definitive cell wall formed, which completely blocked oil-coalescence.

The earliest stage studied by electron microscopy (Steward and Mühlethaler, 1953) is that 15 hours after the initiation of aplanospore formation, when the spore is already enclosed in a well-defined wall. At this time the wall consists of a loose network of interlaced cellulose fibrils (Figure 3.2a) embedded in an abundant amorphous network. Shortly after this stage, a lamella of parallel fibrils is laid down on the inner surface of the first formed random network (Figure 3.2b). This pattern is then followed in all subsequent lamellae, as described for the mature vesicle. By 5 days after the initiation of aplanospore formation, the wall of the young vesicle consists of the original net-like layer and 4 to 6 lamellae of parallel fibers. Amorphous material is interposed not only between the cellulose threads but also, in the early stages, between neighboring lamellae.

An interesting feature is that during this early period, as the wall thickens, the aplanospore is also rapidly expanding in size. Yet cellulose fibers themselves are almost inextensible. Steward and Mühlethaler observed tears in the lamellae where the fibrils had pulled apart (Figure 3.2b). They concluded that the process of expansion occurs largely by an irregular tearing of the previously deposited outer lamellae, while new intact layers are being deposited on the inner wall surface. In addition, the abundant amorphous material in the young wall would permit slipping and readjustment of the inelastic wall components.

GENERAL CHARACTERISTICS OF PLANT CELL WALLS

The structural framework of the walls of most plants consists of a matting of long cellulose fibrils.[3] These fibrils originally form in the cytoplasm and, being relegated to the periphery of the protoplast, become so closely adjacent as finally to form a stiff wall.

The early formed wall of the plant cell (the primary wall of the

[3] The cell walls of many fungi consist largely of chitin with a similar type of structure, except that the basic unit of the macromolecule is not β-glucose, as in cellulose, but a derivative, acetylglucosamine. Chitin is also the principal component of the extraneous coats of many animal cells, such as the chorion of the *Ascaris* egg.

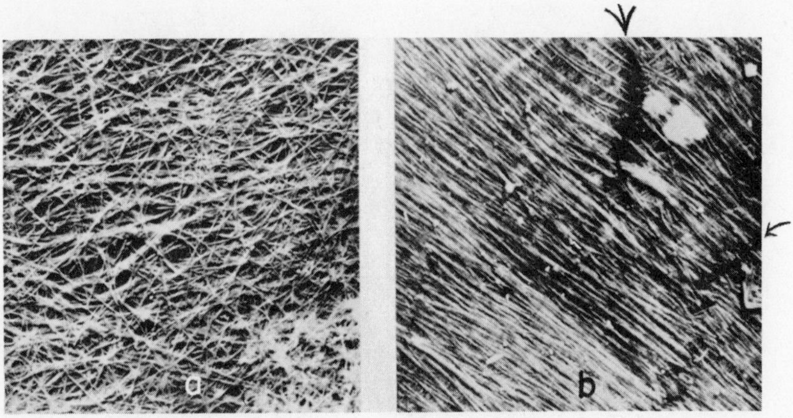

FIGURE 3.2. Electron photomicrographs of surface of wall of aplanospores of *Valonia ventricosa,* 7,200×; amorphorus materials removed. (a) 15-hour sporeling, showing tangled weft of cellulose threads. (b) Older sporeling, parallel arrangement of fibrils well established. Arrows point to tears in wall resulting from tension due to expansion. (From Steward and Mühlethaler, 1953.)

higher plants, Figure 3.4), laid down during cell division and while the cell continues to grow in size, is characterized by a high content of pectic substances, a high state of hydration, a relatively low content of cellulose, and the presence of proteins. In the initial stages of growth the cellulose fibrils are laid down in a relatively loose, interwoven network (Figure 3.3),[4] as described above for the early formed wall of the *Valonia* aplanospore. As growth and enlargement continue successive layers of fibrils are deposited in an oriented parallel pattern. At this early stage, however, the extraordinary regularity of arrangement seen in the later formed lamellae of the *Valonia* cell wall is unusual.

Striking properties of the young plant cell wall are, first, its ability to increase in area during enlargement of the growing protoplast, and, second, its ability to reversibly expand and contract (as much as 30 percent in surface area) without wrinkling during turgor changes. During growth the expanding young cell wall thickens, or at least does not become appreciably thinner, indicating that the synthesis of new cell wall materials keeps pace with the enlargement of the protoplast. In the early phases of cell wall growth, in the interwoven mesh stage, an organized insertion of new materials must occur throughout the volume of the wall,

[4] Mühlethaler (1950); Scott, Hamner, Baker, and Bowler (1956); Frey-Wyssling and Müller (1957).

FIGURE 3.3. Electron photomicrograph of cell wall of very young parenchyma cell; pectic materials removed. Cellulose fibrils form a dispersed network. (From Frey-Wyssling and Müller, 1957.)

while during later phases of growth deposition of cellulose fibers occurs largely at the inner surface of the wall.

The plastic and elastic properties of the young, expanding cell wall are undoubtedly related to its high content of pectic substances, the latter having the property of forming gels. Embedded in this pectic matrix, the cellulose fibrils themselves are rigid, relatively inelastic strands. The ability of the young wall to expand indicates that the constituent cellulose fibrils, embedded in the pectic matrix, must undergo separation. This could occur (1) by a slipping of fibers in an interwoven mesh past one another; (2) by a lateral separation of fibers oriented parallel to each other; and (3) by a rupture of the cellulose fibrils when these are long[5] or continuous, as described for the expanding wall of the *Valonia* aplanospore.

Conceivably, expansion of the wall during growth could occur by a process of simultaneous reabsorption and replacement of cellulose fibrils. We have no indication that a mechanism of this type is involved, but a limited reabsorption to loosen the cellulose fibrils

[5] O'Kelley and Carr (1954) present electron microscopic evidence that in the initially formed walls of pollen tubes the cellulose may occur as short rodlets.

in previously deposited lamellae might be a necessary preliminary to facilitate expansion.

As the rapid phase of cell enlargement comes to a close, thickening of the wall continues, with the laying down of successive lamellae inside the original wall. This later formed part consists principally of cellulose containing little if any pectic material. The cellulose fibrils are closely packed and oriented with great regularity parallel to each other in any given lamella, as we have already seen in the wall of *Valonia*.

In many plant cells enlargement together with thickening of the wall may continue over long periods of time, as in many of the coenocytic species (for example, *Valonia* and *Nitella*). Here no abrupt transition exists between the younger and older parts of the cell walls.

Most of the cells in the flowering plants, however, especially those composing the supporting tissues, undergo two relatively distinct growth phases: an early period of rapid increase in size when the wall thickness does not increase markedly, followed by a later phase of extensive wall thickening with no further expansion of the cell. The parts of the wall formed during each of the two growth phases are here distinguishable as the outermost, thin primary wall formed during the period of size increase, and the inner, later formed, thick secondary wall (Figure 3.4). The primary wall has the characteristics already described for the early formed portion, and the entire thickness of the secondary wall has the structure described for the later formed portion. Pectic compounds are said to be lacking in the secondary wall. Although a switching in the direction of parallel fibrils occurs in the deposited lamellae of the secondary walls, this does not take place with the frequency described for *Valonia*.

The successively deposited secondary lamellae may constitute a relatively thin investment surrounding the cell or a layer of considerable thickness, as in the sclerenchyma; or they may become so thick as to largely replace the cell contents, as in adult cotton fibers. Root hairs, the function of which is to absorb water and salts from the soil, have especially thin walls (Figure 13.4). These resemble primary cell walls in being plastic and in having a high content of pectin and a thin matting of cellulose fibrils (Frey-Wyssling, 1952).

An intimate relationship exists between the surface of the protoplast and the cellulose wall to the extent that the cytoplasm probably permeates between the cellulose fibrils of the wall, at least in the expanding, early formed walls of young plant cells and in the newly formed lamellae of the secondary walls of adult cells. This becomes evident when attempts are made to plasmolyze cells. When a young, growing plant cell is immersed in a concentrated solution of a nonpenetrating sugar, as the protoplast shrinks in size the cell wall is pulled inward and crumbles. No separation occurs between the protoplasmic surface and the wall, due, evidently, to the intimacy of the union between the two. In the older cells, with rigid cell walls, this phenomenon of plasmolysis occurs in either of two ways: the protoplast, in the process of pulling away from the cellulose wall, disintegrates, while the central vacuolar membrane[6] retains its integrity and shrinks in size; or, in other cases, the protoplast itself remains viable and is pulled away from the unyielding cellulose wall as the central vacuole shrinks. What happens in both these types of plasmolysis is that the external portion of the cytoplasm is ripped off. In the first instance repair of the surface does not occur sufficiently rapidly, while in the second a new surface film instantly sweeps around the protoplast, which thus maintains its integrity.

Further indication of the intimacy of union between the protoplast and the cell wall is the fact that the protoplast cannot be drawn away from the cell wall when the fluid contents of the sap vacuole are withdrawn by means of a micropipette inserted into the central vacuole (Martens and Chambers, 1932). When vacuolar fluid is withdrawn from a *Tradescantia* stamen hair, for example, the vacuole shrinks in size while the protoplasm occupies increasingly more space. Instead of a separation developing between the rigid cell wall and the protoplast, water is drawn into the cytoplasm from the external environment to make up for the decreased vacuolar volume. When the suction is released, the protoplast shrinks back to its normal state of being a thin layer of protoplasm surrounding its vacuole.

These experiments on plant cells offer strong evidence that the cell wall is an integral part of the protoplast.

Further evidence for the intimate relation between the protoplast and its cell wall is the interesting fact that once the protoplast has become separated from its original cell wall by plasmolysis, even after deplasmolysis, reintegration of the two components does not occur. The protoplast remains separated, and a new cellulose wall is formed under the old. An example of this is seen in certain red algae which undergo recurring plasmolysis and deplasmolysis

[6] The isolated sap vacuole with its continuous vacuolar membrane, the tonoplast, maintains its differential permeability and will shrink or enlarge according to the concentration of the external sugar solution (Chambers and Höfler, 1931).

in sea water tide pools on rocks which become exposed between tides. Shifts in tonicity of the sea water in these pools occur at every turn of the tide, being specially marked during the summer when there is more sunlight. The protoplast of plasmolysis generates a new cellulose wall each time deplasmolysis occurs. In consequence, some of the older algae can be found choked with many successively formed cellulose walls which have accumulated with time in a single cell.

Quite apart from the behavior of plant cells when plasmolyzed, evidence from a number of sources indicates that at least the growing wall of many species of cells is an integral part of the protoplast,[7] with active structural change and deposition of new materials occurring throughout the volume of the wall. It is difficult, for example, to conceive how the cellulose threads in the interwoven matting of an early formed cell wall could be twisted about one another except by being deposited within the cytoplasmic cortical zone.

INTERCELLULAR CEMENT SUBSTANCES

The intercellular binding materials in plants are polysaccharides. In all the higher plants and in many of the lower the pectic compounds constitute a continuous cementing matrix which intervenes between and pervades the outermost (initially formed, or primary) parts of the walls of adjacent cells (Figure 3.4).

This distribution has been determined largely on the basis of the selective extraction of pectic substances from tissue sections by treating them with acids and soaking them in solutions of calcium-binding agents (ammonium oxalate, sodium versenate, etc.), combined with staining procedures. The latter depend on the ready combination of basic dyes[8] with the free carboxyl groups of pectic substances. When the pectic materials have been extracted, only the cellulose framework of the individual cell walls remains intact, and the tissue falls apart.[9] On the other hand, when cellulose is extracted from plant tissues, there remains a coherent tissue consisting of the pectic substances of the primary walls and the intercellular regions. Young and growing tissues may contain

[7] Frey-Wyssling (1952); Preston (1952); Green and Chapman (1955). Cytoplasm has been described within the cell walls of *Valonia* in studies using electron microscopy, and the high protein content of primary walls may be due to their permeation by cytoplasm (Preston, 1952). The electrical measurements of Bennett and Rideal (1954) also indicate that cytoplasm pervades the interstices of the cellulose wall in *Nitella*.

[8] For example, ruthenium red (Mangin, 1892, 1893).

[9] The softening of over-ripened fruits is due to loosening of their cells, and this occurs coincidentally with a decrease in content and an increase in solubility of the pectic materials (Carré, 1922).

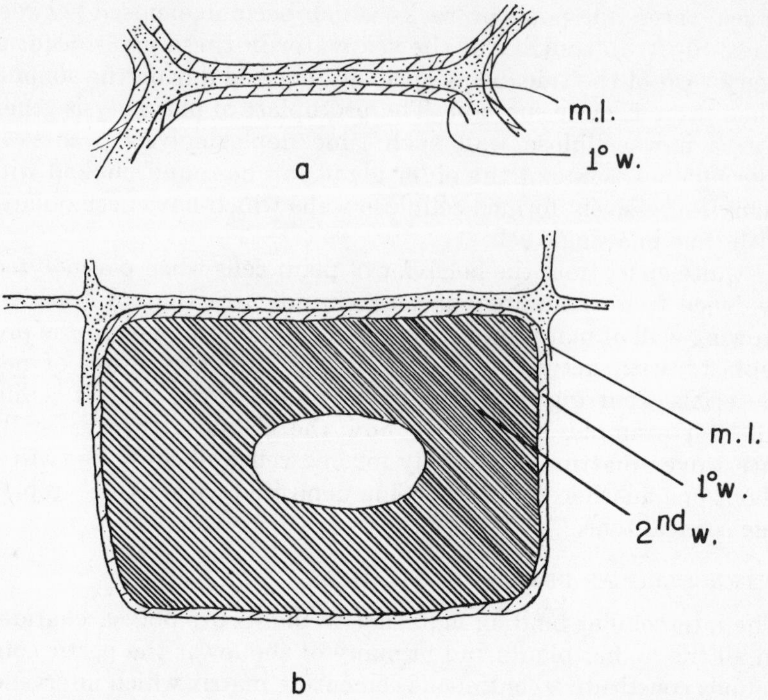

FIGURE 3.4. Diagrams showing transverse sections of cell walls of flowering plants. (a) Wall of young growing cell, with adjacent cell walls. (b) Wall of mature tracheid (adapted from Kerr and Bailey, 1934). Middle lamella (m. l.), primary wall (1° w.), secondary wall (2nd w.). Stippling shows distribution of pectic compounds; diagonal hatching shows distribution of cellulose; relative amounts indicated by density of stippling or hatching. Thickness of middle lamella and primary wall exaggerated.

large amounts of pectic substances (up to 10 percent), while in adult and especially in woody tissues the amount of pectic material may be extremely low (0.5 percent and less). This, however, reflects only the vast increase in amount of cellulose and the very small space occupied by the intercellular regions and primary walls in adult tissues.

The pectic compounds are polysaccharides, chiefly found in the walls of higher plants. The pectic acid molecule is a long chain polymer (Owens and others, 1946) of D-galacturonic acid units linked together in α 1, 4 glycosidic linkages. The aldehyde groups, therefore, are securely blocked, and each unit presents one free carboxyl group and two free —OH groups.

In pectinic acid a substantial proportion of the carboxyl groups of the galacturonic acid units is methylated. Pectin is simply pectinic acid with a high proportion of methylated carboxyl groups. Totally methylated pectin does not occur in nature (Deuel and Solms, 1954).

As extracted from plant tissues, the polygalacturonic acid chains and their methyl esters, although predominant in amount, are found closely associated in an as yet unknown manner with an arabosan and galactosan.

In nature the pectinic and pectic acids occur almost entirely as salts (pectinates and pectates). In accordance with the dissociation "constant" of the —COOH groups of polygalacturonic acids (Deuel and Solms, 1954), at the usual neutral or slightly alkaline environmental pHs, the free carboxyl groups of the galacturonic acid units are dissociated as sodium and potassium salts or undissociated as the calcium complex.

The characteristic features of the undegraded pectic compounds are their high molecular weights, the high viscosity of their colloidal aqueous solutions, and their ability to form gels of varying degrees of rigidity. Gels of the highly methylated pectinic acids (pectin) tend to be relatively soft. Here salt formation plays an unimportant role due to the paucity of free carboxyl groups. The free pectinic acids with low methoxyl content and the free pectic acids are insoluble in water, while as their sodium and potassium salts they form colloidal solutions. Gel formation or precipitation occurs on adding calcium. The divalent ion is believed to form cross linkages between carboxyl groups of adjacent chains. The rigidity of the gel formed can be controlled by varying the ratio of calcium and sodium salts, the greater proportion of sodium favoring a less rigid gel.

In young, growing tissues pectin is the principal pectic compound which fills the intercellular spaces and permeates the primary walls. The predominance of pectin, jelly-like in nature, while serving to hold the cells together in a continuous matrix, also permits the growing, enlarging, and differentiating cells to slip over one another. As the plant tissues approach maturity, the pectin becomes progressively demethylated (splitting of methyl groups from the esterifield carboxyl groups) and converted, therefore, to pectinic and pectic acids (see chemical discussion above). The free carboxyl groups then combine mainly with the calcium, and to some extent with the magnesium, present in the environmental waters, forming the undissociated calcium or calcium-magnesium pectates. Unlike pectin, calcium pectate has the properties of a rigid cementing material. Evidence that pectic acid occurs in the intercellular regions[10] as the calcium salt was provided by Molisch (1913), who treated sections of plant tissues with dilute sulfuric acid and observed the formation of calcium sulfate crystals in the intercellular regions. Furthermore, calcium pectate has been shown to have the same solubility characteristics as the intercellular material (Mangin, 1891–1893; Bonner, 1935).

The role of the calcium ion becomes evident when plants are grown in calcium-free water, especially at low pH. Decomposition of the root tips tends to occur, and separated cells, many of which remain living, may be found. If decomposition has not gone too far, adding calcium and raising the pH corrects the situation.

An interesting observation in relation to the value of calcium for plant growth was made by Reed (1907): Although mitotic figures in cells of the plants *Spirogyra* and *Zea* will form in calcium-free media, the division of the cells is frequently not completed because of their inability to form a new transverse wall. The partitioning of plant cells is achieved, as mitosis is completed, by the deposition of the cell plate in the mid-region of the spindle remnant. Presumably pectic material is an important constituent of the cell plate, although this has not, as yet, been proved.

An excess of calcium ions increases the rigidity of fresh potato and tomato tissues and retards or reverses the loosening of cells and softening of plant tissues which occur during attempts at their preservation. These effects, which occur because calcium converts the pectates to, or keeps them in, their insoluble form,[11] reveal the importance of calcium pectate in maintaining the rigidity of tissues.

While the cementing materials in the higher plants and in many of the lower plants are pectic compounds, among the brown marine algae the salts of alginic acid, especially calcium alginate, serve an analogous role. That this compound has the properties necessary to serve as a strong intercellular binding agent is indicated by the fact that fibers composed of calcium alginate possess considerable tensile strength. Both pectic and alginic acids are polymers of hexuronic acids and, because of their free carboxyl groups, form salts at neutral or alkaline pHs. Characteristically, their sodium and potassium salts are soluble, while their calcium salts are insoluble.

[10] In plant tissues of the woody variety the presence of pectic materials may be masked by heavy deposition of lignin in the intercellular spaces and primary walls. Lignification does not alter the content of pectic materials, which continue to serve in their important role as cementing substances. This can be readily demonstrated by extracting the lignin.

[11] Personius and Sharp (1939); Loconti and Kertesz (1941).

It is interesting that alginic acid is a structural counterpart of pectic acid. In alginic acid β-D mannuronic acid is the repeating unit in the linear chain, instead of α-D galacturonic acid. In both the units are coupled in 1, 4 glycosidic linkages. Like sodium pectate, sodium alginate is a polymer of high molecular weight, dissolving in water at neutral pH and forming solutions of characteristically high viscosities even in low concentration. When calcium is added, and as the concentration is raised, first gel formation occurs, followed by the precipitation of insoluble calcium alginate (Steiner and McNeely, 1954).

Free alginic acid is insoluble in water. Being a relatively strong acid (pK determined at approximately 2.0 to 3.0, Saric and Schofield, 1946), at the nearly neutral pH of fresh seaweed it occurs in the form of its salts, as alginates. The ratio of cations saturating the carboxyl groups depends on the concentration of cations in solution (Mongar and Wassermann, 1952). The specific affinity of algin for calcium is such that in a sea water environment it cannot exist entirely as the insoluble calcium salt, but as the insoluble mixed calcium-magnesium-sodium complex. As is to be expected, the complex is solubilized by calcium-binding agents.

The protoplasts in the tissue of the brown algae are surrounded by substantial cellulose walls, and these in turn are enclosed within an outer coat composed of a transparent material of variable thickness. The outer coats constitute a continuous matrix, composed of calcium alginate, in which the cells are firmly held (Figure 3.5a). When treated with calcium-binding agents, the binding material swells and dissolves, and the individual protoplasts, now without their outermost coats, but within their cellulose walls, are set free (Figure 3.5b).[12]

Polysaccharide sulfates are intercellular constituents in the red and brown marine algae.[13] Their precise role has not been determined, but the polygalactose sulfate esters, agar-agar and carrageenin, which occur in large amounts in certain species of the red algae, are the main constituents of a jelly-like intercellular matrix which is abundant in the medullary regions of the stems and thalli.

In the red alga, *Irideae laminaroides,* the polygalactose sulfate constitutes about 40 percent of the dry weight of this plant, and there is no evidence for the presence of cel-

[12] The transparent binding material is believed to be an alginate on the basis of its solubility, birefringency, and staining characteristics (Thiele and Andersen, 1955).

[13] Fucoidin, a polymer of L-fucose monosulfate, occurs in small amounts in the brown algae; its role is unknown.

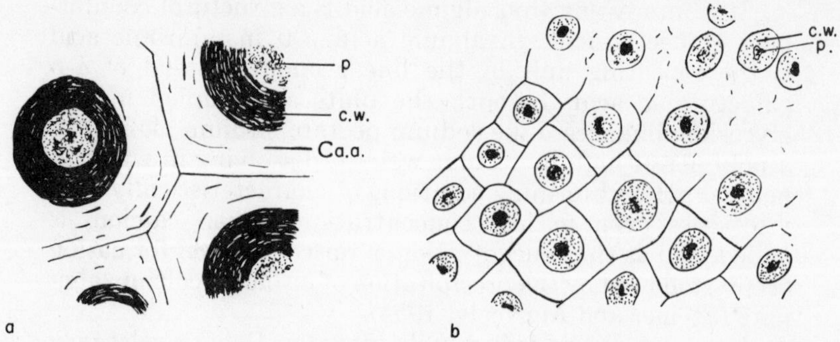

FIGURE 3.5. (a) Section through medullary region of *Laminaria,* showing protoplast (p.) surrounded by heavy cellulose wall (c. w.) and embedded in matrix composed of calcium alginate (Ca. a.). (b) Thin slice of the medulla immersed in sodium citrate. At one edge the protoplasts within their cellulose walls are set free, due to swelling and dispersion of alginate matrix. (From Thiele and Andersen, 1955.)

lulose or any other polysaccharide. An extremely slow rate of incorporation of labelled carbon dioxide into the polygalactose sulfate indicates that it is not a metabolic constituent, but rather a structural compound which, when formed, is not used by the plant (Hassid, 1936 and personal communication). The absence of cellulose in this species is an extraordinary feature, since the cells of algae are generally enclosed within a cellulose wall.

As extracted from the algae, these polysaccharide sulfates occur as the mixed sodium-potassium-magnesium-calcium salts. Most of the salt-forming groups (all in agar-agar and fucoidin) are sulfate groups, and these are highly ionized (Harwood, 1923). Accordingly, their calcium salts, at least at the salt concentrations prevalent in sea water, are largely dissociated.

The cementing materials in plants have generally been shown to be calcium-insoluble, sodium-soluble polysaccharide salts. The highly sulfated polysaccharides, as such, hardly fit into this category. Although their structural function may be that of providing a jelly matrix in which the cells are embedded, their physical properties alone are inconsistent with a functional role of binding cells into close-knit, highly organized units.

4

Extraneous Coats of Animal Cells

VITELLINE AND FERTILIZATION MEMBRANES OF THE ECHINODERM EGG

The vitelline membranes of unfertilized echinoderm ova vary from the tough and readily visible membrane of *Asterias* eggs to the exceedingly thin and tenuous membrane of *Arbacia* and other sea urchin eggs. The latter can be visualized only by lifting it off the surface of an uncentrifuged (or centrifuged) egg using microneedles, or inferred from oil-coalescence measurements.[1] We have already discussed in Chapter 2 (Table 2.1) the evidence from oil-coalescence data for an extraneous coat at the surface of the jelly-free, unfertilized egg. Following insemination elevation of the membrane starts at the site of sperm entry, from where the elevation sweeps over the surface of the egg. In the sea urchin (*Arbacia*) egg the elevation of the membrane is completed within 2 minutes. In the starfish (*Asterias*) egg the elevation is slower and continues progressively for a much longer time and to a greater extent.

That the vitelline membrane becomes transformed into the fertilization membrane is shown by the fact that if the vitelline membrane is removed from the unfertilized sea urchin egg, no fertilization membrane forms after sperm entry.[2] As the membrane rises it becomes increasingly prominent, and its physical properties, such as consistency and solubility, progressively change from those characteristic of the vitelline membrane to those of the fully formed

[1] Chambers (1921e); Hobson (1932); Chase (1935); Chambers and Kopac (1937a); Kopac (1940). Runnström and Monné (1945) state that the vitelline membrane can be readily demonstrated in oocytes by immersing them in hypertonic solutions, but in the mature egg the vitelline membrane adheres to the protoplasmic surface. Afzelius (1956), in electron microscopic studies, described the presence of a delicate vitelline membrane in the sea urchin egg.

[2] Chambers (1921e); Moore and Moore (1931). Electron microscopic studies by Afzelius (1956) indicate that the outer rim of the fertilization membrane is identical in structure with the vitelline membrane.

fertilization membrane (Table 4.1).[3] The fertilization membrane, in contrast to the vitelline or to the elevating membrane, is characterized by its toughness, greater thickness, and greater stability. The calcium ion plays a major role in this transformation, since if the egg is transferred to a calcium-free medium immediately after insemination, the membrane continues to elevate, but the change in physical properties to those characteristic of the fully elevated fertilization membrane does not take place, the membrane remaining soft and tenuous.

Loeb (1908) proposed that the rising of the fertilization membrane is due to secretion by the egg of a colloidal substance to which the membrane is impermeable and consequent intake of water into the space between the egg surface and the fertilization membrane. The lifting of the membrane, however, cannot be due to such mechanism, since Chambers (1942a) showed that the rising membrane can be torn with microneedles without stopping further expansion, as described below.

> While the membrane of an *Arbacia* egg is elevating, it is punctured and torn in several places. At first, the membrane collapses and wrinkles. Several minutes later the collapsed condition still persists, but the wrinkles are more pronounced and more distinct, indicating a progressive increase in surface area of the membrane.
>
> The increase in surface area of fragments of the fertilization membrane can also be demonstrated in the starfish egg by tearing off pieces of the membrane before its elevation has been completed. Carbon particles, which adhere to the torn off fragments, can be observed to move apart.

Furthermore, the progressive expansion of the fertilization membrane can occur independently of the presence of the egg within it. This can be shown by inserting a fine micropipette through the elevating fertilization membrane of an *Asterias* egg and sucking out the entire contents of the egg. After withdrawal of the shaft of the pipette, the diameter of the empty but still distended membrane increases progressively with time. Evidently the elevation of the membrane is due to the intrinsic expanding properties of the membrane material itself, and not to osmotic forces within the space beneath the membrane as postulated by Loeb.

Painstaking microscopic observations of eggs during the first 30 seconds after insemination have revealed that immediately preced-

[3] Hobson (1932); Kopac (1940).

TABLE 4.1. Properties of extraneous coats of mature sea urchin egg, and effects of various agents

	Vitelline membrane	Elevating membrane	Fertilization membrane	Hyaline plasma layer	Jelly coat
When present	Unfertilized egg	Fertilized egg, 15 sec. to 2 min. after fertilization	Fertilized egg (after elevation completed)	Fertilized egg, appears 2 to 3 min. after fertilization	Immature, mature unfertilized and fertilized eggs
Normal properties	Soft, plastic, film-like	Soft, pliable, thin	Tough, rigid, thick	Firm gel, outer part rigid	Weak gel
Effect of excess Ca in sea water	Stiffens, rigidifies	Stiffens	Stiffens, rigidifies	Stiffens, becomes inelastic[e, i]	Stabilizes
Effect of acid sea water, pH 4 to 5	Does not disperse[g]	(No data)	Does not disperse[g]	Disperses[e]	Disperses[d]
Effect of monovalent salts, pH 6	Does not disperse[f]	Partially disperses[b,‡]	Does not disperse[f]	Disperses[a]	Disperses[c]
Effect of non-electrolytes, 1 M, pH 8*	Disperses[f, j]	Disperses[h, j]	Does not disperse[f]	Disperses[h]†	Disperses
Effect of trypsin in sea water, pH 8	Disperses in low concentration[h,k]	Disperses in low concentration[h]	Does not disperse except in high concentration[h]	Disperses[h, l]	Disperses[m]

[a] Herbst (1900); [b] Hobson (1932), Kopac (1940); [c] Lillie (1921); [d] Loeb (1914); [e] Moore (1928); [f] Moore (1930); [g] Moore (1932); [h] Moore (1945, 1952); [i] Moore (1949); [j] Moore and Moore (1931); [k] Runnström (1948); [l] Runnström, Monné, and Broman (1944); [m] Tyler (1941).

* Acidity inhibits the dispersing effect of the non-electrolytes. A complicating feature is the non-electrolyte urea, which induces activation of the unfertilized egg, i.e., breakdown of the cortical granules and elevation of the membrane (Motomura, 1934; Moser, 1940).

† Formation of the hyaline layer is largely suppressed if prior to insemination the eggs are exposed to a solution of a non-electrolyte. The same result is obtained if fertilized eggs are immersed in a solution of a non-electrolyte for a minute or two immediately after insemination (Moore, 1930; Moore and Moore, 1931).

‡ Immersion of eggs in monovalent salt solutions prevents the transformation in physical properties to those characteristic of the fertilization membrane.

ing the rising of the membrane, the contents of specialized granules located in the cortex of the egg are ejected between the vitelline and plasma membranes (Figure 19.1).[4] The breakdown of the cortical granules begins at the point of sperm attachment and then spreads around the egg in a wave-like fashion. Evidently sperm attachment causes a propagated cortical response, which in turn induces breakdown of the cortical granules. In a normal environment the material ejected from the cortical granules adheres to the inner surface of the rising vitelline membrane and becomes incorporated with it to constitute the thick and ultimately tough fertilization membrane. The process has been depicted by Sugiyama (1956) as follows:

sperm attachment → propagated cortical response → breakdown of cortical granules → elevation of fertilization membrane

When calcium is lacking in the environment, the cortical granules can still be ejected but the material is not incorporated with the elevating membrane; instead it is dispersed in the perivitelline space.[5] This should account for the previously noted softness and tenuousness of the membrane when it elevates in the absence of calcium.

The intrinsic expansibility of the rising membrane could be explained by a change in configuration of the constituent molecules (for example, from a globular to a planar shape) or by an increased hydration. The factor which initiates this change in configuration could be the interaction between material liberated from the cortical granules and the vitelline membrane itself. The role of the calcium ion is to achieve insolubilization of the cortical granule material and its incorporation with the vitelline membrane to form the stiff fertilization membrane.

INTERCELLULAR CEMENT SUBSTANCE

The intercellular cementing material is the one investing coat which under normal conditions lies closest to the protoplasmic surface of animal cells in general. It serves to bind together individual cells to form aggregates of cells or cell complexes.[6] An example of

[4] Harvey (1911); Hendee (1931); Moser (1939); Motomura (1941, 1957); Runnström, Monné, and Wicklund (1944); Runnström (1948); Endo (1952).

[5] Motomura (1941); Endo (1952).

[6] We do not claim that the cement substance is the only agency which serves to bind together cells, since specialized cell connectives and interlocking devices occur in many epithelia of the adult organism and in embryonic tissues, even in sea urchin and amphibian blastulae (Balinsky, 1959; Holtfreter, 1943b).

the intercellular cement substance is the hyaline layer of the echinoderm egg.

The intercellular cement is a direct product of the cell and is deposited only when calcium is present in the environment. It can be readily removed or weakened either by rendering the medium free of calcium or by acidifying the medium sufficiently to increase the ionization of the calcium salts constituting the cement.

Hyaline layer of the echinoderm egg

The formation of this layer is to be regarded as a secretion initiated by the process of fertilization. As we have seen, the surface of the unfertilized egg is invested by a delicate vitelline membrane which lies directly on the protoplasmic surface of the egg (Figure 2.1). Immediately after insemination, the vitelline membrane elevates, leaving below it a truly naked protoplasmic surface. Within a few minutes, the hyaline layer appears as an exudation product on the egg surface. The oil-coalescence data described under the application of this method (Chapter 2) clearly reveals this transformation from a naked surface. The hyaline layer material may be derived from the substance of the cortical granules. As has already been described, these are expelled shortly after attachment of the spermatozoon.

The dimensions of the hyaline layer vary considerably in ova of different species. In the *Arbacia* egg it reaches a thickness of 3 to 4 micra within 8 to 10 minutes after fertilization. During cell cleavage some of the hyaline layer material is carried in between the walls of the advancing furrow and keeps the actual surfaces of the furrow apart.

The existence of the hyaline layer was recognized by Selenka (1878) and Fol (1879) when they first described the process of insemination. Herbst (1900) correctly interpreted it as a "Verbindungs-Membran," although later others came to regard it as an integral part of the protoplasm of the egg. This latter view has since been discarded because of the fact that the hyaline layer can be removed entirely without impairing the life of the egg.

Herbst demonstrated that the hyaline layer material always dissipates when the egg is washed in sea water lacking divalent cations, resulting in a falling apart of the blastomeres, although the time needed for its complete removal increases the longer the time the hyaline layer has been in the presence of calcium. Lowering the pH of the medium to 5.0 or 6.0 favors dissipation of the hyaline layer, even in the presence of calcium (Table 4.1).

When excess calcium is added to sea water (at a pH of 8.2 to

8.4) the stiffness of the hyaline layer is markedly increased, while decreasing the calcium content of sea water weakens the cement substance. The minimum concentration of calcium in sea water needed for formation of the hyaline layer at pH 8.0 is 10^{-4} M.

If within a few minutes after fertilization the sea urchin egg is immersed in artificial sea water lacking divalent cations, the hyaline layer never appears. The material for the formation of the hyaline layer continues to be exuded by the egg, but is dissipated. This can be shown by returning the egg, after a half an hour or longer, to sea water containing calcium. The space between the egg surface and the fertilization membrane becomes filled with a cloud of fine granular material. This is the hyaline layer material which had been kept, although in a dissipated state, within the confines of the fertilization membrane.

The ability to exude hyaline layer material is maintained throughout development by the blastomeres, but this material is never secreted in such a large amount as during the first half hour after fertilization. The continued ability of the blastomere to form this material might be explained by the combination of calcium in the sea water with negatively charged proteinates in the naked cell surface. At the relatively alkaline pH of the environment an undissociated complex would be formed, which would be drawn out of the protoplasmic surface film to accumulate as an insoluble layer of material surrounding the egg. Another possible origin of this material is the few cortical granules which are stated to persist even in late blastomeres.

The fact that hyaline layer material is secreted after first cleavage is shown in the following experiment.

> The fertilization membranes were shaken off within a few minutes after fertilization, and the eggs placed in an isosmotic mixture of sodium chloride and potassium chloride in proportions of 19 to 1, the mixture being at a pH of 7.0. In this solution the naked eggs continued normal cleavage for many generations. The resulting blastomeres rounded up, fell apart, and continued to exist as isolated blastomeres.
>
> Some of the eggs were transferred back to sea water a few minutes after first cleavage. The first two blastomeres remained separated, but the blastomeres of the subsequent cleavages were held together by the presence of fresh hyaline layer material. This material was much less in quantity than if the eggs had been in normal sea water from the time of insemination. Normally constituted swimming larvae subsequently developed from the daughter blastomeres, but the

cells, although weakly adherent, could be easily separated with a microneedle.

Secretion of the hyaline layer material can be suppressed by a prior washing of the unfertilized sea urchin eggs in a solution of urea or other non-electrolyte, isosmotic with sea water, at a pH of 7 or 8 (Moore, 1930a). In addition, this treatment causes the dispersal of the vitelline membrane. A possible explanation for the suppression of the hyaline layer is that exposure of eggs to isosmotic urea causes an explosive breakdown of the cortical granules (Moser, 1940). The substance of the granules, having dissipated in the urea solution, would no longer be available for hyaline plasma layer formation upon return of the eggs to the calcium-containing sea water, and subsequent fertilization.

The hyaline layer can also be suppressed by washing the fertilized sand dollar egg (Moore, 1930b) or sea urchin egg (for example, *Lytechinus pictus,* Brooks and Chambers, 1954) in urea shortly after insemination while the fertilization membrane is still rising. The fertilization membrane disperses, and a visible hyaline layer does not form. Cleavage results in the formation of cell plates, the constituent cells of which are loosely adherent, as revealed by testing with the microneedle. Dan and Ono (1952) believe that cell plate formation indicates the presence of a small amount of cementing material.

The hyaline layer of an egg kept continuously in sea water progressively stiffens so that it can be torn with microneedles and pulled out as semi-elastic strands. Its stiffness is greatest on its outermost surface, where it constitutes a stiff cuticular membrane, while the region between it and the protoplamic surface is relatively fluid. During cleavage the material of the soft inner part sinks inward to cover the walls of the advancing furrow, while the stiff outer part of the hyaline layer is not dragged in but bridges the gap between the blastomeres. The softness of the inner layer might be due to a greater acid reaction of the region directly external to the protoplasmic surface. This would promote the dissociation of the calcium salt composing the hyaline layer.

Dan and Ono (1952) have presented good evidence that the surface of the egg is attached to the outer stiff part of the hyaline layer by radial fibers, which extend out from the surface of the egg. The attachment would serve to fix the relative positions of the developing blastomeres.

In the starfish egg the weakness and extensibility of the hyaline layer permit the blastomeres, as they are forming, to exert their inherent property of rounding up as spheres. On the other hand, in the *Arbacia* egg the hyaline layer is sufficiently strong and elastic to prevent the full expression of the mitotic lengthening of the egg. When the mitotic spindle and asters disappear, with a consequent decrease in rigidity of their interior, each blastomere becomes constrained into the shape approaching that of a hemisphere. An intermediate condition is represented by the sand dollar egg, in which the hyaline layer is not strong enough to prevent the mitotic lengthening of the egg. However, when cleavage is completed and the asters fade, the layer is elastically strong enough to pull the blastomeres together. The mitotic lengthening and the maintenance of the rounded shape of the recently formed blastomeres against the pull of the investing hyaline layer is due, in part, to the gelation accompanying the formation of the asters. By agitating the aster with the microneedle the gelated state is reversed. If this is done immediately after cleavage, the blastomere, of which the aster has been destroyed, tends to collapse and is pulled toward the daughter blastomere, against which it presents a concave surface (Figure 4.1). When the interfering needle is

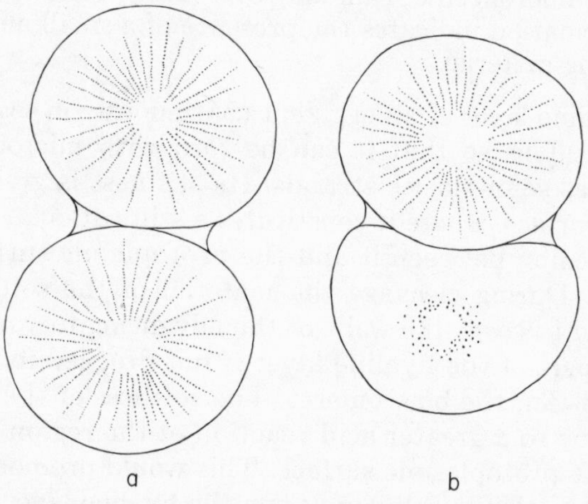

FIGURE 4.1. Fertilized sand dollar egg in the two-blastomere stage divested of its fertilization membrane. Blastomeres are held together by hyaline layer visible as line across borders of the two contiguous blastomeres. (a) Blastomeres, immediately after cleavage, rounded because both contain asters which are in gel state. (b) Aster of one blastomere has been caused to revert to sol state, whereupon other blastomere, which still contains an aster, bulges against its solated neighbor. Elasticity of hyaline layer causes blastomere with intact aster to deform the other.

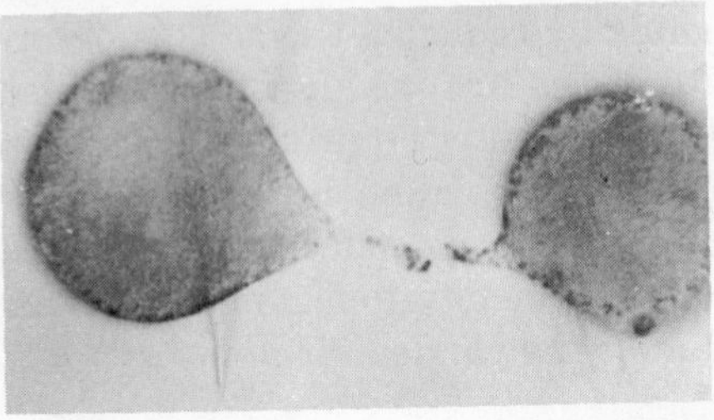

FIGURE 4.2. Connecting stalk between two blastomeres of living, fertilized *Arbacia* egg divested of its fertilization membrane and immersed in monovalent salt solution (NaCl, 17 parts; KCl, 3 parts) isosmotic with sea water. The two blastomeres, after cleavage completed, were mechanically separated by microneedles to show persisting connecting stalk. This stalk is eventually broken by spontaneous movements of subsequently cleaving blastomeres. (After Chambers, 1938b.)

removed so that the aster re-forms, the flattened blastomere rounds up again, pulling itself away from its neighbor. These observations demonstrate the elastic properties of the hyaline layer which holds the blastomeres together and serves to press them against one another.

Even if the egg is allowed to develop in a medium free of divalent cations, the developing blastomeres do not fall completely apart. This is because of the persistence of a stalk holding together the blastomeres at the termination of each cleavage (Moore, 1930a, b). The cleavage furrow is not carried to completion. Normally, in sea water, the blastomeres are so closely appressed that the stalk is not visible. In order to observe the stalk the fertilized egg, with its fertilization membrane shaken off, is allowed to develop for 30 minutes after insemination in an isosmotic calcium-free medium (to remove the hyaline layer). The egg is then returned to sea water before first cleavage occurs. The hyaline layer material, which forms after the return to sea water is insufficient to cause a close adhesion of the blastomeres. After cleavage is completed, moving the two blastomeres apart with microneedles renders the connecting stalk visible (Figure 4.2). Under the influence of the pull, the cytoplasmic part of the stalk breaks along its middle into spherical, presumably cytoplasmic, droplets, while the portions extending from the surface of the blastomeres slowly merge with the main body of each blastomere.

Adequate chemical analyses of the isolated hyaline plasma layer material have not, as yet, been made. Protein is undoubtedly a constituent, in view of the fact that the hyaline layer can be dispersed by trypsin (Table 4.1). The material has all the properties of an organic calcium salt (Table 4.1), presumably a calcium proteinate,[7] with the divalent calcium ion maintaining the firm gel state of the hyaline layer by establishing cross linkages between adjacent macromolecular elements. Recently Mazia (1958) has shown that the hyaline layer is dispersed by mercaptoethanol, a sulfhydryl reagent, which splits disulfide bonds by reducing them. This observation indicates that the gel state of the hyaline layer depends also upon the presence of disulfide linkages between protein molecules. Histochemical studies reveal that the hyaline layer contains a polysaccharide component.[8] A polysaccharide-containing substance is present between the blastomeres of the developing embryo. This is presumably the hyaline layer material carried inward.

Intercellular cement of epithelial cells in tissue culture

The cement substance which binds together the cells of many types of epithelial tissues depends, for its stability, upon its constitution as a reversible calcium salt. Removal of calcium from the external medium causes the epithelial cells to round up and fall apart. Acidification promotes the dispersal of the cells. We have already seen that the blastomeres of developing sea urchin eggs fall apart when exposed to similar conditions.

Investigations of the role of calcium were made on tissue cultures of the embryonic epithelial tissues of mouse and chick (Chambers, Cameron, and Grand, 1949). When a normal medium is used, outgrowths develop from fragments of the tissue, forming extensive coherent sheets. Acinar and cyst-like spaces develop in the sheets which form as a result of the proliferation of such epithelial cells as the proximal tubule cells of the kidney and the secretory cells of the lingual glands. When well-developed sheets had formed, the cultures were washed in calcium-free Tyrode's solution for varying periods (30 minutes to 3 hours). The cultures were finally returned to a calcium-containing culture medium at the normal pH of 7.4 for the purpose of determining the viability and the extent of recovery of the cultures after the experimental procedures. The results of several experiments are shown in Figure 4.3. In each instance, whether the epithelial cells were of visceral or ectodermal origin, malignant or not, the effect of washing coherent sheets and acini which had developed in the cultures with calcium-free media was the rounding up and separation of the cells. The disorganization of the sheets was more complete in the case of epithelial cells of visceral origin (Figure

[7] Moore (1928, 1930a, b); Nakano (1956).
[8] Monné and Slautterback (1950, 1952); Monné and Hårde (1951).

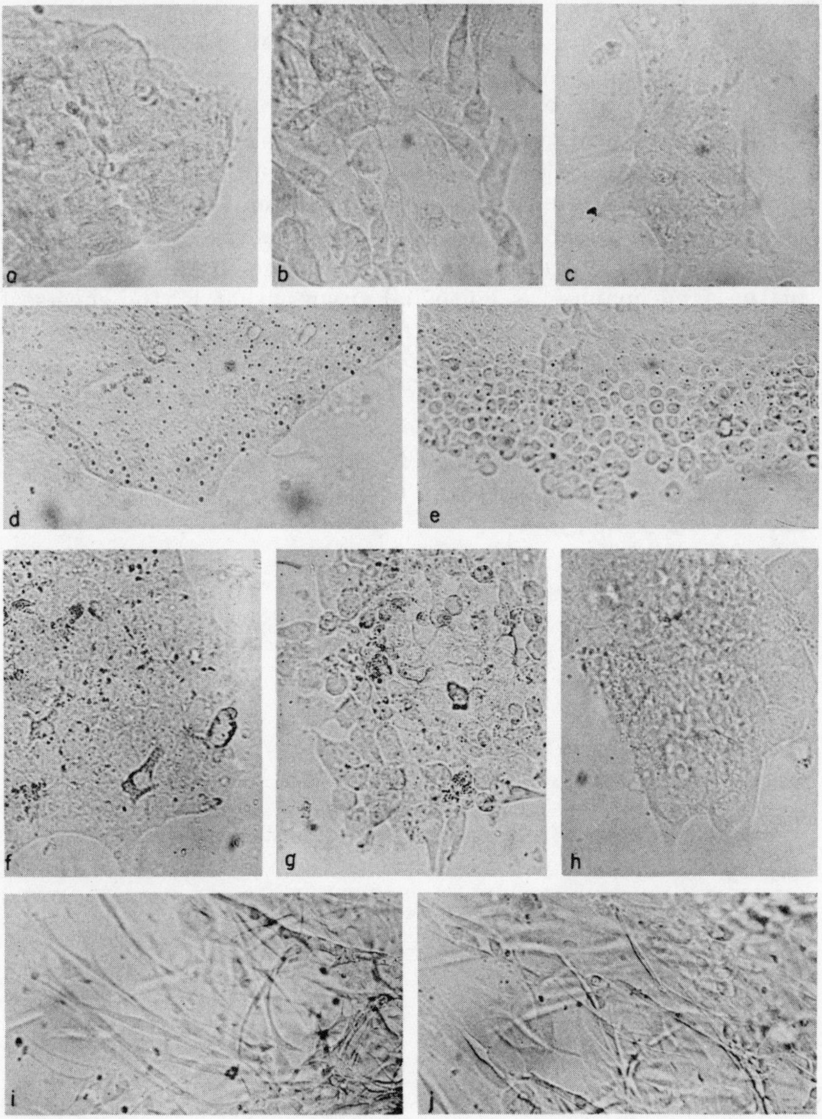

:GURE 4.3. Effect of presence and absence of calcium on sheets of epithelial cells and ›roblasts grown in tissue culture.

(a–c) Tongue epithelium of mouse embryo, 48 hours' growth: (a) epithelial sheet ›own in blood plasma; (b) same, after 85 minutes' exposure to calcium-free Tyrode's lution; (c) same, after recovery of 3 hours in plasma medium.

(d, e) Intestinal mucous epithelium of 9-day chick: (d) epithelial sheet grown in blood asma medium; (e) same, after 3 hours' exposure to calcium-free Tyrode's solution.

(f–h) Carcinoma epithelium of mouse, 72 hours' growth in tissue culture: (f) epithelial eet grown in blood plasma; (g) same, exposed 2 hours in calcium-free Tyrode's solution;) same, after recovery of 3 hours in plasma medium.

(i, j) Human fibroblasts: (i) fibroblasts in blood plasma medium; (j) same, exposed 5 urs in calcium-free Tyrode's solution (no effect).

4.3d–h), since these were not connected by the intercellular bridges characteristic of epidermal cells (Figure 4.3a–c). Returning the noncoherent cells to normal media resulted in the complete reconstruction of the sheets and acini. Recently Grand (personal communication, 1958) has found that sheets of mammalian endothelial cells growing in tissue culture respond to a lack of calcium in the same manner as epithelial tissues; that is, the sheets of endothelium dissociate when perfused with calcium-free media and reconstitute when the calcium is replaced.

Cultures of fibroblasts exposed to calcium-free media offered a striking contrast to those of the epithelial tissues. Exposure for the same or longer periods of time to calcium-free media in no way altered the shapes of the cells or their arrangement in pseudosheets (Figure 4.3i, j).

5

Role of Intercellular Cement in Cell Organization

ROLE IN THE ORGANIZATION AND DIFFERENTIATION OF THE EMBRYO

One of the important functions of the hyaline layer material is preventing fusion of the blastomeres as these are formed. In the dividing *Arbacia* egg the blastomeres are tightly held together by the investing hyaline layer, the outer part of which is strong and serves as a pseudomembrane. The inner part is more fluid; with the sinking in of the cleavage furrow, this hyaline material is carried inward. With the deepening of the furrow, its advancing edge assumes the shape of a tear drop, with the broad part of the drop in advance. The hyaline layer material contained within the drop coats the newly forming surfaces of the blastomeres. Even if the contiguous surfaces of the blastomeres are firmly pressed against one another by compressing the egg, the tear drop shape is maintained, and the furrow continues to advance with the contiguous surfaces kept apart by the intervening cement material (Figure 5.1). Fusion of the surfaces of the forming cleavage furrow, however, can be induced by dissipating the hyaline layer in a calcium-free medium and constraining the dividing egg in a groove in order to force the walls of the furrow into contiguity. Since the surface of the truly naked egg is liquid in nature, fusion between contiguous newly forming surfaces is to be expected in the absence of an intervening material.

Grooves were made on a glass slide 80 micra deep and having the same width, to accommodate fertilized *Arbacia* eggs whose fertilization membranes had been removed by mechanical shaking. The membraneless eggs were transferred through several washings with a mixture of sodium

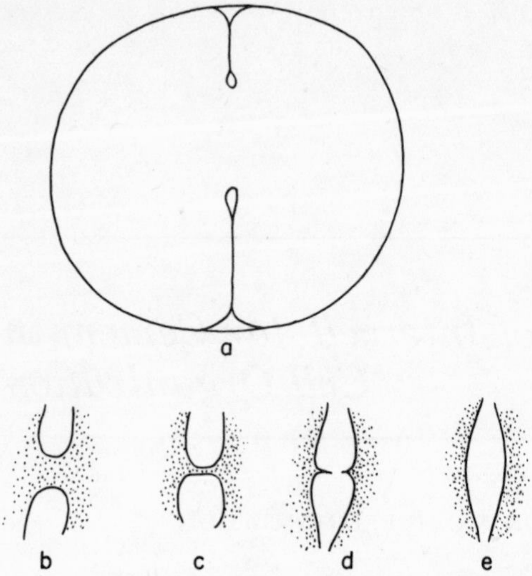

FIGURE 5.1. (a) Sand dollar egg cleaving under compression, showing tear-drop shape at edge of advancing furrow. (b–e) Details of advancing furrow: (b, c) presence of a space at tip of advancing furrow causes connecting bridge to be of considerable length; (d, e) bridge breaks, leaving oblong space. (From Chambers, 1924a.)

chloride (19 parts) and potassium chloride (1 part) isosmotic with sea water. In this calcium-free mixture the eggs survived and underwent normal cleavage. A drop of the mixture containing eggs was placed on the grooved slide and covered with a glass coverslip. The coverslip was tightly pressed on the slide so that those eggs which were in grooves were held firmly in place, with their surfaces in contact with the walls of the grooves. At about 55 minutes after fertilization, karyokinetic lengthening of the eggs occurred preparatory to the first cleavage. The lengthening of the egg was along the axis of the mitotic spindle and at right angles to the direction of the cleavage furrow. Those eggs in which the lengthening was oriented along the long axis of the groove underwent their cleavages normally. Those which were oriented at a right angle to the groove underwent abortive cleavage. In these, the cleavage furrow started to form, but the indentation flattened out as soon as the opposite surfaces of the furrow made contact, and the furrow was obliterated.

Furthermore, since the cement substance holds together the blastomeres of the developing embryo within the limits of a confined space, the cement layer conditions the shape of the cells as they multiply and flatten against each other. These changes in shape play an important role in cellular differentiation.

The mature, unfertilized egg of the starfish (*Asterias*) and of the sea urchin (*Arbacia*) is totipotent. Any fragment, as long as it is not too small,[1] when fertilized is able to develop into a larva normal in every respect except in size. In addition, each of the two sister blastomeres of a sea urchin egg which has undergone its first cleavage, when separated from the other has the potentiality of developing into a complete larva. However, in their normal positions they are pressed against one another by the tightly investing hyaline layer, each thereby tending to be hemispherical. This constrained position conditions their development, and they grow into a single embryo.

A significant operative experiment which reveals the importance of shape and contiguity of cells is the following. Owing to the natural toughness of the vitelline membrane of the unfertilized starfish egg, it was possible to cut the egg into several fragments without rupturing the membrane. This was done by bearing down upon the egg with the shaft of a horizontal microneedle. By performing the operation several times on the same egg, several fragments were obtained, all held within the same vitelline membrane. Upon insemination, every fragment, only one of which contained the egg nucleus, received a spermatozoon. Each of the fragments is potentially a complete ovum and, if the fragments had been isolated by tearing the originally enclosing membrane, they would have developed each into a complete haploid or diploid larva. However, because they were held together by a common extraneous coat, their constrained position while undergoing cleavage resulted in the formation of a single integrated larva.

An interesting example of the effect of mechanical deformation of a cell on differentiation is reported by Whitaker (1940) in his experiments on the egg of *Fucus,* a brown alga. The spherical *Fucus* egg, within 10 to 15 minutes after fertilization, was made to assume a cylindroid shape by drawing it into the bore of a narrow capillary tube. Immediately following fertilization, the naked egg secreted a thin layer of soft gelatinous material, which gradually hardened to form a rigid coat, or cell wall. When the egg was blown out the deformed shape was retained due to hardening of the external cell wall, and the rhizoid which grew out of it always developed at one end of the cylinder.

The hyaline layer also plays a role in the formation of the blastocele. In the echinoderm egg the developing blastula has a wall of

[1] Fragments of unfertilized sea urchin eggs, down to one twentieth of their original size, when inseminated segment regularly and form swimming blastulae. Smaller fragments do not develop beyond this stage. The larger fragments, as long as they are not less than one fourth the diameter of the intact egg, develop through to plutei (Moore, 1912, and personal communication).

cuboidal cells enclosing a primitive blastocele, or segmentation cavity. Blastocele formation results from the fact that the blastomeres, from their very inception, are more firmly attached to the investing hyaline layer than to each other. Dan (1952) ascribes this firm attachment to the presence of radial fibers, which extend from the surface of the blastomeres into the hyaline layer.

With each successive cleavage the blastomeres, attached over their outer surfaces to the investing hyaline layer, attempt to round up and thus draw away from the center where the blastocele is forming. The blastomeres continue proliferating while their external surfaces remain stuck to the investing hyaline layer, with the result that they become pressed in between one another to form a wall of simple cuboidal epithelium surrounding the blastocele.

The fertilization membrane can assume the role of the hyaline layer if the latter is dispersed by immersing the fertilized egg in a solution of sodium and potassium chlorides (19 to 1) isosmotic with sea water. In this solution the hyaline layer disappears, while the fertilization membrane persists but becomes very adhesive. The isolated blastomeres within the membrane are at first distributed haphazardly. However, as they undergo successive cleavages, each daughter cell becomes plastered to the sticky inner surface of the fertilization membrane until they form a normal appearing blastula. When these blastulae are transferred to sea water, the secretion of cement becomes again evident; the fertilization membrane loses its stickiness; and eventually a typically formed blastula is released.

Invagination and exogastrulae

An interesting case is the formation of exogastrulae when a sea urchin egg develops in sea water to which lithium has been added (Herbst, 1895). At the time of gastrulation the endodermal part of the wall bulges out and forms an everted endodermal sac. Runnström (1935) ascribed this abnormality to a specific effect of lithium on the metabolism of the cells involved. Moore and Burt (1939), however, were able to prevent exogastrulation of *Dendraster* embryos even in the presence of lithium by rendering the sea water medium hypertonic with the addition of sucrose, to which the embryos, at the late blastula stage, are impermeable. This indicated that a supposed specific lithium effect was counteracted by a relatively innocuous, nonpenetrating, non-electrolyte. Simply by lowering the internal pressure of the blastocele within the blastula the tendency for exogastrulation could be counteracted.

With the micromanipulative technique it is possible to show

that the presence in sea water of lithium, a monovalent cation, at the concentration used for obtaining exogastrulae, weakens the investing hyaline layer of the blastulae of both *Arbacia* and *Echinarachnius* eggs to the degree that the tensile strength of the layer is greatly diminished. The cells constituting the wall of these blastulae exposed to lithium are so loosely held together that they can readily and individually be picked off with the tip of a microneedle, a condition that is not possible in ordinary sea water or even when excess sodium chloride or potassium chloride is added to sea water. Evidently lithium is a powerful dispersing agent of the hyaline layer and can serve as such even in the presence of the calcium as it is normally found in sea water. Therefore, the formation of exogastrulae in the presence of lithium becomes comprehensible on the ground that the exogastrulation is conditioned by a weakening of the hyaline layer. This interpretation is strengthened by the cited findings of Moore and Burt.

The experimental data indicate that the normal process of invagination is mechanically favored, at least in part, by the restraining effect of the hyaline layer investment.

ROLE IN THE FUNCTIONAL ORGANIZATION OF EPITHELIAL CELLS

A characteristic of epithelial tissues which possess secretory activity is their functional polarity. Substances are taken into the cells at one surface and are secreted at the opposite surface. The process is essentially unidirectional. The secretory activity of the cells depends on their organization into definitive groups, such as tubules in the case of the proximal tubule cells of the kidney.

The walls of the kidney tubules consist of columnar cells, the outer surfaces of which are firmly attached to a basement membrane, or cuticle, investing the tubule. When explanted fragments of chick mesonephros are exposed to a solution of phenol red in a tissue culture medium (Chapter 16), the cells of intact tubules soon take on the yellow color of the acid range of the phenol red as it passes through the cytoplasm of the cells to accumulate in the lumen of the tubule (Chambers and Kempton, 1933). This passage of the phenol red occurs only when the cells are in their proper position in the wall of the tubule (Figure 16.2). If the tubule is cut into segments, the cells in the immediate region of the cut loosen from the basement membranes and become rounded. While the cells are changing from a columnar to a spherical shape, they become colorless and no longer give evidence of picking up the dye. In the tissue culture medium these partly isolated cells undergo proliferation to form sheet-like outgrowths. None of the cells in the flat sheets give any evidence of coloration with phenol red. As growth continues, however, at certain regions in the sheet the cells pile up in several layers, and between them occasional spaces appear. This results in an alveolar-like orientation, following which the cells bordering an enclosed space reacquire their capacity to pick up and secrete phenol red. Ultimately isolated cysts filled with the indicator are formed.

The suspension of secretory activity resulting from separation of the tubule cells can be demonstrated by bathing a tissue culture of the colored tubules with a calcium-free Ringer's solution. This results in a dispersion of the intercellular cement substance, whereupon the cells draw away from one another and round up. As in previous experiments the cells lose their color and their ability to pick up phenol red. When the Ringer's solution is exchanged for one containing calcium, the wall tightens up so that the cells become again closely pressed against one another, and the reappearance of the phenomenon of secreting phenol red is again exhibited. This phenomenon was found to be repeatedly reversible.

The necessity for the organization of cells into multicellular units for secretory function has also been demonstrated in the case of the ependymal epithelium of the chorioid plexus in tissue culture (Cameron, 1953). When the epithelium is organized in the form of alveoli, the epithelial cells are colored by the sulfonated dyes, which accumulate in the alveolar cavities. On the other hand, cells in outgrowing sheets of the ependyma remain colorless when exposed to the same dyes.

Evidently a significant relationship exists between the constrained shape of cells and their functional polarity. In the disorganization of the epithelium at the cut ends of the tubules, or following exposure to calcium-free media, the cells are loosened sufficiently for them to become spherical. This is accompanied by a loss of secretory activity. The reassumption of the columnar shape with the reappearance of the binding cement is accompanied by a return of secretory activity. The inference is that the shape of the cell affects orientation within the cell, resulting in the development of polarity. It may be that the chemical constituents of the protoplasm essential for its polarity develop an anisotropy when under strain.

Another instance of functional polarity is seen in the uptake of trypan blue, a colloidal dye, by the kidney tubule cells. If trypan blue is added to a culture medium containing kidney tubules or cysts, the dye is not picked up by the cells, nor can it enter the lumen. If, however, the dye is injected into the lumen of a tubule, it is ingested by the cells at their luminal surfaces. Isolated kidney epithelial cells phagocytose the dye. In this instance, organization into tubules has caused a restriction of the phagocytic activity to the luminal border of the cells.

PERMEABILITY OF CELLULAR SHEETS

Moore has shown by immersion experiments that sand dollar (*Dendraster*) blastulae with extraneous coats intact, up to about the tenth cleavage, are freely permeable to sucrose.[2] We have confirmed these results by injecting hypertonic sucrose into the early blastulae. No swelling occurs following the microinjections. At later stages, when the number of cells in the wall of the blastula has increased, the wall becomes impermeable to sucrose, as revealed by the fact that late blastulae remain collapsed when immersed in sea water made hypertonic with sucrose.[2] Colloidal substances do not penetrate into the blastocele of either early or late blastulae. However, the salt constituents of sea water and the water-soluble salts of the sulfonphthalein indicators readily penetrate into the interior not only of early and late blastulae, but also of gastrulae and plutei. Late blastulae immersed in hypertonic sea water, for example, do not undergo deformation.

We know that the fertilization membrane and the hyaline layer are freely permeable to salts and sucrose. The fertilization membrane is impermeable to colloidal substances, and the hyaline layer restricts the passage of these substances. The cells themselves are totally impermeable to colloids, sucrose, and the sulfonphthalein indicators. The relative impermeability of the blastomeres to the salt ingredients of sea water[3] is shown by the fact that the cells remain shrunken indefinitely when immersed in hypertonic sea water. An explanation of the development of impermeability to sucrose in the late blastula is that the intercellular spaces, containing hyaline layer material, become narrowed by proliferation of the cells to such an extent that the number of sucrose molecules which can traverse the blastular wall via the intercellular regions is negligible. The intercellular spaces, however, have not been sufficiently narrowed in the late blastula to prevent the rapid passage of salts and indicators. To understand the impermeability of the blastular wall to colloidal substances, at both early and late

[2] Moore and Burt (1939); Moore (1940, 1945); confirmed by Dan (1952).

[3] We know that potassium and sodium readily exchange between the interior of fertilized echinoderm eggs and their environment (E. L. Chambers, 1949; Chambers and Chambers, 1949), but this is relatively a slow rate of passage compared with the extremely rapid rate at which these ions must diffuse across the blastular wall.

stages, we have only to consider that the extraneous coats surrounding the cells of the blastula are the barrier, including hyaline layer material which intervenes as a thin layer between the cells.

In the Metazoa we meet with three types of cellular membranes:

(1) Included in the first type are the leaky membranes. They readily permit the indiscriminate passage by diffusion of substances generally considered incapable of penetrating the protoplasmic surface, but they do offer a barrier to the passage of colloidal particles. The permeability of membranes of this type resembles that of the early echinoderm blastula; that is, other routes are available for the passage of substances across these multicellular membranes in addition to those routes which involve diffusion across the protoplasmic surface.

(2) In the second category are the membranes of intermediate permeability. They are impermeable to colloidal particles and to water-soluble substances of large molecular dimensions, but are permeable to those water-soluble substances of smaller molecular dimensions which, nonetheless, generally do not diffuse across the protoplasmic surface, or do so relatively slowly. The permeability of membranes of this type resembles that of the late echinoderm blastula. Passage of substances can occur to some extent by routes other than across the protoplasmic surface.

(3) Membranes of the third type possess the highly selective permeability characteristics of the protoplasmic surface; that is, the passage of substances across such membranes involves traversal of the protoplasmic surface films of the constituent cells, to the exclusion of other routes.

An example of the first, or leaky, type of membrane, is the endothelial lining of blood vessels (exclusive of those of the vertebrate central nervous system). The endothelium is characterized by the indiscriminate nature of its permeability. For example, the endothelium of even the least permeable of the systemic capillaries is not fully impervious to protein molecules. Large water-soluble molecules, even as large as inulin, pass through mammalian muscle capillaries at a measureable rate, and sucrose passes through quite rapidly (Pappenheimer, 1953), in spite of the fact that muscle capillaries belong in the category of the less permeable of the systemic capillaries. None of these substances is known to be able to diffuse across the protoplasmic surface. Furthermore, for water and the smaller lipid-insoluble molecules, their rates of passage across the endothelium are far in excess of any known rates across the protoplasmic surface. The mere thinness of the endothelial cell cannot account for the wide range of permeability, since very thin layers of living protoplasm are known to possess a high degree of selectivity. An example of this is the protoplast of the marine plant cell *Valonia,* of only several micra in thickness, which maintains an almost pure solution of 0.5 M potassium chloride within its central vacuole. The width of the protoplasmic layer is not a factor because the selective permeability of the cell is a property of the almost infinitesimally thin protoplasmic surface film.

It is virtually impossible for the cell physiologist to conceive how a body of protoplasm could maintain the highly specific attributes of its interior, essential for the living state, and yet possess the indiscriminate type of permeability which characterizes endothelium. Nor can the endothelial cell be considered some type of cell remnant, since it has the characteristic structure[4] and properties of a living cell:[5] It is irritable and reacts to prodding; it may phagocytose particulate matter; it undergoes mitotic division; and its nucleus is a prominent structure which stains with basic dyes only after death of the cell. Accordingly, we should search for means, other than by diffusion across the protoplasmic surface films of the constituent cells, whereby substances can traverse the endothelium.

Passage of those substances which do not generally traverse the protoplasmic surface film, or do so only very slowly, could occur through sites where the protoplasm, with its bounding surface film, is discontinuous: for example, through the intercellular regions[5] or via cytoplasmic discontinuities[6] in the endothelial cell. Passage of substances across the capillary wall at sites where the protoplasmic surface is lacking would involve only passage through the matrix of the intercellular cementing substance and of the basement membrane. These materials could certainly restrict the passage of colloidal particles, especially when the porous matrix of the cement is clogged with the plasma proteins.[5] We have already encountered an example of the imperviousness of extraneous coatings to colloids in the fertilization membrane and hyaline layer of the echinoderm egg.

Since the total surface area of the discontinuities must be small relative to the total surface area of the endothelial cells (with the

[4] Palade (1953, 1956); Moore and Ruska (1957); Buck (1958).

[5] Chambers and Zweifach (1940, 1947a); Zweifach (1954).

[6] Electron microscopic studies have revealed the presence of abundant perforations in the endothelial cell lining of the glomerular and peritubular capillaries of the kidney (Pease, 1955) and intestinal villus (Bennett, Luft, and Hampton, 1959). Cytoplasmic discontinuities, but of a lesser extent, have been described in the endothelial cells of the capillaries of various endocrine organs (Pease, 1955; Ekholm and Sjöstrand, 1957). Such cytoplasmic discontinuities, however, have not been detected in the capillaries of cardiac and skeletal muscle (Moore and Ruska, 1957), lung (Karrer, 1956), and central nervous system (Dempsey and Wislocki, 1955).

exception of the glomerular capillaries), substances which do not diffuse across the protoplasmic surface film (for example, sucrose) or do so relatively slowly (for example, the sodium ion) should pass across the endothelium far more slowly than substances which can readily traverse the protoplasmic surface (for example, lipid-soluble substances). This has actually been found to be the case in the capillaries of skeletal muscle (Pappenheimer, 1953).

It is important to realize that the surface film of the living cell constitutes a barrier to the diffusion across it of all classes of substances, even those which traverse it fairly rapidly, such as water. Accordingly, any influence which increases the relative proportion of "gap" surface area would increase the rate of diffusion and bulk flow of fluid across the endothelium.

Perfusion of blood vessels of the frog with a colloid-containing Ringer's solution lacking calcium, or containing calcium but slightly acidified, markedly increases the bulk loss of the circulating fluid from the capillaries and also causes the extravasation of particulates and cellular elements through the interendothelial cell regions.[5] When the perfusion fluid is changed to one containing calcium at the appropriate pH of 7.4 to 7.8, the normal condition of permeability of the blood vessels is re-established.

A plausible explanation of these results is that a cementing substance, a product of the endothelial cell, plays an important role in capillary permeability, in that it fills the intercellular regions, serves to hold the cells together and to the basement membrane, and serves to restrict the passage of colloidal substances.[5] The cement substance has the properties of an organic calcium salt, presumably a proteinate, analogous to the hyaline layer of the fertilized egg. Accordingly, diminishing the calcium content of the external medium or acidifying the medium softens and weakens the cementing substances by causing dissociation of the calcium salt (Chapter 4). This permits loosening of the endothelial cells, widening of the gaps, and dispersal of the intervening cement materials, and results in the excessive loss of fluids, of colloids, and even of cellular elements.

Electron microscopic studies of endothelium have revealed the presence of numerous small vesicles, or vacuoles, in the cytoplasm, many of which appear to open at both surfaces of the endothelial cell.[4] The suggestion has been advanced that the vesicles represent a mechanism for transferring fluid and dissolved substances across the endothelial cell, involving engulfment of fluid, movement of vacuoles across the cell, and emptying of the contents at the opposite surface. If such an active transfer process played an important role in the bulk movement of fluid across the endothelium, it could account for the indiscriminate nature of the permeability. Obviously such a process could accomplish the transfer of substances, segregated throughout in vacuoles, without involving their diffusion across the highly selective protoplasmic surface film. In effect, a pinocytic process could achieve a bypassing of the surface film as effectively as passage through regions where the surface film is discontinuous. The principal difficulty with such a mechanism, however, is that it must involve the expenditure of energy. The evidence, at least up to the present, indicates that the purely physical processes of diffusion and flow play the major roles in the passage of substances in solution across the endothelium of the systemic blood vessels.[7]

An example of a membrane of intermediate permeability is the intestinal mucosa. Here substances traverse the membrane principally by passage across the protoplasmic surfaces of the constituent cells, but another route is available to a limited extent, presumably involving passage through the intercellular regions. Permeability is considerably more selective than that of the capillary endothelium. Höber (1901) described a striking experiment with methylene blue and ammonium molybdate showing conclusively that these substances can be demonstrated to have diffused into intercellular spaces in the course of their passage across the mucosa. The intestinal mucosa of a frog tadpole was vitally stained with methylene blue, and some pieces were exposed to ammonium molybdate and others to corrosive sublimate. The molybdate did not penetrate the epithelial cells, while methylene blue and sublimate did so immediately. With molybdate he obtained only an intercellular precipitation of methylene blue, while with sublimate he obtained both an intracellular and an intercellular precipitation of methylene blue.

With phenol red, a sulphonphthalein, to which mucosa cells of the intestine are impermeable, it has been possible to demonstrate an intercellular diffusion in both directions through the mucosa of the rabbit. A loop of the intestine of a rabbit was exposed and the two ends ligated while the loop was left attached to the mesentery with a prominent branch of the coeliac artery and vein. Phenol red was injected into the blood circulation, and in a few minutes it could be detected in the lumen of the intestinal loop. In another animal the phenol red injected into the lumen of the intestine was quickly detected in the venous blood. In both cases strips of the

[7] Landis (1946); Wilbrandt (1946); Chambers and Zweifach (1947a).

mucosa were removed, immersed in weakly ammoniated Ringer's solution (to intensify the red color of the phenol red), and examined under the microscope. The dye could be observed by its red color on both sides of the membrane, but not a trace was seen within the cells. Occasionally, color was detected between the cells.

Höber and Höber (1937) presented evidence for the passage of polyhydric alcohols of different molecular sizes in accordance with the existence of a definite pore size in a sieve-like membrane. They ascribed the sieve-like nature of the membrane to a condition of the mucosa cells, but it might also be regarded as a property of the intercellular cement filling the intercellular regions.

Examples of the third type of cellular membrane are the wall of the proximal tubules of the kidney and the epithelium of the chorioid plexus. Under normal conditions the cells fit so closely together that the permeability of the wall is entirely that of the constituent cells. Any cellular membrane which depends for its permeability entirely on the cells constituting the membrane must be highly selective, since the membrane behaves as if it were bounded by a continuous protoplasmic surface. High concentration differentials can be maintained across such membranes even with respect to water-soluble substances of small molecular dimensions. Membranes of this type are secretory epithelia in which work is performed in the transport of certain substances across the cell. One-way permeability and unidirectional transport are outstanding features. As we described previously, when fragments of the proximal tubules of the kidney are grown in tissue culture at 40° C., their cut ends close over; upon the addition of phenol red or other sulfonphthalein indicators to the external medium, the cells pick up the indicators, become diffusely colored, and transfer the indicators into the lumina of the tubules. This process continues until all the indicator from the external medium has been transferred into the lumina. That is to say, transfer can occur against an almost infinite concentration gradient. The selectivity of the tubule cells is so pronounced that they are able to drive water against a concentration gradient (Keosian, 1938), distending the tubule sacs until the constituent cells are flattened to a pavement shape.

When the cells are metabolically inhibited, the indicators are not accumulated in the tubules; in fact the tubules become completely impermeable to them (Chapter 16). This indicates the high selectivity of this type of membrane, whether metabolically active or inactive.

That the high degree of selectivity of the tubules is dependent on the intimacy of union between the cells, to the extent of a virtual elimination of the intercellular regions as a factor in their permeability, is shown by cooling a tubule which has previously accumulated phenol red. The cooling does not in any way disturb the organization of the tubules, but causes a slight loosening of the cells in the wall, as revealed by the increased distinctness of the cell boundaries. Simultaneously the phenol red in the lumen streams out of the cooled tubule, not through the cells but between them, through the intercellular cementing material, which still holds the cells together. When the cultures are returned to normal temperature conditions, the wall tightens up, the prominence of the cell boundaries diminishes, and the tubules reaccumulate the phenol red. Identical results can be obtained by washing tubules which have previously accumulated phenol red in a calcium-free culture medium, but only just sufficiently to increase the distinctness of the cell boundaries. Loss of phenol red results from the weakening of the intercellular cement.

The data presented in this chapter indicate that the intercellular cement serves important functions over and above that of binding cells together. The cement substances are susceptible to variations in the electrolyte content and pH of their medium. A large measure of the known physiologic effects of monovalent and divalent cations on cellular tissues is to be ascribed to the action of the salts on the intercellular cements. The permeability of many cellular membranes is determined by the extent of the intercellular spaces and the cement substances which fill them. The cement substances, by their constraining action, influence cell organization, differentiation, and intracellular polarity.

III

The Protoplasmic Surface Film

INTRODUCTION TO PART III. Throughout this presentation we use the term protoplasmic surface film for the bounding, differentiated surface layer of the protoplasmic unit. This surface film can be destroyed by tearing with a microneedle, and unless the tear is repaired instantly by the formation of a new film, disintegration spreads into the exposed interior, and the life of the protoplasmic unit is destroyed.

The micrurgist pictures the protoplasmic surface as a lipoprotein film, with ubiquitous properties, dynamic and ever changing, conditioning the passage of substances into and out of the protoplasmic unit, and ensuring its physical integrity.

6

The Protoplasmic Surface Film as a Barrier

We are well aware that the protoplasmic body, divested of its extraneous coats, is immiscible with its aqueous environment. We know that many substances dissolved in the aqueous environment are unable to penetrate living cells. Moreover, those substances which do enter, even the rapidly penetrating ones such as water, do so more slowly than would be expected on the basis of free diffusion. These prominent characteristics reveal that the cell unit constitutes a definite barrier. This is exclusively the property of the living protoplasm, since immediately upon death of the cell, the barrier is lost, even though the visible morphological structures persist. Two opposing points of view have existed in regard to the location of this barrier. One side claims that the barrier is a property of the protoplasm as a whole, while the other holds that the barrier is located at the surface of the protoplasmic body.

Bütschli (1894), who advocated the emulsion structure of protoplasm in his theory on the foam structure of protoplasm, was opposed to the idea that the "barrier" characteristics of the living cell could be accounted for by its possession of an external, water-immiscible, selectively permeable layer. He maintained that these were properties of protoplasm as a whole. He showed that the passage of water into or out of cells immersed in aqueous media of different tonicities was compatible with his belief by comparing the protoplasmic system to a droplet of viscid oil containing, in suspension, minute crystals of salt. Immersion of the droplet in an aqueous medium would result in the appearance of vacuoles containing water which had slowly passed through the oil to dissolve the crystals.

Bütschli's claim that protoplasm as a whole is water immiscible is supported principally by the undoubted fact that protoplasm can absorb only a limited amount of water and still survive. This feature was cited in support of Bütschli's thesis by Lepeschkin (1930), who found that the swelling of such cells as the marine protozoa, foraminifera, and many algal cells, upon exposure to diluted sea water, is due to their becoming interpenetrated with aqueous vacuoles, the size and number of which vary with the dilution of the aqueous medium. A similar phenomenon was described for marine eggs by Leitch (1934). In addition, the contractile vacuole, pre-eminent in the fresh water protozoa, may be considered to have a function analogous to that of the vacuoles observed in echinoderm eggs exposed to hypotonic sea water; both are concerned with the segregation of excess water from the protoplasm. The basic necessity for preventing the effects of an excessive influx of water by the active functioning of the contractile vacuole can be shown by slowly microinjecting into an ameba moderate amounts of distilled water (two times the volume of the fully dilated contractile vacuole). Accelerated pulsations of the contractile vacuole rapidly dispose of the introduced water. Furthermore, if that part of an ameba which contains the contractile vacuole is cut away with a microneedle, the remaining nucleated remnant at once generates a fresh contractile vacuole.

A striking demonstration of the intolerance of the protoplasmic matrix to the introduction of excess water, which can almost be interpreted as evidence for its water immiscibility, is observed when distilled water is rapidly microinjected into an ameba. The introduced water "rushes" through the interior to collect at one side as a blister and is discarded by being pinched off (Figure 6.1).

In contradiction to Bütschli's point of view is that of Wilhelm Pfeffer, who believed that the chief permeability barrier of a living cell resides in a differentiated, water-immiscible, peripheral layer —the plasma membrane. Pfeffer considered that he observed

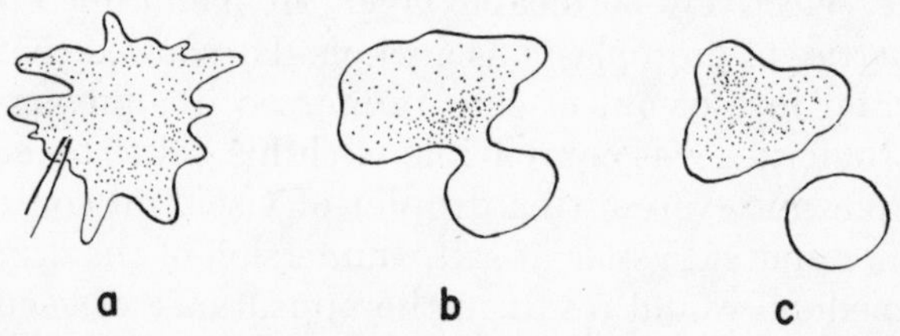

FIGURE 6.1. *Amoeba dubia* injected with large amount of distilled water, followed by pinching off reaction. (a) Before injection. (b) Blister forms following injection. (c) Blister is pinched off, and ameba returns to normal.

freshly formed plasma membranes enclosing vesicles of protoplasmic material which exuded from the tip of an *Hydrocharis* root hair when the latter was crushed. He concluded that the boundaries of these vesicles could not be simply interfaces between two immiscible fluids, the protoplasmic material and the surrounding water, because he also observed that these vesicles swelled in pond water and, when crushed, burst, the boundary disappearing and the contents dispersing in the surrounding aqueous medium.

The experimental evidence obtained using the micromanipulative method supports Pfeffer's concept by demonstrating that the chief barrier which preserves the integrity of the protoplasmic unit resides in a differentiated surface layer.

We can demonstrate that a continuous aqueous phase pervades the protoplasmic matrix of the living cell, since dilute solutions of monovalent salts and of water-soluble acid dyes, when microinjected, diffuse readily through the hyaloplasm (Figures 11.1; 11.4). Hence, the immiscibility of protoplasm with its aqueous environment must be due to the special properties of its surface layer.

During life the surface of the protoplasmic body, denuded of all extraneous coats, presents a sharp, smooth contour to the exterior. This contour represents the protoplasmic surface film. The essentiality of this film can be clearly demonstrated by sharply tearing a naked sea urchin egg at a point on its surface with the microneedle. A wave, or "ripple," of disintegration sweeps around the surface; the sharp contour is destroyed and replaced by a fuzzy border. This process represents dissipation of the surface film. Momentarily the inner part of the egg remains normal in appearance, but the process of disintegration rapidly spreads inward from the surface until the entire mass is transformed into debris.

The protoplasmic surface film serves the strictly mechanical purpose of retaining within its confines the protoplasmic ingredients. This is readily demonstrated by immersing cells in solutions of the noncoagulating salts, sodium and potassium chloride. Upon ripping the protoplasmic surface with a microneedle, complete dispersion of the cell contents occurs. For example, amebae are active for many hours in a mixture of 0.08 M sodium chloride and 0.02 M potassium chloride. If, however, the surface film is disrupted by tearing it with a microneedle, there occurs an outpouring of the contents of the interior, until the protoplasmic interior is fully dispersed (Figure 6.2). In the case of echinoderm eggs divested of their extraneous coats the same phenomenon of dispersal occurs also when the tearing is done in a solution of monovalent salts isosmotic with sea water.

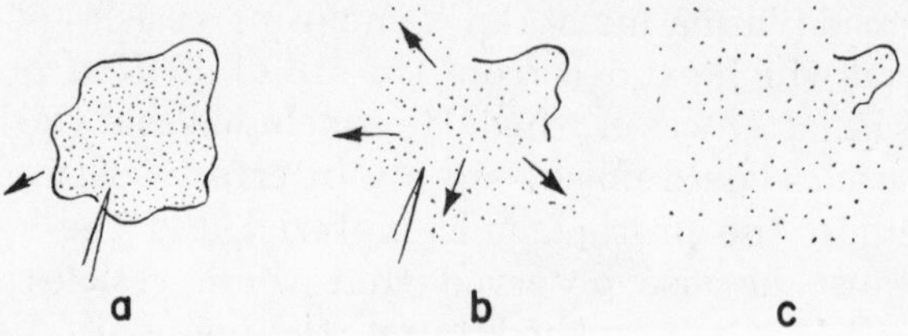

FIGURE 6.2. *Amoeba dubia,* living, torn in solution containing mixture of 0.08 M NaCl and 0.02 M KCl. Surface rapidly dissipates and granular contents disperse.

That the integrity of the protoplasmic unit and its selective permeability are conditioned by the protoplasmic surface film rather than by the underlying gelated cortical layer is indicated by the fact that this cortical gel can be temporarily converted to the sol state by mechanical agitation with the microneedle, with no adverse effect on the cell.

The surface film may elevate from the cortical gel by the collection of a liquid hyaloplasm under the surface. This occurs during the insemination of the echinoderm egg. The insemination cone, which appears on the egg where the head of a spermatozoon has entered, is a minute blister containing a clear liquid invested with a very delicate, film-like membrane. These blisters alternately well up and sink on the surface and persist for 5 to 6 minutes after insemination. Insemination cones form also on denuded, immature eggs, but these cones are unusually large and form rounded blisters. By gentle manipulation with a needle, the blisters can be made to coalesce and spread over the surface so as to present the aspect of a delicate film extending over a considerable portion of the egg (Figure 6.3a, b).

The fact that the surface film, elevated as an insemination blister, is the same as the barrier upon which the integrity of the cell depends is shown by pricking the blister (Figure 6.3c). The effect is the breakdown of the raised film, which spreads in a wave of disruption over the exposed cortical gel layer and leads to destruction of the entire egg. This extreme susceptibility of the protoplasmic surface film to injury is found only when the delicate papilla is pricked. Elsewhere, the part of the egg's surface immediately overlying the cortical gel layer tolerates puncture without any destructive effect.

Micrurgical experiments have unequivocally established that the special properties of the protoplasmic surface are responsible for the inability of many substances to penetrate the protoplasmic body.

Cells, in general, are impermeable to the highly dissociated sulfonated acid dyes. Cells can be immersed in aqueous solutions of the sodium and potassium salts of the Clark and Lubs series of sulfonphthalein pH colorimetric indicators and left indefinitely in the solutions; the living cells give no sign of coloration. When the aqueous solutions of these indicators are microinjected into a great variety of living cells, the indicators spread evenly through the cytoplasm and, on reaching the boundary, do not diffuse out. The injections, unless in large quantities, cause no adverse effect. These experiments indicate that there is a surface bounding layer which is responsible for the impermeability of the cell to these indicators. Since these indicators are soluble only in aqueous solutions, the readiness with which they diffuse through the interior of the cell shows that the immiscibility of the protoplasmic unit with the external environment must be due to an intervening protoplasmic surface film which these indicators do not penetrate.

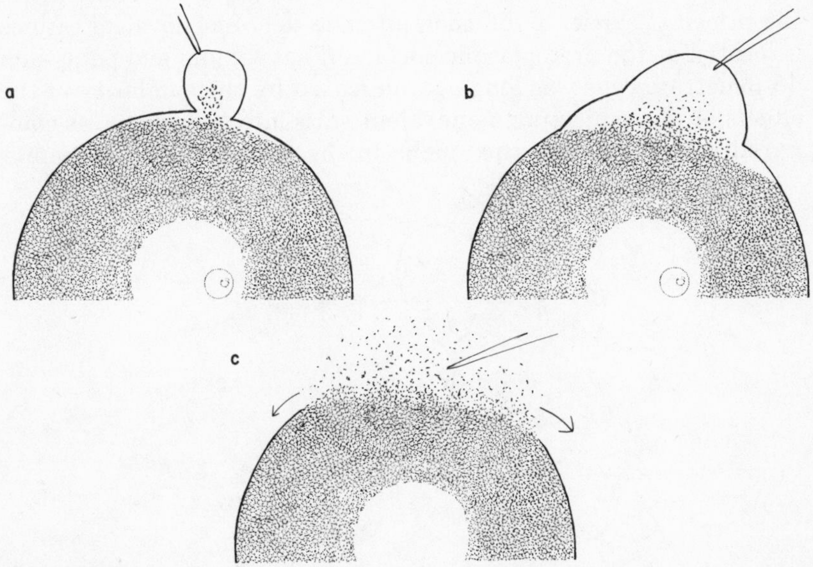

FIGURE 6.3. Enlarged insemination cone, or blister, on immature *Arbacia* egg denuded of its extraneous coats. (a, b) Successive views while blister is manipulated to induce spreading at base. Note partial escape of granular material from underlying cytoplasm into blister. (c) Blister pricked with microneedle. Wave of disruption sweeps around egg and exposed cytoplasm undergoes cytolysis.

Other micromanipulative experiments indicate that the barrier to the penetration of calcium ions into a living cell is located at the protoplasmic surface. When a fresh water ameba is immersed in a solution of 0.1 M calcium chloride, it moves about undisturbed for a day. However, if the surface of the ameba is pricked with a microneedle (Figure 6.4a), the exposed cytoplasm in the vicinity of the tear immediately coagulates (Figure 6.4b). This effect spreads inward as the ameba moves away, while contracting, in successive attempts to close the exposed gap and to pinch off the involved region (Figure 6.4b, c). This reaction results in a steadily lengthening column of solidified material, while the still living part at the end farthest from the tear continues attempting to pinch itself off (Figure 6.4d, e). Finally, this part also succumbs, and the entire ameba becomes converted into a solid column of coagulated material (Figure 6.4f). This experiment shows that only when the surface is torn can calcium ions penetrate the interior in appreciable amounts. Similar experiments on echinoderm eggs also show that the barrier to the penetration of calcium salts is located at the protoplasmic surface.

Evidence obtained using micrugical methods has shown that the principal barrier to diffusion into the cell of monovalent cations is located at the protoplasmic surface. That sodium and potassium do penetrate cells was long ago indicated by the similarity of the effects of microinjecting monovalent salts into the ameba, as compared with immersing the ameba in the same solutions (Chapter

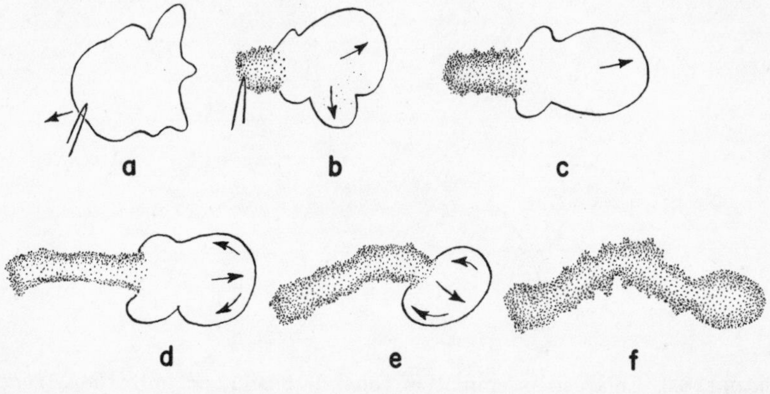

FIGURE 6.4. *Amoeba dubia* torn in 0.08 M $CaCl_2$ solution. (a) Before tearing. (b–e) Ameba flows away from region which has been torn. Inward diffusion of calcium chloride and retreat of ameba from tear converts ameba into ever lengthening column of coagulated material until (f) entire ameba is set and is dead.

11). Upon immersion, the effects appear after considerable delay, in comparison with the immediately observed effects of the microinjections. The rapid rate of diffusion of these salts in the protoplasmic interior is made evident by microinjecting them as salts of colored anions.

The conclusion that the cell surface is the principal barrier to the passage of monovalent cations has been established on a more quantative basis by the use of radioactive sodium and potassium, and by electrical conductivity measurements. The radioactive ions have been found to pass into or out of resting cells, and at rates very much slower than would be expected on the basis of free diffusion.[1] The slow rate of exchange is not due to ion binding in the cell interior. For example, when radioactive potassium is microinjected into single giant nerve fibers, the tagged ions are found to diffuse as rapidly in the axoplasm as in an aqueous salt solution.[2] These data are corroborated by the finding of a very low electrical conductivity across the cell membrane in contrast to the high conductivity of the cell interior (approximating that of a salt solution of similar composition).[3]

From the experiments described in this chapter we may conclude that the principal barrier to the inward diffusion of nonpenetrating or slowly penetrating substances resides at a differentiated protoplasmic surface layer. Bütschli's proposal that the permeability of the cell to all substances is a property of protoplasm as a whole is, therefore, incorrect. Nonetheless, Bütschli deserves credit for having drawn attention to this feature, since in certain instances, in addition to the protoplasmic surface, the internal protoplasm, or intraprotoplasmic components, do restrict diffusion.

We have already mentioned that even those compounds which enter cells rapidly do so at rates considerably slower than would be expected on the basis of free diffusion. We know little about the location of the barrier for such rapidly penetrating substances. Water is an example of a compound which passes into or out of most cells with great rapidity. An analysis of the osmotic properties of cells indicates that in addition to the cell membrane, the internal protoplasm restricts the diffusion of water (Dick, 1959).

[1] Brooks (1937, 1940); Keynes (1951).

[2] Hodgkin and Keynes (1953). Furthermore, radioactive sodium diffuses in axoplasm almost as fast as in free solution (Hodgkin and Keynes, 1956). Vapor pressure measurements in living frog muscle indicate that the soluble constituents of the sarcoplasm are all in the free state (Hill and Kupalov, 1930): the predominant intracellular cation being potassium, "there is no latitude for supposing that an appreciable part of the K is combined."

[3] Cole and Hodgkin (1939); Caldwell (1955); see also values for specific protoplasmic resistance and specific membrane resistance in Shanes (1958), his Table 1.

Possibly the restricted diffusion of water in protoplasm and the limited capacity of protoplasm to absorb water, long ago noted by Bütschli, may have a common basis in the presence of forces which resist excessive separation of submicroscopic cytoplasmic components as water enters the cell.

Protein constitutes 20 to 30 percent of the wet weight of cells. Consequently, the density of fixed charges in protoplasm must be very high. It would hardly be surprising to find an alteration in the rate of diffusion of charged particles through this "framework." As mentioned above, no evidence could be obtained for any limitation in the rate of diffusion of monovalent cations in the axoplasm of giant nerves. However, the possibility cannot be ruled out that the diffusibility of a small fraction of these ions is restricted. A small fraction of the potassium contained in isolated mitochondria is held quite firmly (Bartley and Davies, 1954). Abelson and Duryee (1949) followed the penetration of radioactive sodium into immature frog eggs by making radioautographs of frozen sections at successive intervals. They concluded that the sodium ion diffuses through the cytoplasm of the egg at a markedly slower rate than through Ringer's solution.[4] With regard to the divalent cation calcium, we have already seen that the principal barrier to its inward diffusion is located at the protoplasmic surface. Nonetheless, the diffusibility of calcium in the cell interior is markedly restricted, almost all of it being present in a bound form (Chapter 11; Hodgkin and Keynes, 1957).

We need to keep in mind that suspended in the cytoplasmic matrix are various inclusions and organoids, any of which may serve as subsidiary barriers. In a cell containing vacuoles the rate of interchange of a substance between the protoplasmic interior and the environment may be determined, in part, by the ease with which the substance in question traverses the vacuolar membranes. For example, the tonoplastic membrane enclosing the central vacuole in plant cells possesses permeability properties quite different from those of the external protoplasmic surface (Chapter 16). Brooks (1940) has shown that radioactive potassium diffuses more rapidly from the external environment into the protoplasm of plant cells, than from the protoplasm into the central vacuole. A similar condition may exist in animal cells, in which vacuolar inclusions are smaller, but frequently very numerous. Another example is the mitochondrion, recent evidence indicating that this

[4] The cytoplasm of the immature frog egg contains many yolk globules. Is it possible that the presence of these globules may have influenced the measured diffusion rate of the sodium ion?

organoid is surrounded by a selectively permeable membrane (Palade, 1953). Measurements using radioactive potassium have shown that potassium in the cell interior exchanges with the same cation in the external environment at a nonuniform rate.[5] This behavior is best explained on the basis that the diffusing ions encounter multiple barriers, such as the protoplasmic surface and one or another subsidiary cytoplasmic barrier.[6]

[5] Harris (1957); E. L. Chambers (1949).

[6] An uninvestigated feature concerns the possible influence of the system of membranes associated with the endoplasmic reticulum on the rate of diffusion of substances through the cell interior. This system of membranes is extensive in some cells and sparse in others (Palade, 1956).

7

New Formation of the Protoplasmic Surface Film

DeVries and his followers concluded that the plasma membrane is an organ which, like the cell nucleus, can be derived only from structures already existing. On the other hand, Pfeffer (1899) believed that any part of the protoplasm, freshly exposed to an aqueous environment, immediately acquires a new protoplasmic surface and assumes the function and character of a limiting plasma membrane.

Any part of the protoplasmic matrix is capable of forming new surface films as long as the cell is exposed to the appropriate environment.

The formation of new surface films is readily observed in eggs immersed in sea water after puncture of the germinal vesicle of an immature starfish egg has induced a localized cytolytic reaction in the cytoplasm (Figure 13.2), or after the surface of the egg has been sharply torn with the microneedle (Figure 7.1a). As the exposed cytoplasm starts disintegrating, a succession of new films is seen to sweep across the gap, continually forming and breaking down again (Figure 7.1b). Ultimately a stable film is formed, walling off whatever healthy cytoplasm remains. The process is described more fully below.

> Following breakdown of the surface, in the exposed cytoplasm vacuoles of 2 to 3 micra in size swell and burst. Their fluid contents then merge with the protoplasmic matrix, in which the microsomes and other fine granules present now exhibit active Brownian movement. While this is happening, surface films appear which sweep around masses of the disorganizing region to form spherules of all sizes, the larger ones containing fine particles in active Brownian movement. These spherules swell and ultimately burst until the entire

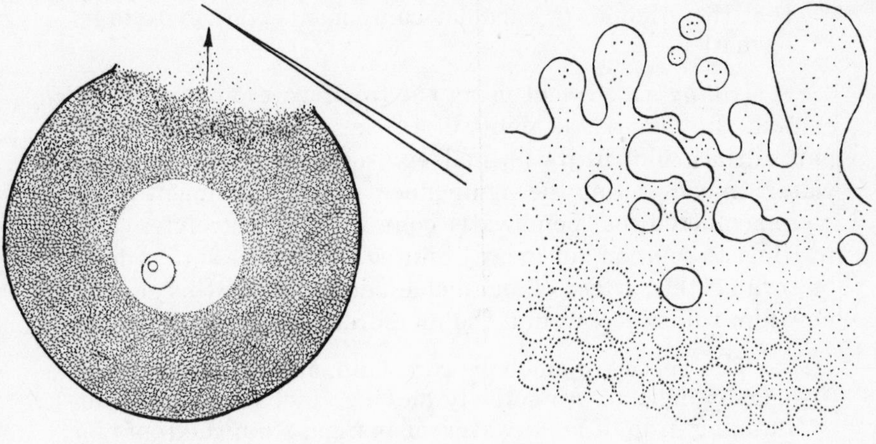

FIGURE 7.1. (a) Living starfish egg disintegrating as result of shooting tear through cortex. (b) Magnified sketch of torn margin, showing sweeping of new surface films over outpouring cytoplasmic material. New films continue to form and to break down until eventually disintegrative process is walled off.

mass is dissipated. In all this commotion the observer is struck by the succession of surface films which form and are as continually broken down (Figure 7.1b). Occasionally a film sweeps around a part of the cytoplasm which is not disorganized, converting it into a globule of apparently healthy cytoplasm. Such a globule maintains its normal appearance as long as it is surrounded by its protective film. It is only the globules of disorganizing material which swell and burst. A surface film may also form a short distance ahead of the boundary of an advancing zone of disintegration. Occasionally a film may form which reaches the margin of the still intact film which surrounds the healthy part of the cytoplasm. Thereupon, further disintegration ceases, and the cell remnant gradually rounds up. The remnant of the mature egg is fertilizable; it will segment and undergo further development. On the other hand, if the extracellular environment has affected the cytoplasm beyond the margin of the advancing film, disorganization starts in again behind it, and the film breaks down and disappears. Another film then forms farther ahead of the advancing wave of destruction. In this way several successive films may form and break down before further destruction is finally stopped by a film which persists. The film apparently is formed from ingredients in the hyaline matrix of the cytoplasm. There is no indication that the granular or vacuolar components contribute to its formation.

A striking experiment using the fresh water ameba, *A. dubia,* exhibits the remarkable ability of a living cell to form a completely new surface film in its normal environment, in this case pond water. In considering the significance of this experiment it should be remembered that pond water contains the electrolytes common to Ringer's solution but greatly diluted. The entire surface film can be torn off the surface of an ameba, and following this an entirely new film will sweep around the membraneless protoplasmic mass.

> This species of ameba is very fluid, and its pellicle is unusually delicate. An actively moving ameba is placed in an hanging drop of pond water suspended from the roof of a microdissection moist chamber. An horizontal microneedle is then inserted into the hanging drop, and the ameba, actively moving on the surface of the coverslip, is gently released from its attachment by means of the needle. The ameba, being heavier than the water, falls through the drop. Usually, when an ameba reaches the bottom of the drop it tends to be caught in the water–air interface, where it is ruptured by the tension forces. The delicate pellicle, with the surface film, is ripped off, the granular contents of the ameba are exposed and scattered over the surface of the drop. However, in this case, just after the pellicle had been ripped off, the still adhering granular mass, including the nucleus and contractile vacuole but lacking a surface film, is seized by the tip of the needle and quickly lifted into the deeper region of the drop. When this is done a fresh protoplasmic film immediately regenerates at the surface. Complete recovery of the ameba ensues.

The readiness of the protoplasm of cells in their normal electrolytic environment to form new surface films subsequent to surface tears varies greatly according to the extent of the trauma and the type of cell operated upon. For example, the immature starfish (*Asterias*) egg will recover readily from large tears, while the mature *Arbacia* egg may disintegrate entirely following a simple puncture with the microneedle.

Conditions for the formation of new surface films are at an optimum when the electrolytes sodium, potassium, and calcium are present in the external medium in the proportional concentrations characteristic of the cell's natural habitat. In the case of the marine starfish eggs, film formation readily occurs following small tears of the surface of eggs immersed in solutions containing the same concentrations of sodium, potassium, and calcium as sea water. If the calcium is omitted from the mixture, repair from a tear is definitely impeded. The granular contents of the egg then simply disperse rapidly out through the gap (Chapters 6, 11). However, when an egg is torn in a solution of calcium chloride so strong as to be isosmotic with sea water, new surface film formation does not take place due to a too rapid, inwardly progressing clotting reaction, which converts the interior of the cell into a frothy coagulum.

The action of calcium and of sodium in relation to surface film repair seems to be related to the opposing effects of the two ions on the cytoplasm's ability to generate new surface films. These two reactions are the coagulating action of calcium and the dispersing action of the monovalent salts (Chapter 11). The needed balance between the two opposing tendencies is obtained in physiological salt solutions. In these the sodium and calcium chlorides present are mixed in proportions of 50 or 75 to 1. The antagonistic action of the two salts in this mixture is such that the dispersing action of the sodium is almost negated, while the coagulating effect of the calcium is only weakly evident. When the torn cytoplasm is exposed to this combination of sodium and calcium, the rate and extent of dispersion, and the inactivation by coagulation of the cytoplasmic components, are sufficiently gradual and limited to allow the occurrence of just those specific reactions which are needed for surface film formation. Presumably these reactions involve an orientation of surface film ingredients at the boundary between the healthy cytoplasmic material and the aqueous environment. The newly formed surface films are as impermeable to the sulfonphthalein indicators (for example, phenol red) as is the original.

An increase in area of the protoplasmic surface film, which occurs when cells divide or when cells are stretched or cut into fragments with microneedles, is explained not only by the stretching of a pre-existing surface layer, but also by the addition of surface layer material from the underlying cytoplasm. As long as there is no rupture of the surface, this type of surface formation occurs readily in isosmotic solutions of calcium chloride or in isosmotic solutions of monovalent salts in the complete absence of calcium. This is shown, for example, by the continued survival and repeated cleavages of fertilized sea urchin eggs completely divested of their extraneous coats and transferred through several washings into a mixture at pH 7.0 of sodium and potassium chlorides (in proportions of 19 to 1) isosmotic with sea water and containing sodium oxalate or citrate to remove traces of calcium which might be derived from the few eggs which undergo cytolysis. The absence of divalent salts prevents the formation of cement for binding the cells together but does not prevent the great increase in surface area of the surface films of the many dozens of blastomeres which form.

8

Physical Properties of the Protoplasmic Surface Film

Before the properties of the protoplasmic surface film can be discussed, it is necessary to differentiate clearly between the film and extraneous coats applied to the surface of the cell. In nature the protoplasmic surface film is seldom, if ever, exposed. Extraneous material adheres so closely to the external surface of the surface film that it is frequently difficult to differentiate the one from the other. However, the contrasting effects of calcium on the two can be used for their differentiation by immersing cells in an isosmotic solution of the chloride salt. The calcium ion stiffens the extraneous coats; in contrast, the characteristic fluidity of the surface film is maintained. Such effects are clearly demonstrated in the following experiment.

> Unfertilized *Arbacia* eggs, which live in a marine environment, are transferred to a drop of calcium chloride isosmotic with sea water and are mounted on a coverslip. Through the microscope many of the eggs are seen to be stuck to the surface of the coverslip. The end of a blunt-tipped microneedle is then brought against one of the eggs and moved so as to push the egg ahead of the needle. Extraneous material, principally the so-called vitelline membrane still surrounding the egg, stiffens because of the calcium present and adheres firmly to the coverslip. The pushing action of the needle causes this investing material to break in places, whereupon the enclosed egg can be made to slip out of a relatively rigid shell. The naked egg rounds up immediately. It is no longer sticky and can be rolled about freely. When pressed against the coverslip with the horizontal shaft of a microneedle, the denuded egg can be pinched into two portions, each of which immediately assumes a spherical form. The egg behaves very much like

an oil drop and can be divided repeatedly to form smaller, spherical droplets. However, the operation is unsuccessful if the egg surface is even slightly torn with the tip of the needle. The tear then immediately opens up, and the interior of the egg, now exposed to the calcium of the medium, becomes converted into a hard, frothy coagulum.

The continued existence of the surface film as an integral layer is essential to the life of the protoplasm it encloses. Immediately upon the death of the cell, whether this is caused by a tearing of the surface film or by the induction of general coagulation, as occurs when fixatives are used, the surface film dissipates, and the selectively permeable properties of the protoplasmic unit are lost. Extraneous coats can be removed without detriment to the life of the protoplasmic body. These are highly permeable structures and persist intact after the death of the cell.

The intimate relation between extraneous coats and the protoplasmic surface has caused many investigators to reach erroneous conclusions regarding what is or what is not the actual protoplasmic surface film. Animal cytologists, particularly, have loosely used the term "cell membrane" and have identified this as being identical with the plasma membrane, the selectively permeable barrier of the cell, even after the protoplasm has been destroyed by fixation and dehydration. In view of the extraordinarily labile nature of the surface film, caution is indicated in attempting to localize this structure in sections of cells fixed in osmic acid[1] or other preparatives for electron microscopy. Furthermore, unless measures are taken to remove all extraneous material, it is likely that whatever appears as a "cell membrane" is, in fact, an extraneous coat.[2]

[1] In studies made with the electron microscope Robertson (1957, 1959) described at the cytoplasmic surface of many different cells a three-layered membrane about 75 Å thick. In Schwann cells the ~75 Å unit composes half of each myelin lamella. The electron microscopic picture and x-ray diffraction data suggest that the ~75 Å unit represents a bimolecular leaflet of lipid, the polar surfaces of which are covered by monolayers of nonlipid material. Whether or not this membrane represents the semipermeable barrier of the protoplasmic unit is undetermined, but the membrane structure visualized and its dimensions are consistent with a great deal of indirect evidence concerning the structure of the protoplasmic surface film. For example, a ~50 Å thick bimolecular layer of lipid has been estimated, from electrical measurements, for the semipermeable barrier of the giant axon of the squid (Cole and Hodgkin, 1939; Curtis and Cole, 1938).

[2] Electron microscopic studies on the ghosts of erythrocytes (Hillier and Hoffman, 1953) have been carried out to provide information about the structure of the plasma membrane. The ghosts were obtained by exposing red cells to relatively large volumes of hypotonic salt solutions. We believe such studies may not have relevance to the surface film, since the hypotonic treatment should have caused its disintegration. There is no indication that the ghosts retained the selectively permeable properties of intact red cells, or even of red cells lysed by brief exposure to a small volume of water (Teorell, 1952). The structure studied may represent rather the stroma of the cell or a differentiated extraneous coat which, when the cell was living, was applied to the external surface of the surface film. (See also Chapter 2, footnote 10.)

Mechanical and chemical means may be used to denude cells of their extraneous coats. By mechanical means coatings which are sufficiently brittle may be broken off and eliminated by shaking. By chemical means the investing cement-like substance covering the protoplasmic surface can be dispersed by excess monovalent cations or by chelating agents which bind divalent ions. Moreover, enzymes (for example, trypsin) which are active within the pH range of the normal environment of the cell have been used to digest off extraneous coatings.

In experiments on living cells in their natural environment, only freshly exposed surfaces can be considered naked, since extraneous material is constantly being generated by the protoplasm. In the presence of divalent salts, which are always present normally, this material may be deposited as insoluble salts over the cell surface. In many cases the naked surface can be maintained by keeping the cell in a solution of monovalent salts, which have a dispersive action on this secreted material.

The best criterion found thus far for testing the nakedness of the protoplasmic surface is the oil-coalescence method (Chapter 2). This involves the engulfment by the protoplasm of a liquid, water-immiscible drop, such as an oil drop, with a surface tension higher than that of the cell (Figure 8.5). An oil drop exuding from the tip of a micropipette is brought into contact with the surface of the cell, as described later in this chapter. Whether or not coalescence occurs depends upon (1) the extent to which the protoplasmic surface is exposed by removal of extraneous coatings, (2) the surface tension of the oil being used, and (3) the size of the oil drop applied. Eggs possessing a jelly layer or a tough extraneous coat do not coalesce with oil drops. Coats impervious to the oil prevent the oil from making contact with the egg surface. However, coalescence may occur when a structurally weak extraneous coat, such as the tenuous vitelline membrane of an unfertilized *Arbacia* egg, is present. The greater the rigidity of the coatings, the higher must be the surface tension of the oil and the larger the size of the oil drop applied, to achieve coalescence. Removal or weakening of such coatings permits coalescence of the cell with a small oil drop of relatively low surface tension.

Specific illustrations of the methods used to obtain denuded plant and animal cells are described below.

Naked, living protoplasts can be obtained from plant cells with rigid walls by using the method described in Chapter 21.

Plant materials which yield spherical cells entirely free of extraneous coats are the aplanospores of the coenocyte *Valo-*

nia ventricosa, a subtropical marine plant cell. When a *Valonia,* which is usually turgid in its natural condition, is punctured and caused to collapse, the multinucleated protoplasmic layer underlying the cellulose wall breaks into minute mono- or binucleated ameboid bodies, the aplanospores. During the first hour they are essentially naked bodies of protoplasm. This is revealed by their readiness to coalesce with small drops of relatively low surface tension oils. Subsequently their surfaces become coated with a submicroscopic layer of extraneous material (Chapter 3).

A close approach to a naked protoplasmic surface film is a mechanically induced exovate of an unfertilized starfish egg. The egg, enclosed in its closely fitting vitelline membrane, is seized by a needle and dragged to the edge of a shallow drop of sea water. The vitelline membrane including the margin of the egg directed toward the deeper part of the drop is slightly torn, and, by means of the other needle, the egg is slowly pulled further toward the edge of the drop. The resulting compression of the egg causes the formation of an exovate enclosed within a surface film. The cytoplasm of the *Asterias* egg has an unusually ready capacity to form surface films, so that the exuding material forms itself into an exovate of appreciable size.

A freshly fertilized sea urchin egg can be denuded by shaking off its fertilization membrane and immersing the egg in a monovalent salt solution, as described in Chapter 2.

Our knowledge of the characteristics of the naked protoplasmic surface has been derived chiefly from marine eggs, denuded as above described. Similar studies have been made on the isolated protoplasts of plant cells and on the aplanospores of *Valonia.*

A striking property of the naked protoplasmic surface is the absence of any sign of wrinkling when the surface is mechanically deformed or is caused to diminish in area. This is well shown by tearing the polar surface of a naked, dividing *Arbacia* egg in a solution of potassium chloride isosmotic with sea water. The absence of divalent cations in the medium permits the continued outflow and dissipation of the granular interior of the egg for an appreciable length of time before the disintegration of the cortex. In the case illustrated in Figure 8.1, the incipient blastomeres were initially connected by a wide stalk. After tearing, the torn blastomere dissipated (Figure 8.1a), and the contents of the intact blastomere flowed out through the persisting stalk (Figure 8.1b, c). The intact blastomere shrank in size with no wrinkling of the surface, like a gradually deflating balloon.

The surface area of the protoplasmic surface film of a naked

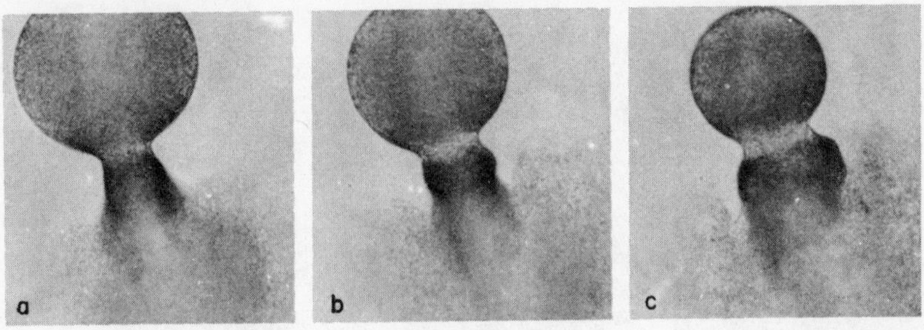

FIGURE 8.1. Polar end of one of two incipient blastomeres of naked living *Arbacia* egg torn in potassium chloride solution isosmotic with sea water. (a) Contents of torn blastomere dissipate while furrow persists. Outflow through open connection causes other blastomere to shrink. (b, c) Continued outflow, with smooth contour of shrinking blastomere maintained. (From Chambers, 1938a.)

cell can be greatly increased without causing the film to rupture. Denuded sea urchin eggs, mounted in a drop of a mixture of sodium and potassium chlorides (19 to 1) isosmotic with sea water, may be compressed between the coverslip of the moist chamber and a fragment of coverslip cemented to the tip of a microneedle to many times their original diameters without showing injury. The limiting factor in this experiment is the necessity of avoiding shearing during the process of compression.

The extent to which the surface may be deformed and its area increased can also be well demonstrated by drawing slender strands out from the surface of a denuded sea urchin egg, an artificially produced exovate of a starfish egg, or the isolated protoplast of a plant cell. These strands of protoplasm, when pulled out with a needle, are highly fluid; they either pinch off and round up immediately or they can be drawn into a lengthening cylinder which breaks into spherical beads joined together by a delicate filament (Figures 8.2, 8.3). As long as this cohering filament persists, a release of the pull causes the beads to merge with one another and with the protoplasmic body from which the material originated (Figure 8.2). No coalescence occurs if the cohering filament breaks apart.

At the naked surface of the cytoplasm of unfertilized *Arbacia* eggs movements resembling currents of flow can be induced by churning (Figure 8.4). In this experiment unfertilized *Arbacia* eggs, freed mechanically from their jelly coats, were mounted in a broad, shallow hanging drop of a solution of calcium chloride isosmotic with sea water (0.34 M). An egg was pushed with the side of

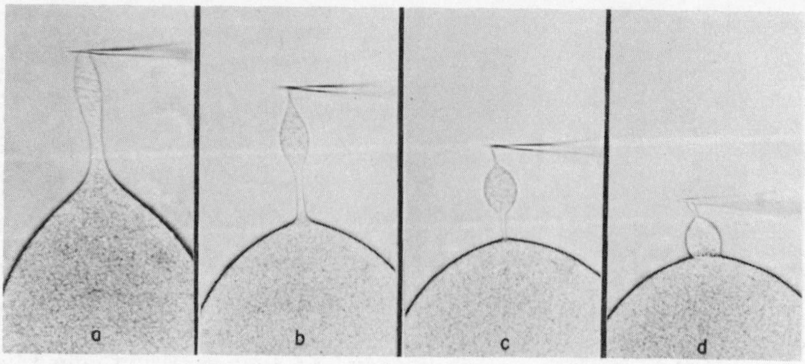

FIGURE 8.2. Four successive views of strand, drawn out from surface of naked exovate of living starfish egg, tending to bead up (b, c), as strand is allowed to return and merge (d) with cytoplasm of egg. (From Chambers, 1938a.)

a blunt microneedle which had been already coated with proteinaceous material. By gentle manipulation the egg could be shelled out of its vitelline membrane, now become rigid by the stiffening action of the calcium salt in the medium. The shelled egg, now naked, was pushed along as it was held against the coverslip by the shallowness of the hanging drop. While the egg was being progressively moved in this fashion, peripheral currents appeared in the cytoplasm, streaming forward along its two sides and backward along the top and bottom of the egg. Minute extraneous particles adhering to the surface showed by their movement that

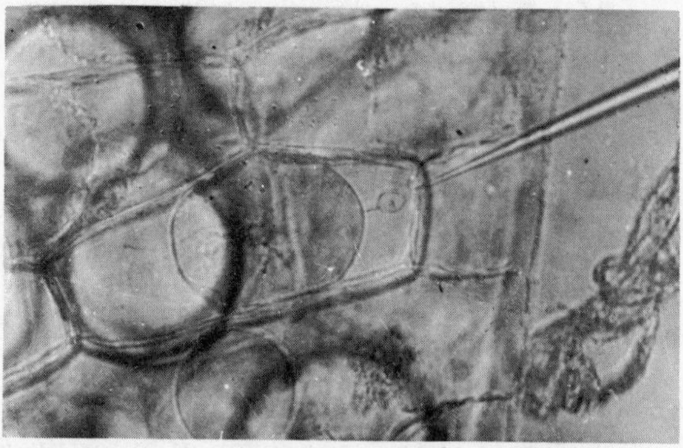

FIGURE 8.3. Plasmolyzed living protoplast with microneedle inserted through end wall of cell into protoplast and then withdrawn. Note protoplasmic strand extending from surface of protoplast. Greater portion of the material of strand has collected into a bead.

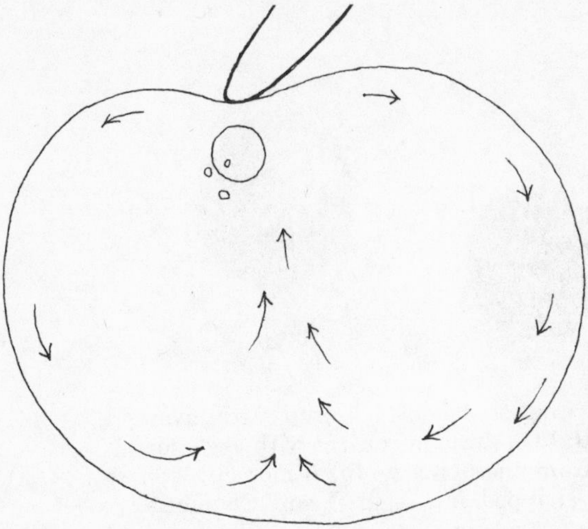

FIGURE 8.4. Naked living *Arbacia* egg being pushed (in a direction toward bottom of page) with blunt needle. Arrows show currents of flow.

the streaming involved the surface. A drop of olive oil was applied with a micropipette to the surface of the cytoplasm. The drop adhered as a cap. When the egg was churned, the oil cap was carried forward along the side of the egg and backward along its undersurface. This movement followed the same direction as the streaming movements in the peripheral cytoplasm.

A striking physical feature of the naked protoplasmic surface is that it permits the denuded cell to coalesce with an oil drop. Provided the protoplasmic surface has been sufficiently exposed by the removal of its extraneous coatings, as has been discussed previously, the determining factor is the magnitude of the tension at the surface of the oil in contact with the aqueous medium. When the tension of the oil is high, the oil snaps in and is engulfed by the naked protoplasm (Figure 8.5). The experimental procedure is described below.

Mature *Arbacia* eggs (75 micra in diameter) were carefully washed free of extraneous material, suspended in a mixture of sodium and potassium chlorides (Chapter 2), and mounted in an hanging drop of sea water in a moist chamber of the micromanipulator. A micropipette containing pure mineral oil (a nonpolar oil with a relatively high surface tension at an oil–water interface) was raised and the tip brought to a distance of 3 to 4 micra from the margin of an egg. A drop of oil was caused to exude from the tip. When the enlarging drop made intimate contact with the surface of the egg, the oil instantly snapped in. In this case, at the

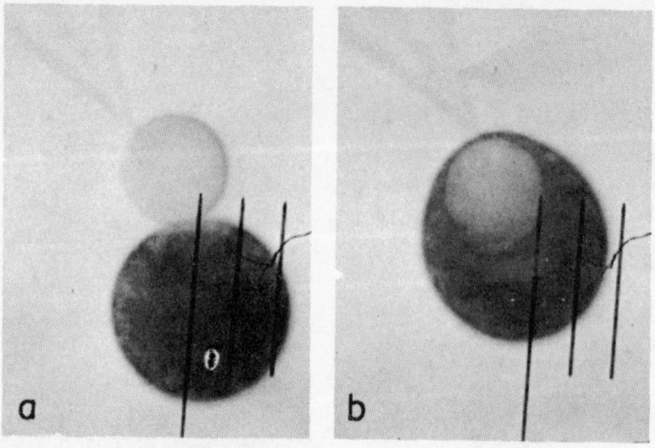

FIGURE 8.5. Coalescence of oil drop with living *Arbacia* egg. (a) Oil drop, in contact with egg's surface, exudes from micropipette. (b) A moment later oil drop has snapped into egg. (From Chambers, 1938a.)

time of coalescence, the oil drop was 10 to 15 micra in diameter. A drop as large as half the diameter of the egg, or larger, will coalesce without causing rupture of the naked egg when this is immersed in sea water. The penetration of the oil is sudden (less than 1/32 second), and motion pictures taken at the rate of 72 frames per second during the penetration revealed no deformation of the oil drop during its passage into the egg.

A recital of what happens when a cell coalesces with an oil drop is as follows. As the drop penetrates, the underlying, gelated cortical layer of the cytoplasm is pushed inwardly and laterally. During this time the gap, being created by the penetrating oil, enlarges by the diameter of the entering oil. As the penetration is being completed, the gap narrows, the cytoplasm with its surface film flows over the oil drop, and the engulfment is completed. The phenomenon occurs only under conditions which determine the coalescence of two liquid spheres in which the drop with the lower surface tension engulfs the drop with the higher tension irrespective of their composition. Thus, a droplet of mercury will be engulfed by a drop of any other liquid with a lower surface tension, such as water or a low surface tension oil. The snapping of oil drops into naked cells indicates only that the protoplasmic surface is liquid and possesses an interfacial tension lower than that of the droplet engulfed.

The lack of wrinkling following induced shrinkage of any naked cell and the ability of the naked *Arbacia* egg to be greatly flattened by compression can be explained on the basis that the surface film has the properties either of an elastic solid or of a liquid. However, the last three experiments involving the pulling out of strands of the surface, the churning, and the coalescence of cells with oil drops indicate the fluid nature of the protoplasmic surface film.

The fluid nature of the surface film is maintained irrespective of the electrolytic composition of the external medium. Churning movements at the surface of eggs can be induced, and coalescence with oil drops occurs, regardless of whether the naked eggs are immersed in a solution of calcium chloride isosmotic with sea water or in isosmotic solutions of sodium or potassium chloride. The maintenance of the surface film in its liquid state, even in the presence of high concentrations of calcium chloride, is in marked contrast to the stiffening action of calcium chloride on the extraneous coats and on the protoplasmic interior (Chapters 2, 11).

A significant feature is that the surface film is rendered unstable and dissipates when it becomes surrounded on both sides by protoplasm. *Amoeba dubia* can be caused, so to speak, to flow over itself, by repeatedly prodding the tip of an advancing pseudopodium with a microneedle. The surface of this species is invested with an extremely delicate pellicle. Occasionally a forming pseudopodium can be caused to roll over to one side and flow over the body of the ameba, bringing about a situation in which the original surface of the ameba is surrounded on both sides with protoplasm. This condition results in progressive erosion of the surface film, disappearance of the contiguous surfaces, and fusion of the backwardly directed pseudopodium with the body of the ameba.

The interposition of the protoplasmic surface film between two masses of protoplasm, and the resulting disappearance of the film, can be demonstrated in the naked, transected protoplasts of plant cells. With microneedles inserted horizontally into the cut-open end of a plant cell (Chapter 21), it is possible to divide the protoplast within its cellulose wall into two sections, each surrounding a segment of the central vacuole. After separation, the two intact segments of the protoplasts are pushed into contact. The protoplasmic surface film between the two disappears, and there results immediately a complete fusion of the two segments.

9

Nature of the Protoplasmic Surface Film An Hypothesis

As we have seen, the protoplasmic surface film is liquid and water immiscible. Of the organic compounds, only lipids possess these two properties, and lipids, therefore, should enter into the composition of the surface film.

Since a surface film can form over any part of the cytoplasm when this is exposed to the external medium, as by a tear, the components which constitute the surface film must be distributed throughout the cytoplasmic matrix. The very low interfacial tension at the surface (Harvey, 1954) reveals that at least the superficial molecular components of the protoplasmic surface contain many hydrophilic groups. We conclude from this information that proteins also enter into the constitution of the protoplasmic surface. Since enzymes are present in the cytoplasmic matrix, we should expect to find them at the surface also; evidence for this has recently been presented (Rothstein, 1954).

In considering the structure of the protoplasmic surface, we need to take into account certain characteristic features of the protoplasmic interior. One of these is the absence of any appreciable surface activity on the part of the native protoplasmic proteins. This is indicated by the fact that when a drop of an inert oil is microinjected into the cytoplasm of a living starfish egg there is no evidence of adsorption of proteins at the cytoplasm–oil interface. Only when cytolysis sets in do the proteins become surface active and form adsorption films at oil–cytolyzing protoplasm interfaces. The native proteins presumably form aggregates of molecules constituting a continuous network dispersed through the continuous aqueous phase. Presumably, interchain forces prevent the protein molecules from unfolding and collecting at oil–cytoplasm interfaces. However, upon cytolysis, the protein molecules of the protoplasmic matrix disaggregate. They become, therefore, surface active and collect at oil–cytolyzing protoplasm interfaces to form palisade-like films, their hydrophilic groups directed outward and their hydrophobic groups inward toward the oil.

As we have seen, when the protoplasmic interior is exposed to the external environment, as by a tear, surface films sweep back and forth at the margin between healthy and cytolyzing cytoplasm. It is possible that the tendency of the protoplasmic proteins to become surface active when exposed to the external environment plays a role in orienting the lipoprotein molecules during formation of new surface films. We have already drawn attention to the fact that no orientation is possible, and the protoplasmic surface film disperses, when it is bounded on both sides by the protoplasmic interior.

As an hypothesis we conjecture that certain lipoproteins at an exposed surface of the protoplasm react with the external aqueous medium by orienting their hydrophilic groups toward the external aqueous medium in a palisade arrangement while the hydrophobic groups of the lipid side chains are directed inward.[1] Escape of the lipoprotein molecules from the film to the exterior would be prevented by the attachment of their protein chains to other protein molecules, which presumably form a network of aggregates extending throughout the protoplasmic interior. The outwardly directed hydrophilic groups would be predominantly anionic, since the cytoplasmic proteins, as we have previously stressed, have isoelectric points well to the acid side of the pH of the environing media. The exposed negative groups would be free to combine with the cations in the aqueous media.

The oriented lipoprotein molecules in the surface film would be constantly shifting and being replaced by other molecules in the protoplasmic interior. The extent to which this is a simple interchange, or removal of molecules in the surface film to the exterior and their replacement from the interior, would depend upon the ions in the external medium with which the hydrophilic groups can combine.

When the divalent calcium ion makes contact with the exposed negative groups of the molecules in the film, cross linkages would be formed which would act to stabilize the palisade arrangement and hold the component molecules in place. This would account for the stability of the surface film in the presence of calcium and

[1] The lipid chains may comprise a double layer (Chapter 8, footnote 1), with their hydrophobic groups in apposition and the hydrophilic groups projecting in opposite directions into the two aqueous phases: the external medium and the protoplasmic interior.

the ability of this ion to promote new surface film formation. Calcium, in forming undissociated complexes and cross linkages with proteins, would tend to prevent shifting movements of molecules between the surface and the protoplasmic interior. The calcium ion, therefore, would penetrate the cell very slowly if at all. There would also be a tendency for insoluble calcium proteinate molecules to be drawn out of the film and to accumulate at the interface between the protoplasmic surface and the external medium. Here the calcium proteinate molecules would take part in the formation of an extraneous coat on the outer surface of the protoplasm.

Sodium chloride is the most abundant salt of the extracellular fluids. When the sodium ion in the medium makes contact with the protoplasmic surface film, the sodium proteinate remains dissociated. This would not prevent shifting movements of the molecules, which would continue to be readily displaceable by underlying molecules. Thus sodium would gain ready access to the interior. The same would be true for the potassium salts.

An interchange of lipoprotein molecules between the protoplasmic surface and the interior would account for decreases and increases in its surface area. The fact that reduction in surface area of the naked protoplasmic surface of a shrinking cell occurs without wrinkling indicates that this must occur by intussusception of the film substance, with lipoprotein molecules, arranged in palisade fashion at the surface, passing into the interior of the protoplasm in proportion to the reduction of the surface area. On the other hand, when the surface area is increased (by compressing the cell, by forming exovates, or by drawing out filaments), additional lipoprotein molecules from the protoplasmic interior would enter into the palisade arrangement at the surface.

As we have seen, the surface film is maintained intact, although fragile and unstable, in the absence of the calcium ion or for that matter in the absence of any ions, as in solutions of nonelectrolytes. However, repair or replacement of the film (for example, following a sharp tear) requires the presence of calcium, and conditions for repair are optimal when sodium, potassium, and calcium are present in the proportional amounts generally characteristic for environmental fluids. Evidently the presence of calcium is required to achieve an orientation of the lipoproteins at the surface, but if calcium is present in excessive amounts relative to the monovalent salts, orientation and repair will not occur. This is because the excess calcium combines with too large a proportion of the negative groups on the protein chains, converting the exposed protoplasm into an insoluble coagulum.

The evidence indicates that the newly formed surfaces have the same permeability characteristics as the original (more data concerning this feature would be highly desirable). This would be because the orientation and spacing of lipoprotein molecules constituting both the newly formed and the original films are identical.

A palisade arrangement of lipoprotein molecules similar to that at the protoplasmic surface should constitute the films which bound vacuoles lying within the protoplasm. Here the hydrophilic groups of the lipoprotein molecules would project into the aqueous contents of the vacuoles. The close relationship between vacuolar membranes and the protoplasmic surface is indicated by the fact that vacuoles can both originate from and fuse with the protoplasmic surface film. For example, the formation of vacuoles during the process of phagocytosis or pinocytosis involves invagination of the surface and pinching off of the invagination. Conversely, fusion of vacuolar membranes with the protoplasmic surface film is seen in the recently fertilized *Arbacia* egg when echinochrome-containing pigment vacuoles empty their contents into the exterior. Afzelius (1956) has provided evidence from electron microscopy that in sea urchin eggs during expulsion of cortical "granules" immediately after fertilization, the membranes of the "granules" fuse with and become a part of the protoplasmic surface. Vacuoles can also form de novo at any region in the cytoplasm; an example of this is the formation of vacuoles in marine eggs exposed to hypotonic sea water. Thus, the constituents required for the formation of both new surface films (like those that form following a tear) and new vacuolar membranes are distributed throughout the cytoplasmic matrix.[2]

Vacuolar membranes are known to have permeability characteristics widely different from those of the protoplasmic surface.[3] For example, in plant cells the permeability of the surface film to ions (Brooks, 1940) and to sulfonphthaleins (Chapter 16) differs prominently from that of the tonoplastic,[4] or central vacuolar, membrane. Differences in permeability are to be expected, since

[2] The membranes of the endoplasmic reticulum (Palade and Porter, 1954; Palade, 1956; Fawcett and Ito, 1958) might be the constituents of the cytoplasmic matrix responsible for new surface film or new vacuole formation.

[3] The surface of the ameba *Chaos chaos* is remarkably impermeable to glucose. However, when the ameba is induced to ingest glucose by pinocytosis, a process in which vacuoles are formed by invagination of the surface, the glucose incorporated within the vacuoles subsequently passes into the cytoplasm (Chapman-Andresen and Holter, 1955). Whether the glucose diffuses unchanged across the vacuolar membranes or a metabolic absorption occurs involving newly formed enzymes at the membrane is not known.

[4] The tonoplast, like the vacuoles of animal cells, can be isolated intact, but if pricked with the microneedle, it dissipates and fades from view (Chambers and Höfler, 1931). The tonoplastic membrane has the properties of a liquid film, stabilized by an absorbed protein layer. In this respect the tonoplastic membrane resembles the protoplasmic surface film in physical properties, but not the nuclear membrane.

the concentrations of the ionic constituents in the vacuolar fluids are know to differ widely from the concentrations present in the protoplasmic matrix and environmental media. This means entirely different physical conditions prevail at the inner and outer interfaces of a vacuolar membrane, as compared with the membrane enclosing the protoplast. Of particular importance would be the differences in calcium- and hydrogen-ion concentrations (Chapters 11, 13). Such differences should alter the orientation and spacing of the palisaded lipoprotein molecules and, consequently, the permeability characteristics of the membranes.

IV

The Action of Salts on Protoplasm

INTRODUCTION TO PART IV. The role of electrolytes in the maintenance of protoplasmic structure and function is one of the most important aspects of cell physiology. The methods usually employed in studying this question have involved chiefly the immersion of cells and tissues in solutions of various electrolytes. By these methods much has been learned, and it is now well recognized that the presence of the cations sodium, potassium, calcium, and magnesium, and their maintenance in definite proportions, is essential to the cell for its specific activities.

However, immersion methods offer us no means of knowing whether the changes observed within the cell are caused by the actual entrance of the electrolytes in question, by their effect on the surface, or by their action in causing the diffusion of substances from the cell. An analysis of this problem is possible by means of the micrurgical apparatus, which permits not only the manipulation and dissection of the living cell, but also the introduction of electrolytes directly into the internal protoplasm.

10

Critique of Immersion Studies of the Effects of Salts on Protoplasm

"Protoplasm," a broad term, refers to the living part of a cell, a morphological entity possessing a selectively permeable barrier enclosing a matrix containing various structural and chemical elements, together composing a complex which is endowed with the property of being alive. Micromanipulative experiments have indicated that the matrix of protoplasm is a continuous phase and that this phase is aqueous. In the matrix are numerous inclusions, such as membraned vacuoles containing watery fluids, oil droplets, crystals, as well as chloroplasts, mitochondria, and the cell nucleus. The inclusions are either inert or metaplastic and can be mechanically removed with little or no detriment to the life of the cell. The matrix may be quite fluid, as in the case of the streaming protoplasm in the interior of an ameboid cell, the astral rays of the division figure, or the fluid coursing through tortuous channels in the plasmodium of the Myxomycetes. On the other hand, the matrix may have the firm consistency of a gel, forming reversibly gelated structures in the protoplasm, such as the plasmagel of an ameba, the cortical region of the protoplast of certain plant cells, the cell aster with its extended rays, and the channel borders in the plasmodium of the Myxomycetes. In the gelated regions the inclusions are held in place.

The term "protoplasmic viscosity" has been used widely in describing the action of salts on the physical state of protoplasm. Viscosity has significance only in relation to homogeneous fluid systems. Use of the term can be misleading in connection with a polydispersed system, such as protoplasm, consisting of many components with widely differing physical properties, especially when some parts are in a gel state and others in a sol state.

Of principal interest is the effect of the salts on the consistency, or degree of rigidity, of the matrix of protoplasm. In order to evaluate the results which have been obtained, we must also take into consideration the action of salts on the various other components of the protoplasm. For example, when interpreting measurements of protoplasmic consistency obtained by the method of centrifugation, possible alteration of the physical properties of the inclusions must be taken into account. The centrifugation method measures only the ease with which granules or chloroplasts are moved through the protoplasmic interior.

The problem has aroused considerable controversy. Those who have microinjected solutions of salts directly into the protoplasm of cells claim that the calcium ion solidifies or coagulates, and that sodium and potassium either have no effect or tend to liquefy, the hyaline matrix. Other investigators have used the immersion method and they claim the very opposite. The cells which have been most widely investigated are fresh water amebae, echinoderm eggs, and plant cells.

The interpretation of the results obtained from immersion experiments presents difficulties. Questions arise concerning to what extent the findings actually indicate the purported changes in protoplasmic consistency. The fact that the various salts enter cells at different rates is an additional complicating factor.

Several of the earlier investigators[1] drew conclusions about the consistency of protoplasm from the shape of plant cell protoplasts plasmolyzed in the various salt solutions. The smooth oval forms assumed by protoplasts when plasmolyzed in solutions of calcium salts were taken as an indication of decreased protoplasmic "viscosity," while the irregular, sharp, angular shapes observed in solutions of the monovalent salts were thought to represent increased "viscosity." The more probable explanation is that the shapes of the plasmolyzed protoplasts are related to the different surface effects of the mono- and divalent cations.

Under normal conditions extraneous coats invariably surround the protoplasmic unit. The cell may be denuded of these coatings without injury to the protoplasm. Immersion of cells in solutions of the monovalent salts, in the absence of calcium, generally softens such extraneous coatings as the cement substances (Chapter 2) causing them to become adhesive and, in certain instances, to undergo dispersal, while the integrity of the protoplasmic surface film is maintained. Calcium stiffens these coatings without affecting the fluid nature of the protoplasmic surface. As an explanation of these findings, we suggest that when the plant cell is immersed in a calcium salt solution, extraneous material on the

[1] Cholodny (1923, 1924); Weber (1924b).

surface of the protoplast and within the cellulose wall hardens. During plasmolysis, therefore, the fluid protoplast readily separates from the cell wall without exhibiting adhesiveness, and the protoplast rounds up. The monovalent cations, on the other hand, have a softening effect on the extraneous coatings, causing them to become sticky. Thus, when plasmolysis occurs in solutions of these salts, the extraneous material remains adherent to the surface of the protoplast, causing it to assume irregular shapes and to remain adherent at certain points to the inner surface of the cellulose wall.

In numerous investigations[2] cells were immersed in solutions of different salts and centrifuged; the rate of sedimentation of the protoplasmic inclusions of these cells was then compared with that of cells centrifuged in their normal environment. Greater difficulty in sedimentation was taken to indicate increased "viscosity" of the protoplasmic interior of the cell, and greater ease in sedimentation, decreased "viscosity." Heilbrunn and his co-workers[3] differentiated between the cortical and endoplasmic regions of the cell. The firmness of the cortex, normally in the gel state, was estimated by determining the amount of centrifugal force required to drag out the embedded inclusions. The force required to dislodge granules from the cortical gel is considerably greater than the force required to drive granules through the endoplasmic, or interior, region of the protoplasmic body. On the basis of their immersion studies, these investigators concluded that sodium and potassium increase, while calcium decreases, the "viscosity" of the protoplasmic interior, or endoplasm, of all the cells studied.[4] Conversely, they concluded that sodium and potassium liquefy or decrease the firmness, while calcium increases the firmness, of the cortical protoplasmic gel in those cells whose cortical region is sufficiently pronounced to permit study (for example, *Amoeba proteus* and *Chaetopterus* eggs).

For a proper evaluation of the above conclusions, we must raise the two following questions. (1) Is the rate of sedimentation of inclusions in cells immersed in various salt solutions and subjected to a centrifugal force a valid measure of the consistency of the protoplasmic matrix? (2) To what extent do the salts penetrate the cells from the external environment?

The sedimentation rate would be a valid measure of the con-

[2] Weber (1924a); Heilbrunn (1923, 1926a, b); Heilbrunn and Daugherty (1931); Wilson and Heilbrunn (1952); Northen and Northen (1939); Runnström and Kriszat (1950); Kriszat (1950).

[3] Heilbrunn and Daugherty (1932, 1933); Wilson and Heilbrunn (1952).

[4] The cells studied included *Amoeba dubia, Chaos chaos, Stentor,* unfertilized sea urchin eggs, and *Spirogyra.* Heilbrunn (1952, 1956) makes the qualification that calcium decreases "viscosity" of the protoplasmic interior only when it penetrates slowly into cells, but that when it penetrates rapidly or in large amounts, gelation of the interior protoplasm occurs.

sistency of the protoplasmic interior if it provided a comparative measure of the consistency of the matrices of cells immersed in various salt solutions. This would be true only if the inclusions themselves persisted as inert bodies and were not affected by exposure to the different salts. If the density of the inclusions or the interfaces between the inclusions and the surrounding protoplasm were affected, the sedimentation rate would be altered quite apart from any change in the consistency of the matrix. Under such circumstances results obtained by centrifugation would prove difficult, if not impossible, to interpret.

The monovalent salts, in fact, have profound effects on cell inclusions,[5] causing the granules to become sticky and adherent to one another, and eroding vacuolar membranes. An ameba immersed in a solution of sodium or potassium chloride becomes quiescent and rounds up; the granules fall to the bottom of the cell. Their stickiness is revealed by rolling the ameba over with a needle; the granules adhere as a clump and remain attached to the originally dependent surface. Microinjections of the monovalent salts have the same effects (Figure 11.1). The other observable effect of immersing amebae in solutions of the monovalent salts is the loss of vacuolar membranes which Mast (1926) has shown surround the cytoplasmic inclusions, notably the crystals with which the cytoplasm of the ameba is frequently filled. When the vacuolar membranes have disappeared, the inclusions, which are free in the cytoplasm and of varied shapes, show much less tendency to slip by one another. They tend to aggregate, as they do not when each crystal is encapsulated in a vacuole. Such factors could profoundly influence the time needed for the cytoplasmic inclusions to sediment under the influence of a centrifugal force.

The erosive action of monovalent salts on vacuolar membanes is strikingly shown by microinjecting *Actinosphaerium* (Figure 10.1) or by immersing the organism in the salt solutions.

The cytoplasmic vacuoles of sea urchin eggs immersed in solutions of sodium or potassium chloride also become sticky and agglutinate, especially when driven together by centrifugation. This is well illustrated in a micromanipulative experiment on centrifuged, unfertilized *Arbacia* eggs:

> Normally, *Arbacia* eggs are filled with minute spherical vacuoles (generally referred to as granules) some of which contain the echinochrome that gives the eggs their characteristic brown color. The eggs are centrifuged in their normal

[5] Chambers and Reznikoff (1926); R. Chambers (1949a).

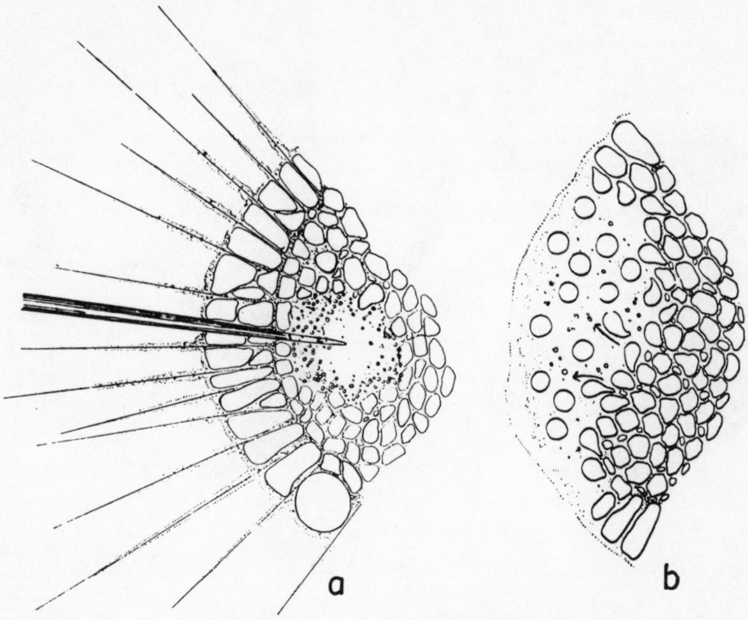

FIGURE 10.1. Vacuolated cytoplasm of living *Actinosphaerium eichhorni* injected with small amount of 1.0 M NaCl solution. (a) Immediately after injection. Vacuolar walls in region of injection have undergone disruption. (b) Pipette removed, several minutes later. Axopodia have disappeared. Cortical vacuoles have disrupted and disappeared. Arrows show vacuoles in endoplasm moving into liquefied region, where they become spherical. (After Chambers and Howland, 1930.)

medium of sea water to a stage in which three zones appear: a lower, granular zone occupying about half of the volume of each egg; a middle, clear zone of a transparent hyaloplasm occupying most of the lighter half; and a dimunitive uppermost zone containing a cluster of oil globules (Figure 10.2a). The nucleus lies below the oil cap. After the eggs are transferred to solutions of sodium or of potassium chloride isosmotic with sea water, some are mounted in the moist chamber to be operated upon with microdissecting needles. An egg is seized with a blunt-ended needle to hold it in place, and with a fine tipped needle a tear is made in the upper, clear zone. Immediately the hyaline material of the clear zone exhibits the fluidity of this region by pouring, together with minute oil globules constituting the oil cap, through the tear and spreading out in the external medium (Figure 10.2b). In another egg, the heavy granular zone is torn, with the

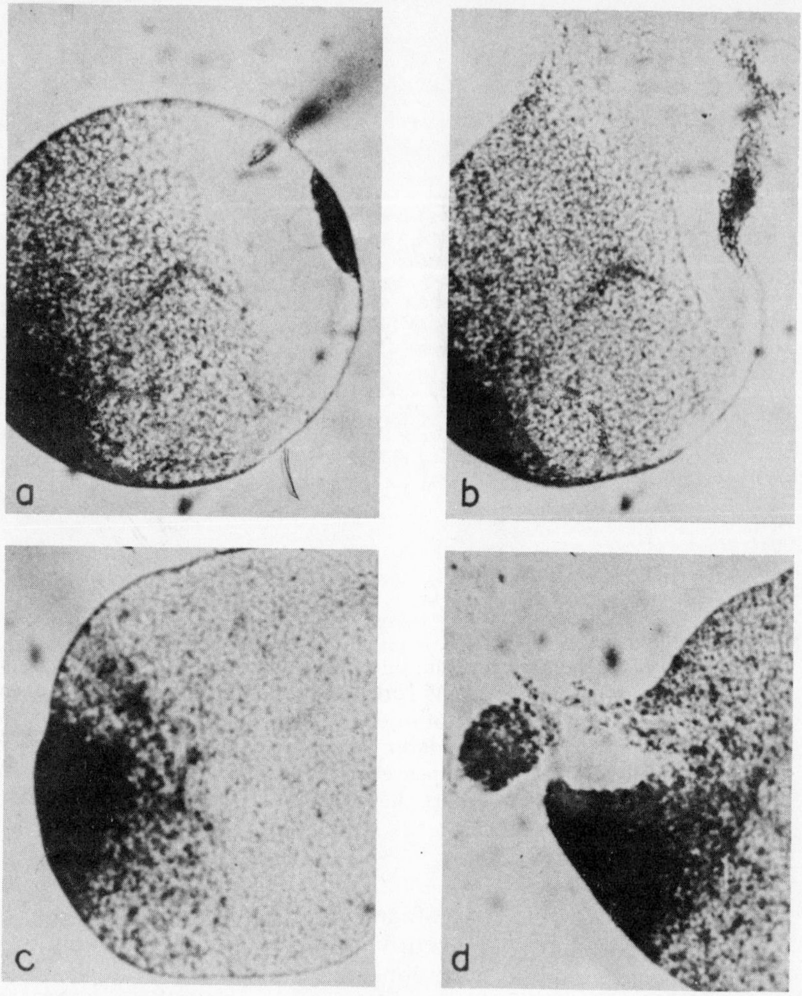

FIGURE 10.2. Mature, unfertilized living *Arbacia* egg centrifuged in sea water into three zones: upper right, hyaline zone with oil cap; middle, colorless granular zone; lower left, pigmented granular zone. Egg then transferred to monovalent salt solution (17 parts of 0.5 M NaCl, 3 parts 0.5 M KCl) isosmotic with sea water. (a, b) Surface of egg in region of hyaline zone torn with tip of microneedle. This results in dispersion of entire hyaline zone, scattering of fatty globules in oil cap and of clear egg pronucleus lying just below oil cap. (c) Lower granular zone closely packed with nonpigmented and pigmented granules which constitute heaviest granules in egg. (d) This zone torn with a microneedle. Note adhesiveness of granules, which remain as a clump. If this egg, uncentrifuged, had been torn in monovalent salt solution, the cytoplasmic granules, not being packed together, would have dispersed in surrounding medium.

result that this zone is found to be a viscid, solid, mass of agglutinated granules (Figure 10.2c, d).[6]

Runnström and Kriszat (1950), using the centrifugation method, concluded that the "rigidity" of the cytoplasmic matrix of sea urchin eggs is increased in the absence of calcium. This conclusion was based on the observation that, when a given centrifugal acceleration (4,350 × g) was continued until maximal stratification had been attained, eggs in sea water containing calcium were highly stratified, while eggs in calcium-free sea water were only partially stratified. Our explanation of these findings is that, quite irrespective of changes which may occur in the matrix, alterations in the physical properties of the inclusions, such as their adhesiveness or density, may determine not only the rate of sedimentation of the inclusions, but also the maximum degree of stratification obtainable in the eggs.

The interpretation of data obtained by measuring the ease of sedimentation of chloroplasts in plant cells[7] presents a different situation. Exposure of the protoplast to different salts may alter not only the physical properties of the chloroplasts themselves (Scarth, 1924), but also the nature and extent of attachment of the chloroplasts to the cortical layer of the protoplast.

The second question we raised regarding the interpretation of results obtained from immersion experiments concerns the extent to which the salts penetrate from the external medium. The changes noted cannot be ascribed to a specific interaction between the salt of the environing medium and the protoplasmic matrix unless penetration of the salt through the cell surface can be demonstrated.

In viability studies in which amebae were immersed in various salt solutions at neutral pH, the monovalent cations were found to be decidedly more toxic than the divalent cations.[8] The results are shown in Figure 10.3. Immersion in solutions of the monovalent salts, sodium chloride and potassium chloride, at concentrations above 0.01 M, causes the cessation of streaming movements and induces liquefaction of the interior of amebae. At a concentration of 0.02 M the amebae die within a short time in the solutions of the monovalent salts, while in solutions of the divalent salts at the same concentration the amebae move about actively for many

[6] When the centrifuged eggs are operated upon in a solution of calcium chloride isosmotic with sea water, when the upper, clear zone is torn the oil cap and hyaline zone become set as a rigid coagulum. A similar effect is noted when the heavy granular zone is torn open: the massed granules undergo a frothy coagulation.

[7] Weber (1924a, b); Northen and Northen (1939).

[8] Chambers and Reznikoff (1926); Mast (1931).

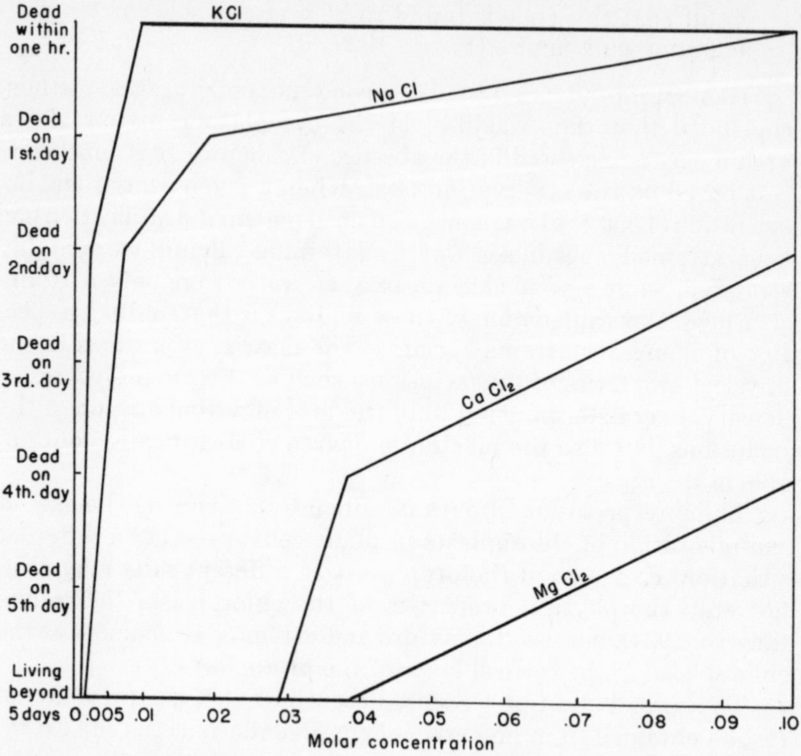

FIGURE 10.3. Chart showing viability of *Amoeba dubia* immersed in increasingly concentrated solutions of potassium chloride, calcium chloride, sodium chloride, and magnesium chloride at pH 7.0. In solutions of the monovalent salts viability is lower than in solutions of the divalent salts.

days. Evidently, the divalent salt solutions do not appreciably affect the viability of amebae until the concentration is increased to the point where the high tonicity becomes a complicating factor.

The toxicity of the monovalent salt solutions must indicate that the sodium and potassium ions enter the amebae readily from the external medium. This conclusion is substantiated by the similarity in the liquefying effects of immersing amebae in solutions of sodium and potassium salts, as compared with directly introducing the same solutions by microinjection (Figure 11.1). In addition, in cells immersed in solutions of monovalent salts a substantial exchange between ions in the cell interior and ions in the external medium must be taking place. An outward diffusion of organic substances may also occur (True, 1922).

It is problematical whether the calcium ion penetrates to any appreciable extent into amebae immersed in calcium chloride solutions. As long as the tonicity of the solution is not too great, the ionic balance in the protoplasm is preserved relatively unchanged. The lack of penetration of calcium from the external medium is substantiated by the dissimilarity between the effects of immersion and those of microinjection. For example, amebae in a solution of 0.02 M calcium chloride stream about actively for long periods of time. However, microinjection with even minute amounts of a concentration as low as 0.005 M induces coagulation of the cytoplasmic matrix.

The same considerations in regard to differential permeability and toxicity apply to cells in general.

A further example of this feature is the plant root hair cell of *Limnobium* (Figure 10.4a). Root hairs immersed in 0.05 M sodium chloride show a cessation of streaming and accumulation of the cytoplasm at the free tip of the root hair (Figure 10.4b). Microinjection of solutions of both sodium chloride and potassium chloride produces the same effect. However, the root hairs immersed in solutions of calcium chloride, up to 0.15 M and above, show no reaction to the salt and continue to live normally. Microinjection of

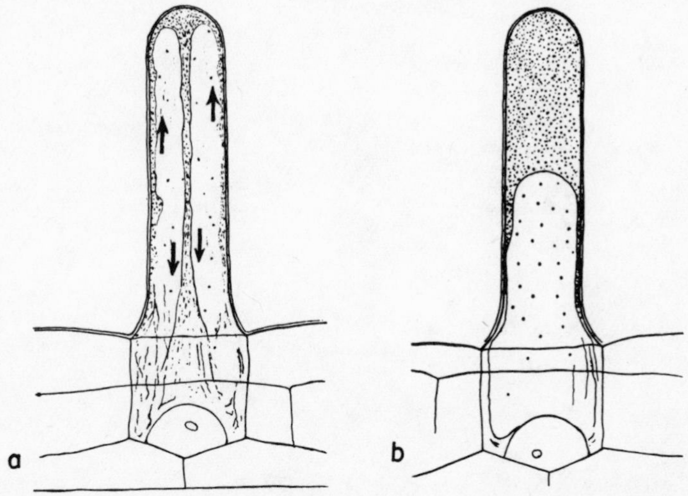

FIGURE 10.4. (a) Normal growing root hair of *Limnobium*. Nucleus at base of cell. Cytoplasm flows peripherally toward tip, where it reverses its direction and flows toward base in axial stream which traverses sap vacuole. (b) Root hair of *Limnobium*, 2 minutes after immersion in 0.2 M NaCl solution. Cytoplasm accumulates at tip of cell as streaming progressively slackens. (From Kerr, 1933.)

calcium chloride at a concentration as low as 0.005 M induces immediate solidification of the cytoplasmic matrix at the site of injection.

The immersion of cells in solutions of monovalent and divalent salts may give rise to such different conditions in the interior that it is hardly justified, on the basis of such experiments alone, to ascribe the changes noted to the specific effect on the protoplasm of the cations penetrating from the external medium. The changes observed in the immersion experiments may result from ionic effects exerted at almost any locus of the interdependent metabolic and enzymatic chains of events occurring in protoplasm. Certainly the alterations noted are not necessarily related to an interaction between the cation in which the cell is immersed and the structural components of the protoplasmic matrix.

11

Effects of Salts Microinjected into Protoplasm

The advantage of the microinjection method lies in the fact that a salt solution may be directly introduced into a healthy cell in its normal environment and the action of the salt on the various components of the protoplasm observed. The introduction of the micropipette into the protoplasm in itself causes no change if done with care and if the investigator has chosen for study a suitable cell type. Characteristically, the microinjection of solutions of divalent salts induces solidification or coagulation of the protoplasmic matrix, while microinjection of solutions of monovalent salts has a dispersing effect on the protoplasmic matrix and tends to liquefy gelated regions. Extensive studies have been carried out on *Amoeba dubia*.[1]

MONOVALENT SALTS INJECTED INTO AMEBAE

The dispersing action of the monovalent salts on the cytoplasm of the ameba is observed following the microinjection of all concentrations ranging from 2.0 M to 0.01 M. Large amounts of the lower concentrations or small amounts of the higher concentrations result in a temporary retraction of the pseudopodia, quiescence of streaming movements, and sedimentation of the granules (Figure 11.1). The retraction of pseudopodia and the rounding up of the ameba following injections of the monovalent salts result from the liquefaction of the plasmagel. These salts also liquefy the gelled borders of the channels seen in plasmodia (Chambers, 1943a).

It is significant that immersion of amebae in solutions of the monovalent salts causes liquefaction of the interior, producing the same effects as microinjection of these salts. This is good evidence

[1] Chambers and Reznikoff (1926); Reznikoff and Chambers (1927); Reznikoff (1928); Pollack (1928a).

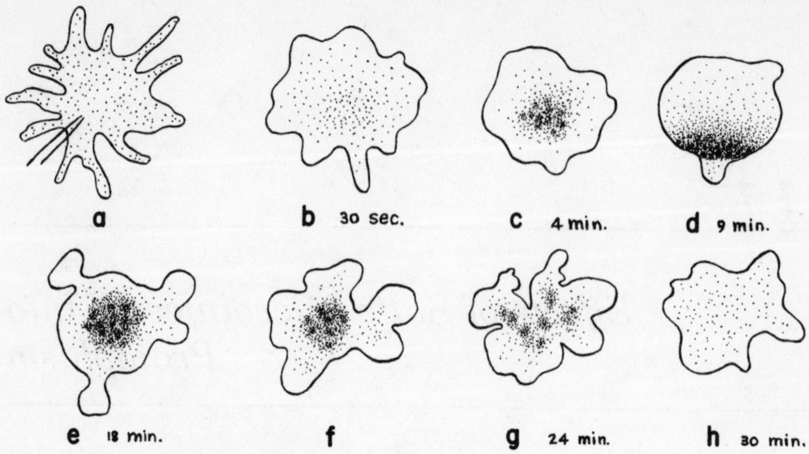

FIGURE 11.1. *Amoeba dubia* injected with 2.0 M NaCl solution, followed by recovery. (a) Before injection. (b) Pseudopodia being withdrawn. (c, d) Assumption of rounded shape and quiescence, with sinking of cytoplasmic granules. (e) Re-formation of pseudopodia after period of quiescence. (f–h) Gradual resumption of former appearance and return of normal streaming.

that sodium and potassium ions enter readily from the external medium.

DIVALENT SALTS INJECTED INTO AMEBAE

Injections of all strengths of calcium chloride from 2.0 M to 0.005 M cause coagulation or solidification of the matrix. In concentrations above 0.005 M the cytoplasm is precipitated as a coagulum, and the coagulated region is pinched off from the healthy and still active part of the ameba. Figure 11.2 shows an interesting case of

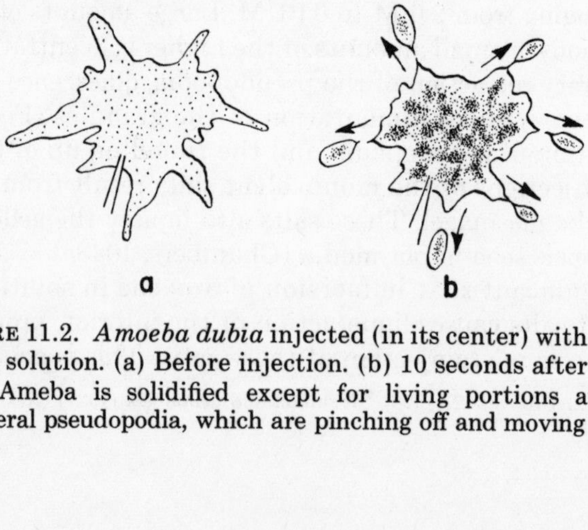

FIGURE 11.2. *Amoeba dubia* injected (in its center) with 1.0 M $CaCl_2$ solution. (a) Before injection. (b) 10 seconds after injection. Ameba is solidified except for living portions at tips of several pseudopodia, which are pinching off and moving away.

an ameba with extended pseudopodia at the moment when 1.0 M calcium chloride is injected into its center. The greater part of the ameba immediately coagulates, while the tips of the pseudopodia farthest from the site of the injection pinch themselves off and move away. A typical pinching off reaction with coagulation of the protoplasmic interior in the immediate vicinity of the microinjected part is shown in Figure 11.3. After an injection of 0.005 M calcium chloride, solidification or gelation of the cytoplasm occurs. The solidified region may be resorbed after an initial attempt at pinching off. At lower concentrations, between 0.005 M and 0.0025 M, injections of calcium chloride cause a temporary, localized solidification of the matrix without other visible change. The solidified portion may be moved about in the streaming cytoplasm with the microneedle and is eventually dispersed.

An important conclusion to be drawn from these microinjection

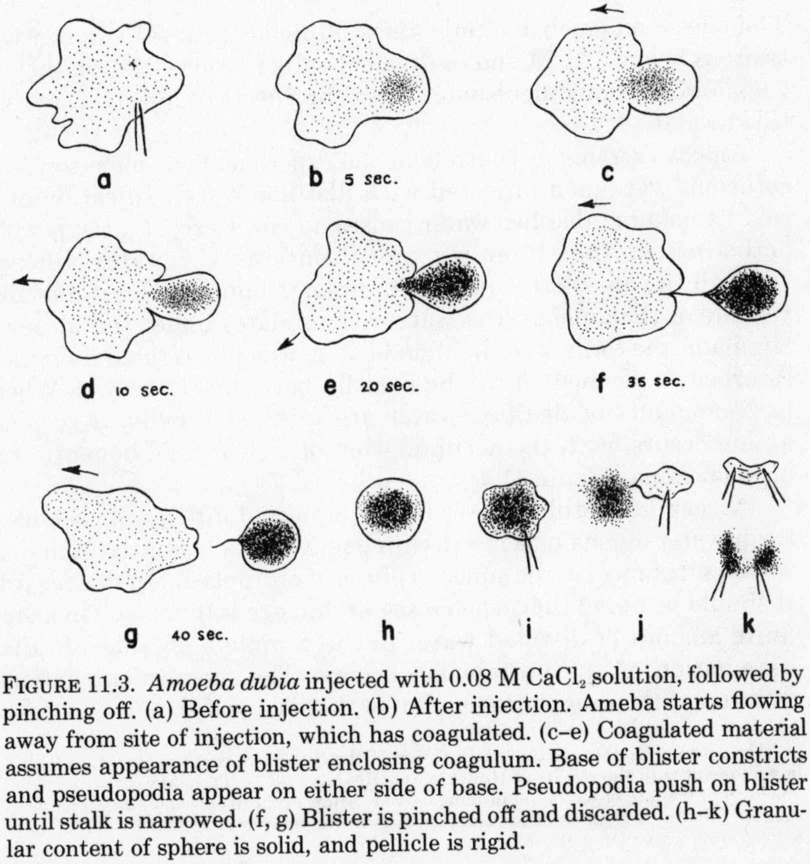

FIGURE 11.3. *Amoeba dubia* injected with 0.08 M $CaCl_2$ solution, followed by pinching off. (a) Before injection. (b) After injection. Ameba starts flowing away from site of injection, which has coagulated. (c–e) Coagulated material assumes appearance of blister enclosing coagulum. Base of blister constricts and pseudopodia appear on either side of base. Pseudopodia push on blister until stalk is narrowed. (f, g) Blister is pinched off and discarded. (h–k) Granular content of sphere is solid, and pellicle is rigid.

experiments is that the concentration of free calcium in the fluid endoplasm of *Amoeba dubia* cannot exceed 0.002 M.[2] Furthermore, these experiments reveal the remarkable impermeability of the protoplasmic surface to calcium ions, since amebae stream about and maintain the fluidity of their endoplasmic contents for many hours after immersion in a solution of calcium chloride even as concentrated as 0.1 M. If the ion penetrated sufficiently to raise the internal concentration of free calcium above 0.002 M, solidification of the ameba would necessarily have to occur! This is in marked contrast to the relatively high permeability of the protoplasmic surface to the monovalent salts.

The microinjection of magnesium chloride causes solidification of the cytoplasm. A pinching off reaction does not occur. The solidification process is not localized but gradually spreads throughout the ameba.

DISTILLED WATER INJECTED INTO AMEBAE

The effects on amebae of injections of monovalent salts in concentrations below 0.01 M and of divalent salts in concentrations below 0.0025 M are indistinguishable from the effects of injections of distilled water.

Amebae serving as controls in the experiments on effects of salt solutions were microinjected with distilled water. Injection of a small amount of distilled water causes no effect except a temporary activation of the streaming movements as the water diffuses through the ameba. If injected in large amounts (at least half the volume of the ameba), the water accumulates under the plasmalemma in the form of an hyaline blister, which is ultimately either resorbed or pinched off by the healthy portion (Figure 6.1). When large amounts of distilled water are injected rapidly, a rushing action occurs, with the accumulation of a clear fluid beneath the plasmalemma (Figure 11.4).

As soon as distilled water is microinjected into a cell such as a fresh water ameba or a sea urchin egg it mixes instantly with dissolved substances in the aqueous phase of protoplasm. In this regard, it should be noted that when a sea urchin egg is immersed in an infinite amount of distilled water in the complete absence of salts, coagulation of the protoplasm occurs. The monovalent cations counteract the coagulating action of distilled water.

[2] Only a small fraction of the total calcium in the cytoplasm is in the free state. This is in agreement with the results of Hodgkin and Keynes (1957). These investigators determined from the rate of migration of radioactive calcium in the axoplasm of the squid nerve that only a minute fraction of the total calcium in the interior of the fiber could be in an ionized state (less than 0.01 mM).

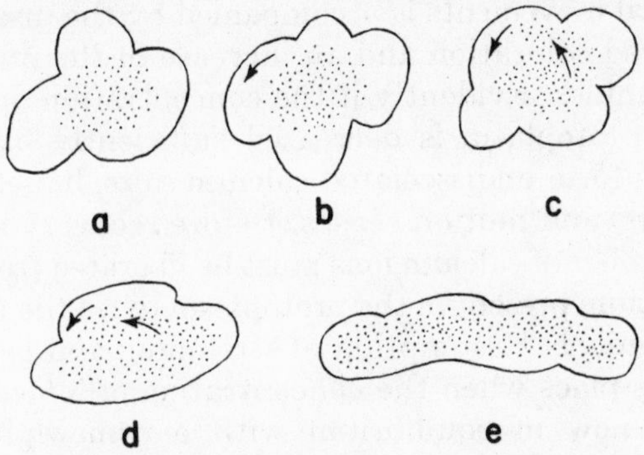

FIGURE 11.4. *Amoeba dubia* injected with large amount of distilled water, followed by recovery without pinching off. (a) Plasmalemma is lifted in several places and granuloplasm clumped in center of ameba. (b) Sudden bulging of plasmalemma in direction of arrow, followed by (c) pouring of granules into hyaline region. (d) Continued scattering of granules throughout ameba and cessation of "rushing effect." (e) Return to normal several minutes after injection.

The coagulating action of distilled water can be demonstrated by plunging a drop of sea water containing unfertilized *Arbacia* eggs into the upper layer of a tall cylinder of distilled water. Because of the high density of the eggs, they fall rapidly through the distilled water. While this is occurring, cytolysis ensues, and all water-soluble substances diffuse out from the interior of the eggs. The eggs are recovered at the bottom of the cylinder as dense solid balls, in which the cell inclusions are embedded in the coagulated matrix.

CALCIUM PRECIPITANTS INJECTED INTO AMEBAE

The microinjection of calcium precipitants such as the sodium salts of alizarin sulfonate or oxalate (Pollack, 1928a) initially causes effects similar to those seen after injection of the monovalent cations, that is, cessation of movement and rounding up of the ameba. The effective concentrations of the calcium-precipitating anions are many times less than that of sodium chloride, showing that the effects observed are not due to the injected sodium. The injection of sodium alizarin sulfonate causes fine, purplish-red granules, representing the precipitated calcium salt, to appear in the cytoplasm. The hyaloplasm is diffusely colored pale red by the excess alizarin. If the quantity of precipitant injected has not been excessive, the recovery of the ameba and the resumption of normal movements is accompanied by the disappearance of the diffuse red coloration and an increase in the number of purplish-red granules. Evidently if the concentration of free calcium ions in the protoplasm is decreased sufficiently by introducing anions which form undissociated calcium salts, liquefaction of the ameba occurs, and motion ceases. Before recovery can occur, an adequate number of calcium ions must be liberated from the supply of bound calcium present in the protoplasm to rid the interior of the introduced anions. Resumption of streaming and pseudopod formation takes place when the concentration of ionized calcium in the matrix, now in equilibrium with a somewhat diminished reserve of bound calcium, is re-established.

Striking evidence for the importance of free intraprotoplasmic calcium ions to ameboid movement is the following observation. After the injection of moderate amounts of sodium alizarin sulfonate, the ameba is found occasionally to elevate a pseudopod, as evidenced by a slight local lifting of the plasmalemma. Thereupon, a shower of the purplish-red granules appear in this region, and the pseudopodial elevation immediately stops.

TEARING AMEBAE IN SALT SOLUTIONS

The liquefying action of the monovalent salts and the coagulating effect of calcium, observed in the microinjection experiments, are also well demonstrated when cells immersed in solutions of these salts are torn with the microneedle (Figures 11.5, 11.6).

> When an ameba is torn in a solution of sodium chloride or potassium chloride (0.08 M and stronger), the contents pour out, the granular cytoplasm scatters, and the pellicle, in which the protoplasmic surface film is incorporated, disintegrates (Figure 11.5). The tearing of an ameba in a solution of calcium chloride (0.1 M to 0.2 M) results in immediate coagulation of the cytoplasm in the region of the tear (Fig-

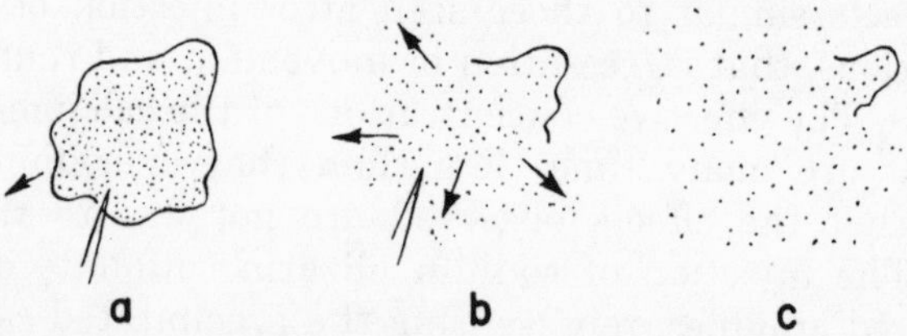

FIGURE 11.5. *Amoeba dubia,* living, torn in solution containing mixture of 0.08 M NaCl and 0.02 M KCl. Surface rapidly dissipates and granular contents disperse.

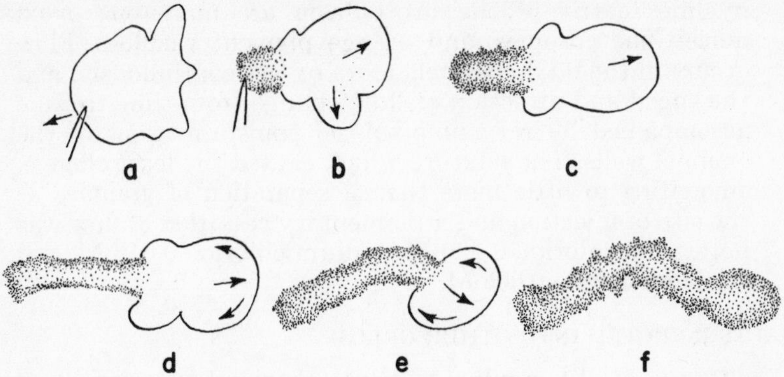

FIGURE 11.6. *Amoeba dubia* torn in 0.08 M $CaCl_2$ solution. (a) Before tearing. (b–e) Ameba flows away from region which has been torn. Inward diffusion of $CaCl_2$ and retreat of ameba from tear converts ameba into ever lengthening column of coagulated material until (f) entire ameba is set and is dead.

ure 11.6a, b). The rest of the ameba begins to flow away and attempts to pinch off the involved region while the solidifying process rapidly spreads inward (Figure 11.6c–e). This reaction produces a steadily lengthening column of solidified material, with the living part at the end farthest from the tear actively flowing and attempting to pinch itself off. Finally, this part also succumbs, and the ameba is converted into a solid column of material (Figure 11.6f). When amebae are torn in lower concentrations of calcium chloride, the solidifying regions are rapidly pinched off.

SALT MIXTURES INJECTED INTO PLASMODIA

The solidifying action of the divalent salts can be neutralized by the liquefying effects of the monovalent salts. A mixture of salts can be prepared which, when microinjected into the ameba even in large amounts, is innocuous and does not alter its physical state. Chambers (1943a) carried out a series of tests with a variety of mixtures using the streaming plasmodium of the myxomycete *Physarum*. This organism was selected, rather than the ameba, because the phenomena of changes in flow and gelation together with the stability of the pigment globules offer highly sensitive criteria for determining the effects of the injections.

The part of the plasmodium studied was the thin, fan-like expansion where the gelated protoplasm is interpenetrated with numerous flowing channels. Suspended in the hyaline matrix of the protoplasm are numerous ovoid nuclei, fine granules, and orange pigment vaculoes. Flow occurs in the fluid channels more or less continuously, and the speed and direction of flow change from time to time accompanied by reversible sol–gel transformations of the channel walls. The mixture which caused the least change, amounting to little more than a separation of granules at the site of injection and a momentary cessation of flow was potassium chloride 0.120 M, sodium chloride 0.013 M, and calcium chloride 0.003 M.

SALTS INJECTED INTO OTHER CELLS

Results essentially similar[3] to those obtained in an ameba with regard to the monovalent and divalent salts have been observed in microinjection experiments using the protozoa *Actinosphaerium, Spirostomum,* and *Stentor,* the plasmodia of *Physarum,* the root hair cells of *Limnobium,* and echinoderm eggs. In the highly vacuolated cytoplasm of *Actinosphaerium* (Figure 21.11) and *Spirostomum* the cytoplasm is reduced to thin strands between the vacuoles. Calcium in concentrations as low as 0.3 M stiffens and coagulates the strands and also ruptures the vacuoles. The rupturing occurs in a peculiar jerky manner, as if the vacuolar walls set before they broke down.[4] The monovalent salts liquefy the intervacuolar cytoplasm and in high concentrations erode the vacuolar membranes (Chapter 10).

The injection of calcium into muscle cells causes coagulation with marked contraction of the clot and its separation from the surrounding sarcolemma of the fiber. Very low concentrations of calcium, however, induce an immediate, reversible, localized contraction, leaving a small temporary scar, or coagulum, at the site of injection (Figure 11.7). An interesting feature concerns the application of calcium chloride to the external surface of a frog muscle fiber. As is well known, spraying calcium chloride from a

[3] Kerr (1933); Chambers and Howland (1930); Chambers (1943a); Chambers and Kao (1952).

[4] The calcium ion may "precipitate" a constituent of the vacuolar walls, resulting in disorientation and even disintegration of the membrane.

Heilbrunn (1930) drew attention to the breakdown of the pigment vacuoles of the *Arbacia* egg in a salt solution containing calcium and the preservation of the vacuoles in solutions of salts of monovalent cations isosmotic with sea water. Harris (1943) observed that in an isosmotic salt solution calcium causes the pigment (echinochrome) to diffuse out through the walls of isolated pigment vacuoles; these persist as "ghosts." Hypotonic salt solutions lacking calcium lyse the vacuoles in a similar manner. These effects on vacuolar membranes may be related to the action of calcium in promoting the transformation of isolated mitochondria suspended in isosmotic media from rod-shaped bodies to highly swollen vesicles (Cleland and Slater, 1953) and to the action of calcium in promoting the liberation of acetylcholine in "quantal" amounts at the neuromuscular junction (Castillo and Katz, 1956).

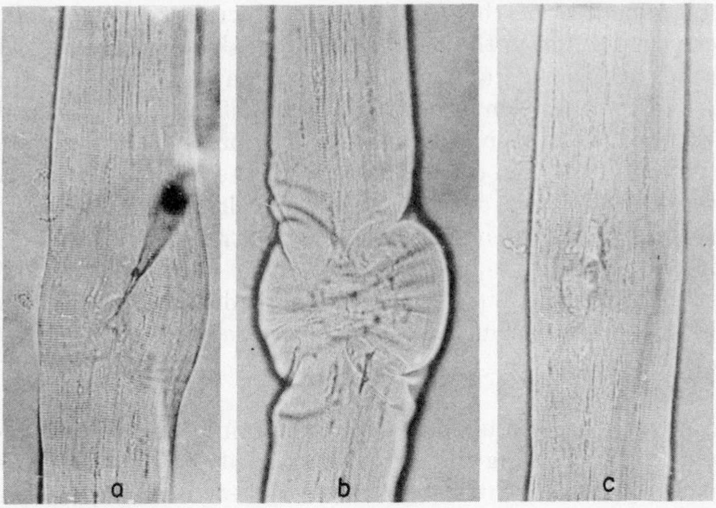

FIGURE 11.7. Sarcolemma of living frog muscle fiber pricked with micropipette containing 0.01 M $CaCl_2$ solution. (a) At moment of insertion of tip of micropipette. (b) Pricking permits introduction of enough salt through sarcolemma to induce sudden effect of localized contracture. (c) Fiber returns to normal shape. Small hyaline scar which formed in sarcolemma at site of puncture disappeared after several minutes.

micropipette onto the surface is without effect. However, if the mouth of a fine-tipped micropipette is pushed against the fiber sufficiently to indent the sarcolemma, but not to penetrate it, with each application of calcium a localized twitch occurs, and there is no scar formation. When very small amounts are applied, the twitch is restricted to the region of several sarcomeres. Evidently the indentation stretches or weakens the sarcolemma, increasing its permeability and enabling calcium ions to enter.

A special case is the squid giant nerve fiber.[5] The microinjection of sodium or potassium chloride (0.5 M) causes no change in the normal gelated state of the axoplasm, while liquefaction follows injection of calcium chloride (0.34 M to 2.7 M). Liquefaction is also observed when the axoplasm is extruded into mixtures of sodium and calcium chloride solutions isosmotic with sea water (calcium concentration varied from 1 to 390 mM). The normally gelated state of the axoplasm is consistent with the very low concentration of calcium present within the squid nerve fiber (total 0.4 mM; ionized less than 0.01 mM).[6]

[5] Hodgkin and Katz (1949); Chambers and Kao (1952).
[6] Keynes and Lewis (1956); Hodgkin and Keynes (1957).

Liquefaction of the axoplasm by microinjection of calcium blocks conduction in the nerve fiber (Hodgkin and Keynes, 1956); whether or not the liquefied axoplasm can be reversed to the gel state with resumption of normal functional capacity has not been determined. For this reason we should at present avoid comparing the action of calcium on axoplasm with the fully reversible solidifying effects of low concentrations of calcium on ameba protoplasm. Certainly the irreversible, coagulative (resulting in death) action of relatively high concentrations of calcium on the protoplasm of a great variety of cells is in marked contrast to its liquefying action on axoplasm.[7] In this respect the axoplasm belongs in a special category.

In brief, the results obtained by microinjecting salts into cells reveal that increasing the concentration of the monovalent salts relative to the concentration of free calcium already present promotes the fluid state of the matrix, whereas increasing the concentration of free calcium relative to the concentration of the monovalent salts promotes gelation or solidification. As long as the ions are introduced in small amounts the effects are reversible and compatible with the living state of the protoplasm. In larger amounts, the effects are irreversible. The injured region is discarded or the entire cell succumbs, the cytoplasmic matrix being dispersed by the monovalent salts or precipitated and coagulated by calcium (an exceptional case is the axoplasm of squid nerve). Findings concerning the effects of salts, obtained by immersing cells in different salt solutions and measuring the relative ease of sedimentation of inclusions in the protoplamic interior, may be misleading. The physical state of the inclusions themselves may be affected by the salts. In addition, the salts penetrate the cells in varying, and undetermined, amounts.

[7] Conceivably, the liquefying action of the calcium on axoplasm might result from the precipitation of the constituents of the axoplasmic gel, the mesh being broken down into discrete precipitation granules, which disperse. An analogous situation is seen in pectate gels. At low concentrations of calcium, gel formation is promoted, but at higher concentrations precipitation occurs (Chapter 3). In the ameba, however, precipitation by excess calcium results in a coagulum.

V

Hydrogen-Ion Concentration of Cell Components

INTRODUCTION TO PART V. Interest in the hydrogen-ion concentration of living tissues, particularly of the protoplasm of cells, arises from two sources: first, the physical state and chemical reactivity of many organic substances, particularly the proteins, depend on the pH; and second, enzymic reactions in general show a sharp dependence on the pH of the solution in which the reactions take place.

When considering pH in relation to protoplasm, the term can refer only to an aqueous phase. Experimental evidence has shown that there is an aqueous phase which is continuous throughout the ground substance of the protoplasm and of the nucleus. In spite of a vast literature concerning the general problem of protoplasmic pH, only a few of the measurements which have been made have reference to the actual living protoplasmic substance.

12

Critique of Various Methods of Determining Intracellular pH

Methods which have been used for measuring intracellular hydrogen-ion concentration may be summarized in seven categories: (1) measurements on aqueous extracts of crushed tissue; (2) measurements on fluid material removed from single cells; (3) potentiometric determinations in single cells; (4) estimations based on indicators naturally occurring as pigments in cells; (5) estimations based on vitally stained cells; (6) measurements of the amounts of dissociated and associated forms of weak electrolytes in cells; (7) microinjections of pH indicators.

MEASUREMENTS ON AQUEOUS EXTRACTS OF CRUSHED TISSUES, PLANT OR ANIMAL

Results obtained with crushed material are always open to question. Precautionary measures are essential to minimize the injurious effect. Michaelis and Kramsztyk (1914) were the most successful of those who measured the pH of tissue extracts. They observed that acid is liberated when fresh tissue is crushed, but that boiling the tissue before preparing the extract prevents the acid formation. They concluded that the hydrogen-ion concentration of the tissue juices during life must be somewhere between those of the boiled and unboiled extracts. By eliminating from consideration tissues in which a large amount of acid formed after crushing, they estimated the true physiological value at pH 6.8.

The production of an acid of injury is not the only complication arising from crushing. Crushing cells disintegrates the structural components and causes an indiscriminate mixing of the protoplasmic ground substance with the nuclear contents and with vacuolar fluids. The microinjection method has demonstrated

(Chapter 13) that each of these three components has a different pH value. Moreover, it is practically impossible to secure a piece of tissue uncontaminated with varying amounts of extracellular fluids. The pH of these fluids is normally on the alkaline side of neutrality. Finally, the acid of injury is about pH 5.2. The pH values of fluids extracted from crushed tissues, therefore, can at best represent a mean of values which is irrelevant to the true reaction of normal cells.

Vlès (1924) and Reiss (1926) used an ingenious but faulty method termed by them "la méthode d'écrasement." They attempted to introduce into cells colorimetric pH indicators to which cells in general are impermeable. They used chiefly echinoderm eggs mounted between slide and coverslip in a drop of sea water containing a given pH indicator. A sudden, momentary pressure on the egg was produced by bearing down on the coverslip. The effect was a bruising (écrasement) of the egg which induced color to appear in the cell. Release of the pressure brought the egg back to its normal spherical shape. Unfortunately, no notice was taken of the fact that the regions which were colored were sharply segregated and gave evidence of being colored because of pronounced cytolysis of the egg cytoplasm. The values they obtained lay between pH 5.0 and 5.3, indicating, as we know now, an acid of injury following cytolysis.

MEASUREMENTS ON FLUID MATERIAL REMOVED FROM SINGLE CELLS

A method also open to criticism is that of testing material sucked out of a cell using a micropipette. Such operations have been performed on large, yolk-laden cells such as fish and frog eggs.

Bodine (1926), using the mature, unfertilized *Fundulus* egg, obtained potentiometrically a pH value of 6.39 for the extracted material. Unfortunately, his results can refer only to the inert yolk, which occupies a large part of the egg. The protoplasm in the unfertilized egg is an extremely thin layer spread over the surface of the yolk. Conceivably, a better way of determining the pH of the protoplasmic portion of the egg might have been to use material from the polar lobe, which forms after the egg has been in sea water for about 50 minutes. Peripheral streaming movements induce the accumulation of the protoplasm at one pole, where it forms the polar lobe. It is possible to suck out an appreciable amount of this material; however, the mechanical insult of doing so introduces an acid of injury. Chambers (1932) microinjected sulfonphthalein indicators into the polar lobe of living *Fundulus* eggs and estimated the pH of the cytoplasmic matrix at 6.6 to 6.8, a value which agrees with determinations on many other types of cells.

POTENTIOMETRIC DETERMINATIONS IN SINGLE CELLS

Microelectrodes are used for securing the pH of protoplasm. Although many different types of electrodes respond to pH differences, the generated potential is not necessarily specific to changes in hydrogen-ion concentration. For example, the hydrogen-platinum microelectrode described by Taylor (1925) accurately registered the pH of simple aqueous solutions and could be used for measuring the pH of sap in the central vacuole of *Nitella;* but when the microelectrodes were inserted into *Nitella* protoplasm, impossible values were obtained.[1] These values were related rather to the oxidation–reduction potential.

The glass electrode, however, responds specifically to pH in the range 0.0 to 9.0 (Dole, 1941), and a microelectrode modification has recently been devised by Caldwell (1954).

The principal problem which faces the investigator who attempts to use a microelectrode for pH measurements is the construction of one fine enough and sharp enough to penetrate into the protoplasm of the living cell without causing injury. Furthermore, when attempts are made to introduce the electrode into the cell, careful observation will usually disclose that the microelectrode tip is lying either outside the protoplasm in a pocket of the surface pushed in by the electrode or else in a vacuole caused by disintegration and denaturation of proteins as a result of the injury produced by the attempt to insert a blunt electrode into the cell. Even if the electrode has been successfully inserted, the protein films that tend to form over the tip set up spurious potential differences.

In spite of the difficulties every effort should be made to construct suitable microelectrodes, since this is, at least theoretically, the only method which will permit the detection of small but important changes in intracellular pH. Although the glass microelectrodes used by Caldwell are of large size, his are the only data of significance which have been obtained by the electrometric method.

In order to make his measurements Caldwell inserted a glass electrode, 100 micra in diameter, for a distance of 2 mm. or more down the length of a crustacean muscle fiber, 600 micra in diameter.

[1] Taylor and Whitaker (1927). The same criticism of nonspecificity to pH changes also applies to the hydrogen-platinum electrode used by Dorfman (1936) and Dorfman and Grodsensky (1937), and to the antimony microelectrode used by Buytendyk and Woerdeman (1927) for measurements of pH in the frog egg.

He states that the insertion definitely injured the fiber. The pH value he obtained for the interior was 6.9. This value is in agreement with that obtained by the microinjection method for the hyaloplasm of many different types of cells. Caldwell made the interesting observations that neither injury nor depolarization of the membrane changes the intracellular pH. This is maintained at 6.9 even when the resting potential across the cell membrane had declined to 0.

Our experience is that the skeletal muscle of the frog (100 micra in diameter) is extremely sensitive to insertion of microinstruments and that it is difficult to insert into the sarcoplasm even a fine micropipette, 0.5 to 1 micra in diameter, without causing injury. Injury to mammalian skeletal muscle (Michaelis and Kramsztyk, 1914) involves the liberation of a considerable amount of acid, with a lowering of the pH to about 5.0. Agreement between Caldwell's measurements and the results obtained by the microinjection method may be due to the fortuitous circumstance that, unlike the vast majority of cells, injury to crustacean muscle is not accompanied by an acid of injury.

ESTIMATIONS BASED ON NATURALLY OCCURRING INDICATORS

Various pigments naturally present in many cells are pH indicators. Since these pigments normally reside in intracytoplasmic vacuoles, their color may serve as a measure of the vacuolar pH but not of the pH of the cytoplasmic matrix.

In the plant cell the protoplasmic portion, the protoplast, is a thin layer investing the relatively large sap vacuole. The pH of the vacuolar sap varies in different plants and in different types of cells of the same plant. In contrast to the protoplast, the sap is very weakly buffered so that penetrating acids and bases readily shift the vacuolar pH. On the basis of the coloration of plant cells immersed in solutions of different pH indicators, Small (1929, 1955) classified the cells of different plants and plant tissues according to the degree of acidity of their cell sap. Small recognized that the protoplast has its own pH independently of the vacuolar sap, but many investigators have failed to notice that the variable values found for different plant cells relate not to the protoplasm but to the vacuolar contents.

To the animal cytologist the problem is somewhat more complicated. The protoplasm of an animal cell constitutes the major part of the cell, and the vacuoles and so-called granules are small, numerous, and distributed throughout the protoplasm. These vacuoles contain fluid, presumably analogous to the sap in the vacuoles of plant cells. The vacuolar fluid is only weakly buffered, and if the vacuoles contain colored compounds which shift their color with pH changes, they can be seen to do so when they are exposed to penetrating acids and bases. The error made by many has been to consider the pH of these distributed vacuoles as that of the protoplasmic matrix.

Vlès and Vellinger (1928) determined the absorption spectrum of the echinochrome as it exists in living *Arbacia* eggs with the idea that they could in this way estimate the pH of the egg cytoplasm. They compared the spectrum[2] obtained from living eggs with that of the alcoholic extracts of the echinochrome at various buffered pH values. They concluded that there exist in the living *Arbacia* egg "points" which have a pH of 5.5 $\pm$ 0.3. It is of interest to note that this value approximates that which they found by their "bruising" method (see above). In this case, their error was that the color of the echinochrome has nothing to do with the pH of the egg cytoplasmic matrix. The value obtained by Vlès and Vellinger refers to the pH of the relatively acid vacuoles containing echinochrome. Chambers (1938d), using the tip of a very finely pointed microneedle, was able to prick one of the echinochrome-containing vacuoles as it lay in its natural position in the cytoplasm of an unfertilized *Arbacia* egg. At once the escaping echinochrome pigment changed color from a ruddy brown to a greyish brown, conforming to the higher pH of 6.6 to 6.8 for the more alkaline cytoplasmic matrix.

Seifriz and Zetzmann (1935) have reported the occurrence of marked hydrogen-ion concentration changes in the protoplasm of the plasmodium *Physarum polycephalum,* ranging from pH 8.0 when fruiting to pH 1.6 when a sclerotium is formed. These determinations were based on the color changes of a naturally occurring pH indicator. The changes in color observed undoubtedly refer to the contents of vacuoles distributed throughout the cytoplasm, in which the pigment is known to be segregated.

ESTIMATIONS BASED ON VITALLY STAINED CELLS

A widely used method of pH determination is the immersion of pieces of tissue or isolated cells in physiological fluids containing in solution pH indicators.[3] A variation of this method is the injec-

[2] The absorption spectrum of echinochrome changes markedly in the pH range from 4.0 to 6.0.

[3] Atkins (1922a,b); Carnot, Glénard, and Gruzewska (1925); Fauré-Fremiet (1923), (1925); Schaede (1924); Rous (1925).

tion of indicators into the circulating blood with the idea that the cells exposed to the blood will be vitally stained in situ.

Most of the indicators used for this purpose have been basic dyes, which readily penetrate cells from the external medium. This method is open to the serious objection that the basic dyes which penetrate cells become quickly segregated in cytoplasmic vacuoles. The color assumed by the dye in the vacuoles may serve as a measure of the pH of the vacuolar fluids but not of the pH of the protoplasmic matrix. Another disadvantage of the basic dyes is their tendency to form undissociated complexes with proteins within the cell, exerting a coagulating action. This renders the coloration imparted by the dye meaningless as an index of pH of the protoplasmic matrix.

Furthermore, many of the basic dyes have a pronounced lipoid solubility. Fauré-Fremiet (1924) immersed *Sabellaria* eggs in solutions of basic dyes (neutral red, nile blue chloral hydrate, and brilliant cresyl blue) in order to determine the hydrogen-ion concentration of the interior of the eggs. He found that the color assumed by the cytoplasm of unfertilized eggs indicated a progressive increase of alkalinity which, by the time the eggs had undergone maturation and were in a fertilizable state, exceeded the seemingly impossible pH of 12.0. Using the same basic dyes he found that the color tint began to fall after sperm insemination to a pH value between 10.4 and 11.0, and progressively fell to 7.0. In a footnote he remarked that the colors assumed by these indicators offer two interpretations, according to whether they are dissolved in an aqueous phase, which would indicate pH, or in lipoids, in which case the dye would show the color of the undissociated base. A third possible interpretation is that the basic dyes form complexes with the cytoplasmic proteins. It may be this latter feature which accounts for the color shifts observed during the changing physiological state of the eggs.

An alkaline reaction for the large refractile granules of the eosinophilic leukocyte has been reported on the basis of experiments using neutral red (Sabin, 1937). The alkalinity of these granules has been inferred because of the yellow color of the neutral red which had penetrated from the medium. As described in Chapter 15, neutral red is a weak basic dye and should pass readily into the cytoplasm from the more alkaline medium of the blood plasma. It should accumulate in regions more acid than the cytoplasm, but not in regions more alkaline. The yellow, alkaline coloration of the eosinophilic granules is not indicative of high pH; it may be due to the formation of a protein complex or to the presence of neutral red as the yellow colored free base dissolved in a lipid component.

Acid dyes, in general, color the cytoplasm diffusely (Part VI). At least this holds true for the fully dissociated acid dyes. In this category virtually the only ones which change color in the pH range of the protoplasmic interior are the sulfonated indicators of the Clark and Lubs series. These, pre-eminently the most suitable for colorimetric measurements (Chapter 13), do not penetrate the vast majority of cells and hence must be microinjected. To this rule of nonpenetrability there is, however, a striking exception: the epithelial cells in situ of the proximal convoluted tubules of the kidney. When segments are immersed in a solution of a sulfonphthalein indicator, the indicator penetrates into the cells by a secretory process and passes into the lumen of the tubule. As the indicator passes through the epithelial cells, it imparts a diffuse, even coloration to the cytoplasm, indicating that it is in solution in the continuous aqueous phase of the cytoplasmic matrix. No segregation in cytoplasmic vacuoles occurs, nor is there any detectable combination with intraprotoplasmic components. The colors assumed by the full overlapping series of indicators (Figure 13.1) are consistent with a pH of 6.8 $\pm$ 0.2 for the cytoplasmic matrix. The following is a description of a typical experiment.

> When the excised mesonephros of an 8- to 10-day-old chick is mounted for culture in a mixture of chicken plasma and embryonic juice, brought to a pH of 7.6 to 7.8 with $NaHCO_3$, and kept at a temperature of 39°C., the tubules remain active for days (Chambers and Cameron, 1932). When phenol red is added to the medium, the cells become yellow in color. Within one to several hours of incubation the lumina of the tubules dilate and become filled with a red fluid, while the walls remain yellow (Figure 16.2). The color of the lumina progressively deepens, but that of the walls remains constant. Examination under high magnification reveals only a diffuse coloration of the cytoplasm of the cells; the granules and the nucleus are not stained.
>
> All the sulfonphthalein indicators color the cytoplasm of the cells in the walls of the tubules (Table 13.1; Figure 13.1). *Cresol red and phenol red:* These indicators are nontoxic even when the plasma medium is deeply colored with the dyes. While in the cytoplasm, they assume the yellow color of their acid ranges. *Brom thymol blue:* In the culture medium this rather toxic indicator assumes a green color at variance with the real pH of 7.6 to 7.8. However, the cells of the few secreting tubules which survive become pale greenish blue. *Brom cresol purple:* This is also toxic, but to

a lesser degree than the previous indicator. While in the tubule cells the indicator is blue in color. *Chlor phenol red:* This is similar to phenol red in being relatively nontoxic. While in the cells it is purplish red in color.

The estimated pH for the matrix of the secreting renal tubule cell is identical with that obtained by the microinjection technique for many living cells.

MEASUREMENTS OF THE AMOUNTS OF THE DISSOCIATED AND ASSOCIATED FORMS OF WEAK ELECTROLYTES

The weak electrolyte most frequently used is carbonic acid. Measurement of the bicarbonate content of an aqueous solution in equilibrium with a known tension of carbon dioxide in an overlying gas phase permits the determination of the hydrogen-ion concentration using the Henderson-Hasselbalch equation.

This principle has been applied to the determination of the intracellular pH of muscle fibers by measuring the total carbon dioxide content of a muscle and the carbon dioxide tension of the blood plasma or surrounding medium. The calculations involve several assumptions, including (1) that the carbon dioxide tension inside the muscle is the same as in the external medium, and (2) that all the acid-labile combined carbon dioxide in the muscle is bicarbonate. The most reliable of these determinations indicate a pH value of 6.9 ± 0.1 for frog and cat skeletal muscle freshly removed from the body.[4] Correction for the amount of extracellular fluid in the muscle tissue was based on the assumption that all the chloride in fresh muscle lies external to the muscle cells.

Many attempts have been made to study the relationship between the internal pH of muscle, estimated from the carbon dioxide/bicarbonate ratio, and the pH of the external medium by equilibrating isolated muscle in Ringer's solution containing different proportional amounts of bicarbonate and dissolved carbon dioxide.[5] Interpretation of the results of such experiments is difficult because the volume of extracellular fluid in isolated muscle varies with time and also because the chloride ion passes into the muscle cells. The most significant experiments in this group were carried out by Wallace and Lowry (1942). They corrected their data for the carbon dioxide and bicarbonate content of the extracellular spaces in

[4] Fenn and Maurer (1935); Wallace and Hastings (1942). It should be noted that Conway and Fearon (1944) calculated a pH of 6.0 for the intercellular pH of rabbit muscle on the basis of carbon dioxide measurements. Hill (1955), however, recalculated their data and obtained a pH value of 6.75.

[5] Stella (1929); Cowan (1933); Fenn and Cobb (1934); Wallace and Lowry (1942); see also Hill (1955).

the isolated muscles by the chloride-ion distribution method. The carbon dioxide tension of the environment was maintained constant, and by changing the bicarbonate concentration the external pH could be varied over a wide range. When the bicarbonate content of the extracellular fluid was high, the data had to be eliminated due to discrepancies. These undoubtedly resulted from error in the chloride method for the determination of the tissue spaces. Nevertheless, in spite of wide changes in the environmental pH, the concentration of bicarbonate within the muscle fibers remained unchanged, and the intracellular pH, therefore, was constant at about 6.9 (since carbon dioxide tension held constant).

The value 6.9 for the intracellular pH of muscle obtained by the carbon dioxide method is identical with that obtained for the cytoplasmic matrix of a large variety of cells by the microinjection of pH indicators (Table 14.1).

The estimation of pH from the carbon dioxide content of a cell represents an average hydrogen-ion concentration of all the intracellular components. In the case of muscle, the pH calculated, assuming that estimation of the tissue space is correct, would essentially represent that of the sarcoplasm, since the aggregate volume of other elements, such as nuclei and connective tissue cells, is proportionately small, and since vacuoles are not present in healthy muscle fibers.

An example of an attempt to use the carbon dioxide method for pH determination in cells with vacuoles is that of Conway and Downey (1950), who calculated the intracellular pH of yeast cells at 5.6 to 6.0. Such values may be more closely related to the hydrogen-ion concentration of the vacuolar fluid than to cytoplasmic pH.

MICROINJECTION OF INDICATORS

The important advantages of this method are that the most useful pH indicators, the sulfonphthaleins, can be introduced into the protoplasmic interior, and the pH of the various components estimated, without injuring the cell. This method, and the results which have been obtained by its use, are discussed in the next chapter.

13

Measurements of pH by Microinjecting Indicators

For the vast majority of cells the micromanipulative method provides the investigator with the only means at his disposal for making suitable intracellular pH determinations. The most useful pH indicators are the sulfonated acid dyes listed by Clark and Lubs, and none of these penetrate living cells in general. Only by microinjection can an adequate series of overlapping sulfonphthalein indicators be introduced into the cytoplasmic matrix and nucleus. This holds true also for the basic dye indicators such as neutral red. As we have seen in the previous chapter, when cells are immersed in solutions of basic dyes, these readily penetrate; but because the dyes segregate in vacuoles, their colors serve to measure only the vacuolar pH.

Another outstanding feature of the micromanipulative method is that it permits determination of the pH of the different intracellular components. Each of these has a pH of its own as well as a different buffering capacity. Results obtained by the microinjection technique are therefore far more meaningful than measurements obtained by other methods which purport to provide values for the average intracellular hydrogen-ion concentration.

A "suitable intracellular determination" can refer only to one carried out on the normal cell, without inducing injury. This is achieved in the micromanipulative method by using a micropipette 0.2 to 0.5 micron in diameter at its tip and by selecting cell types for their resistance to microtrauma.

The usefulness of the Clark and Lubs indicators (Table 13.1) for intracellular pH determinations lies in the fact that they possess sulfonic acid groups, which are fully dissociated throughout the physiological pH range. Accordingly, they are insoluble in lipid deposits and do not combine to any appreciable extent with the

TABLE 13.1. Indicators of useful range used for colorimetric determinations of intracellular pH

Indicator (and abbreviation)	Color and useful pH range
Neutral red (N.R.)	(red) 6.8–8.0 (yellow)
Cresol red (C.R.)	(yellow) 7.2–8.8 (red)
Phenol red (P.R.)	(yellow) 6.8–8.4 (red)
Brom thymol blue (B.T.B.)	(yellow) 6.0–7.6 (blue)
Brom cresol purple (B.C.P.)	(yellow) 5.2–6.8 (blue)
Chlor phenol red (C.P.R.)	(yellow) 4.8–6.4 (red)
Methyl red (M.R.)	(red) 4.4–6.3 (yellow)

cytoplasmic proteins which exist, in the living cell, well to the alkaline side of their isoelectric points. Only damaged, coagulated protoplasm combines with these indicators. When microinjected as the sodium or potassium salts, they promote the fluid state of the protoplasmic interior and diffuse evenly throughout the aqueous phase of the cytoplasmic matrix and of the nucleus.

The indicators may eventually pass into vacuoles, where the colors imparted to the fluid contents bear no relation to the pH of the protoplasmic matrix. Ordinarily this is not a complicating feature, since the segregation process occurs relatively slowly. Furthermore, if a cell type which contains abundant vacuolar inclusions is used, these can be centrifuged to one end of the cell and the pH for the cytoplasmic matrix estimated from the colors in the hyaline zone (see below under "Amoeba dubia").

The relative nontoxicity of these indicators is undoubtedly related to the fact that they do not combine with the protoplasmic constituents. This permits the microinjection of concentrated (0.4 percent) solutions in the amounts needed to achieve sufficient coloration of the cytoplasm without inducing injury. Microinjection of such concentrated solutions is an essential feature, since protoplasm can tolerate the introduction of only limited volumes of fluids.

The validity of the color tints of the sulfonphthaleins as a measure of intraprotoplasmic pH is indicated by the consistency of the results obtained using overlapping indicators. Inconsistencies would be expected if the determinations were subject to considerable colloid or protein error. The absence of significant error from these sources is undoubtedly related both to the noncombining of the indicator anions with the intracellular proteinates and to the inertness of the cytoplasmic (living) proteins with respect to their behavior at interfaces. Furthermore, colorimetric estimations using the basic dye neutral red, which when microinjected in sufficiently dilute concentration also diffuses evenly through cytoplasm and nucleus, agree with those obtained using the sulfonphthaleins. If protein error were considerable, the results would deviate in opposite directions for the basic as compared with the acidic indicators.

The objection has been raised that the micrurgist may be "swamping out" the available hydrogen ions in the cytoplasm with the microinjected indicators. The cytoplasmic matrix and nucleoplasm, however, are highly buffered, and any alteration of their pH values is accompanied by disintegration and death. With respect to the vacuolar inclusions, the characteristic finding is that their contents are frequently unbuffered, and here introduction of indicators could shift the hydrogen-ion concentration.

The technique of determining pH by the method of microinjection is described in the Appendix.

ECHINODERM EGGS

The microinjection method of determining the pH of living cells has probably most often been applied to echinoderm eggs. This was initiated by J. and D. M. Needham (1925, 1926a, b),[1] who microinjected different species of relatively colorless echinoderm, annelid, and ascidian eggs with aqueous solutions of the Clark and Lubs pH indicators. They obtained a uniform pH value of 6.6 to 6.7 for the cytoplasm of the eggs.

Shortly thereafter Chambers and Pollack (1927a) extended this work to the egg of the starfish *Asterias forbesii.* They used both the immature egg, which is characterized by its relatively large vesicular nucleus, the germinal vesicle, and the mature egg both before and after fertilization.

The results of the several injections are graphically presented in Figure 13.1. The colors imparted by the indicators to the cytoplasmic matrix of the starfish egg (unfertilized, fertilized, and in the first and second cleavage periods) are consistent with a pH of 6.6 to 6.8. This closely approximates the value obtained by the Needhams.[2] Mechanical damage of the eggs to the extent of causing cytolysis lowers the pH to that of the acid of injury, 5.5 ± 0.1

[1] Solid particles of the sulfonphthalein indicators were inserted into cells by means of microneedles by Schmidtmann (1924, 1925) and Ogawa (1929). The color tints were observed as the particles underwent dissolution. Reliable results are not obtainable by this method, due to injury of the cell and difficulties in interpretation of the color tints.

[2] Needham and Needham (1926a) state that the eggs of *Ophiura lacertosa* withstood the microinjection of pH indicators far better than all the other eggs they studied. For this egg they determined the cytoplasmic pH at 6.75 ± 0.1. They also described a lowering of pH to between 4.0 and 5.0 following the cytolysis of *Paracentrotus* eggs.

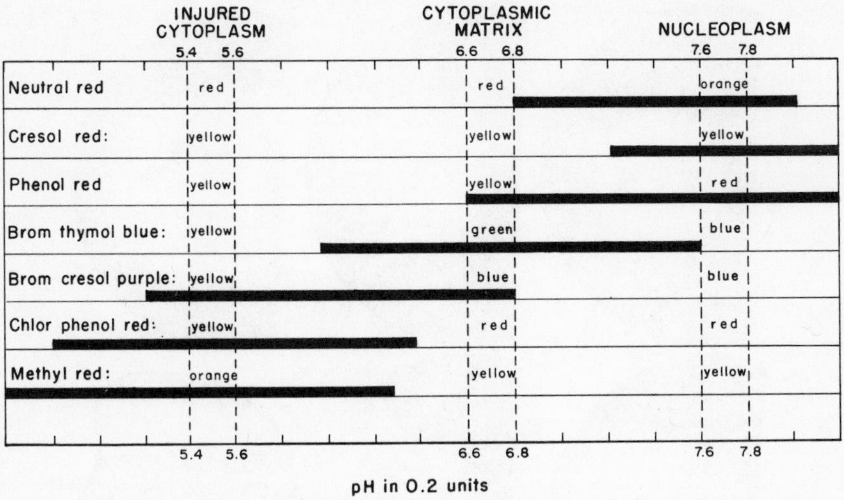

FIGURE 13.1. Tabular representation of pH color indicators observed when microinjected into living cells. Abscissae marked off at regular intervals in 0.2 pH units. Heavy part on each horizontal line marks useful pH range for each indicator. Vertical columns drawn with interrupted lines designate pH color values of each indicator for normal cytoplasmic matrix, nucleoplasm of nucleus, and cytoplasm following injury.

(Figure 13.1), which persists until the encroachment of the surrounding sea water.

It is of interest that the acid due to mechanical injury can also be detected outside the egg in its immediate environment just prior to cytolysis.

The egg reacts in a different manner to a slow tear with a microneedle than to a rapid tear. The egg tolerates a slow tear with impunity; this is probably due to the slight amount of acid produced, which seems to be neutralized as fast as it is formed. With a rapid tear, a localized acid of injury is produced. If sufficient injury is imposed, general cytolysis sets in, made evident by a change to pH 5.5 ± 0.2.

The nucleus, or germinal vesicle, of the immature egg is a highly fluid, optically homogeneous vesicle except for the prominent nucleolus, which, because of its specific gravity, normally rests on the bottom of the vesicle (Gray, 1927). The nucleus is highly susceptible to injury. A mere puncture of it tends to induce irreversible cytolysis, which spreads through the rest of the egg (Figure 13.2). By taking extra precautions it is possible to secure successful nuclear injections and to have the injected eggs continue to

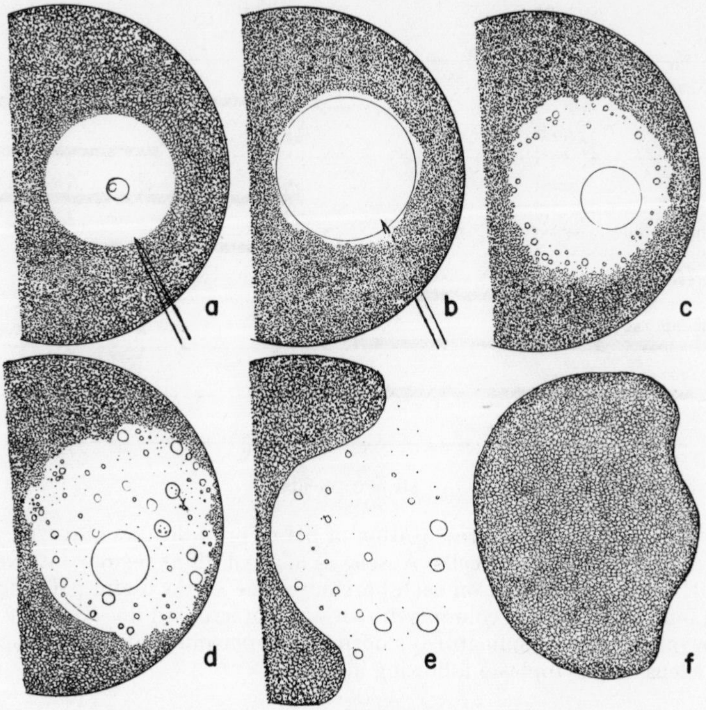

FIGURE 13.2. Successive stages of disintegration in immature starfish egg following puncture of nucleus. (a) Normal living egg as needle is being inserted into nucleus. (b) Following puncture, nucleolus fades, cytolytic reaction in cytoplasm surrounding nucleus spreads at about the same time from entire surface of injured nucleus. (c, d) Disintegration spreads; nuclear remnant shrinks. (e, f) Cytolytic debris walled off from healthy cytoplasm by formation of new protoplasmic surface film. (From Chambers, 1924a.)

undergo maturation (Figure 13.3). The colors imparted to the injected indicators (Figure 13.1) are consistent with a pH of 7.6 to 7.8 for the nuclear fluid.

Maturation of an egg, the cytoplasm and nucleus of which have been previously microinjected with phenol red (Figure 13.3a), is accompanied by wrinkling and fading of the nuclear membrane and a streaming of the red nuclear fluid in all directions from the region of the fading nucleus throughout the surrounding yellow cytoplasm. The diffusion of the red colored nucleoplasm converts the yellow of the cytoplasm to an ever deepening orange tint until all the nucleoplasm has diffused through the egg. Several minutes later the orange color of the mixing nucleocytoplasm changes back to the original yellow color of the cytoplasm (Figures 13.3b, c).

Evidently, the maturation procedure is accompanied by an admixture of the alkaline nuclear fluid (pH 7.6 to 7.8) with the more acid cytoplasmic matrix (pH 6.6 to 6.8). After completion of this mixing, the cytoplasm of the immature egg is converted into what may be termed the nucleocytoplasm of the mature egg. Only then is the egg capable of giving off its polar bodies and of being fertilized, either by insemination or by induced parthenogenesis.

A peculiarity of the nucleus, or germinal vesicle, is its failure to show an acid reaction following injury. When the egg is ruptured,

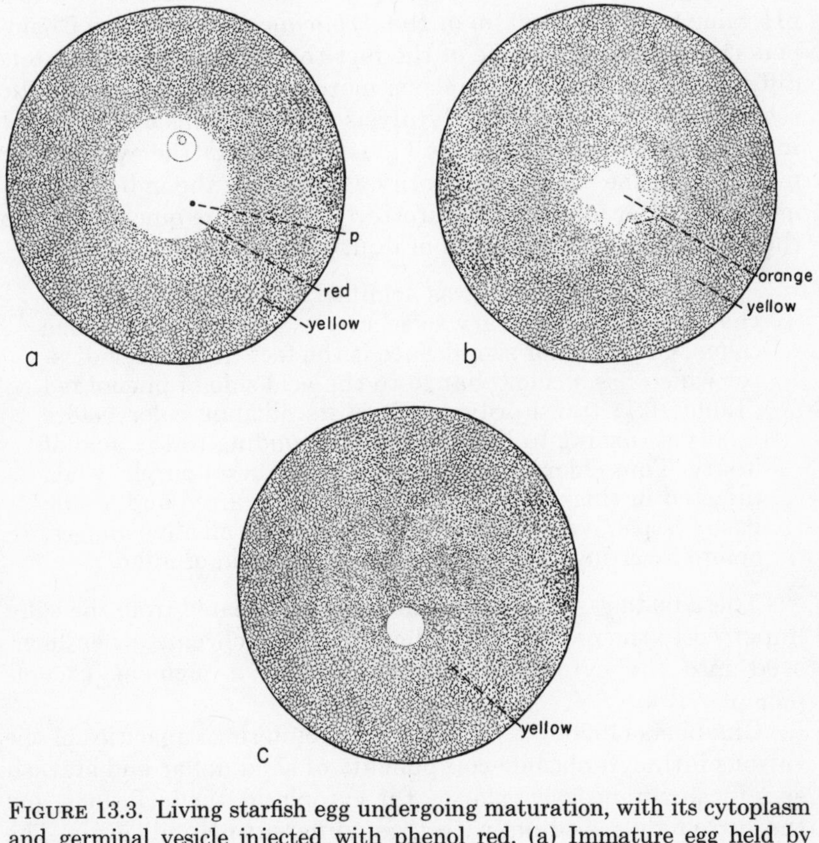

FIGURE 13.3. Living starfish egg undergoing maturation, with its cytoplasm and germinal vesicle injected with phenol red. (a) Immature egg held by needle in shallow region of hanging drop with micropipette (p) vertically inserted into germinal vesicle. Cytoplasm is yellow; germinal vesicle is red. (b) Egg 1 hour later, moved to deeper region of hanging drop. Germinal vesicle has begun to shrivel, and nucleoplasm is mixing with cytoplasm, which is takng on an orange tint. (c) Egg 10 minutes later, with yellow cytoplasm (nucleocytoplasm). Diminutive egg nucleus prior to polar body formation can be seen in egg.

the nucleus becomes converted into an optically homogeneous, spherical nuclear remnant (Figure 13.2) which temporarily maintains its original alkalinity surrounded by the acid reacting cytolytic debris of the rest of the egg. It was this alkaline nuclear remnant that Reiss (1926) probably mistook for the normal nucleus which he observed in sea urchin eggs crushed in solutions of various pH indicators.

The eggs of the sea urchin *Arbacia punctulata* are unusually susceptible to injury when punctured with microneedles. This would account for the low pH value obtained by Wiercinski (1944) for the hyaline zone of the centrifuged eggs. In order to secure a pH value for the cytoplasm of the *Arbacia* egg, Pandit and Chambers (1932) took advantage of the fact that the injected indicators diffuse into the healthy cytoplasm more rapidly than the cytolytic reaction spreads inwardly. Cytolysis always begins at the site of injection. A pH value of 6.6 to 6.8 is inferred for the cytoplasmic matrix from the colors transitorily assumed by the indicators immediately after injection (Figure 13.1). The colors quickly shift to those characteristic of an acid of injury of about pH 5.3.

> Each indicator used was acidified and also alkalinized to the extent just necessary to secure its acid and its alkaline colors. Of particular significance is the fact that each indicator which has a color change to the acid side of phenol red (Table 13.1) transitorily assumes its alkaline color before finally returning to the color corresponding to the acid of injury. Thus chlor phenol red and brom cresol purple, when injected in their acid (yellow) state, give a red and a blue flash, respectively, characteristic of their alkaline ranges, before reverting to the yellow due to the acid of injury.

These findings corroborate the values estimated from the color imparted to the natural pH indicator, echinochrome, when liberated into the cytoplasm by puncture of a pigment vacuole (Chapter 12).

Chambers (1928b) investigated the buffering capacities of the various intracytoplasmic components of sand dollar and starfish eggs by simultaneously coloring the cytoplasm, the inclusions, and the surrounding medium with three different pH indicators. The protoplasm of these eggs is uniformly crowded with granules (or vacuoles). When the eggs are immersed in a solution of neutral red for a short period, the vacuoles are stained with the dye, but no other part of the protoplasm is colored. Phenol red or brom cresol purple is then microinjected into the cytoplasm, and the eggs are suspended in sea water containing cresol red. When the egg suspension was exposed to the weak base ammonia or the weak acid carbon dioxide, the effect of alkalinizing or acidifying the various components could be observed. Both ammonia and carbon dioxide readily penetrate the protoplasm to shift the pH of the intracellular vacuoles, but no shift occurs in the pH of the hyaloplasm, at least none sufficient to change the color of microinjected indicators. These striking experiments are described below. (Refer to Table 13.1 for color shifts.)

The eggs are allowed to remain in sea water containing neutral red long enough to stain only a small percentage of the vacuoles. The eggs are then washed, transferred to an hanging drop of sea water colored with cresol red, and injected with phenol red. It should be remembered that none of the sulfonphthalein pH indicators can penetrate the protoplasmic surface either from within or from without. The result is a striking picture of yellow eggs (acid range of phenol red) containing scattered red granules (acid range of neutral red) and surrounded by a yellow medium of sea water (acid color of cresol red). Ammonia gas is then passed through the chamber until the cresol red in the sea water changes from yellow (acid) to red (alkaline). As soon as this occurrs the cytoplasmic granules, stained with the neutral red, turn yellow (alkaline), while the hyaloplasm maintains the original yellow (acid) color of the phenol red. The result is now a picture of uniformly yellow eggs standing out against a background of red colored sea water (cresol red plus ammonia). Carbonic acid gas is then passed through the chamber until it displaces the ammonia in the hanging drop. As a result the original colors return: The sea water again becomes yellow (acid color of cresol red); the cytoplasmic granules turn from yellow to red (acid color of neutral red); and the cytoplasm itself remains yellow (acid color of phenol red).

Since the cytoplasm has a pH of 6.8 ± 0.1 which is in the acid range of phenol red (yellow), the above experiment is not suited for detecting a possible effect of carbon dioxide on the cytoplasmic pH. For this purpose it is necessary to use brom cresol purple (yellow at a pH more acid than 5.8 and blue at a pH more alkaline than 6.4) which, upon injection, colors the hyaloplasm blue. These microinjected eggs are immersed in an hanging drop of sea water colored blue with the same dye (brom cresol purple). The hanging drop is suspended in an hermetic chamber (Figure A.3) through which moist carbonic acid gas is made to stream until the sea water becomes sufficiently charged with carbon dioxide to change its color to the yellow of the acid range of brom cresol purple. The cytoplasm of the eggs keeps its original blue color in the yellow sea water.

The considerable buffering capacity of the nuclei of immature starfish eggs can be demonstrated by microinjecting nuclei of different eggs with cresol red, neutral red, and phenol red, and then exposing the eggs to an atmosphere containing carbon dioxide or ammonia. In every case the alkaline color within the nuclei of the eggs remains constant irrespective of the color changes of the neutral red stained granules in the surrounding cytoplasm. In other words, the nucleus is sufficiently buffered so that the intranuclear pH of 7.6 to 7.8 is maintained unchanged. When the egg is disintegrated by crushing or tearing, the nucleus loses its capacity to resist changes in pH. The persisting spherical nuclear remnant is then immediately susceptible to the acidity or alkalinity of the environment.

AMOEBA DUBIA

The use of the highly dissociated sulfonated indicators of Clark and Lubs, eminently fitted for intracellular pH determinations, is complicated in the ameba because of the relative rapidity with which the indicators become segregated in vacuoles. This may account for the variable results of the earlier microinjection work on the ameba,[3] in which attempts were made to estimate the pH of the cytoplasm by noting the color of the ameba as a whole. Accurate results can be obtained only by the special method of centrifuging the ameba and noting the color imparted to the hyaline zone, free of inclusions.

Morphologically, the interior of the ameba consists of an optically homogeneous fluid matrix in which are distributed numerous vacuoles. Even the crystals may be enveloped by vacuolar membranes. Along with the crystal-containing vacuoles, which are relatively large, are numerous, much smaller, granule-free vacuoles, difficult to observe as separate bodies except by special illumination. The pH indicators tend to segregate into these vacuoles, all of which give a reaction on the alkaline side of neutrality. These vacuoles are not always present in appreciable numbers. On the other hand, they may be so numerous as to occupy a considerable part of the protoplasm. This variability and

[3] On the basis of an orange-yellow coloration imparted to *Amoeba dubia* injected with phenol red, Chambers, Pollack, and Hiller (1927) reported a pH value of 6.9 ± 0.1 for the cytoplasm. Spek (Spek and Chambers, 1933) described a brick red coloration with phenol red in the cytoplasm of the same species of *Amoeba* and reported a pH of 7.2 to 7.3. Needham and Needham (1925) obtained values which were even more alkaline, reporting a pH of 7.6 for *Amoeba proteus* cytoplasm.

the fact that the penetration into them of injected indicators varies considerably render inconsistent the colorimetric determinations of the pH value of the aqueous phase of the ameba interior.

To overcome this difficulty, amebae, already injected with indicators, were centrifuged in order to ascertain the color tints present in the hyaline zone. These colors were found to be consistent with a pH of 6.6 to 6.8 for the cytoplasmic matrix (Figure 13.1).[4] The experiment is carried out as follows.

After an *Amoeba dubia* is injected with an indicator, the animal is sucked into a capillary tube about 6 cm. long and 0.5 mm. in diameter. One end of the capillary, where it is free of water, is drawn in a microflame to a long cylindroid taper averaging 100 micra in diameter. The capillary is then sealed at both ends. After the capillary has been centrifuged with the tapering tip directed centrifugally, the ameba is found in the cylindroid portion of the taper, where it assumes a long, cylindrical form. The part of the capillary containing the ameba can be observed under the microscope with the tube immersed in cedar oil having the same refractive index as glass. With appropriate centrifugation, the interior of the ameba becomes stratified into three distinct zones: a *vacuolar zone* at the light pole containing the contractile vacuole and surrounded by a halo of minute vacuoles; *an intermediate hyaline zone* consisting of a clear, fluid region free from visible structure; and a heavy *granular zone* containing the gastric vacuoles, some acid and others alkaline, the nucleus, and granular material with crystals in vacuoles. In every case the coloration of the hyaline zone indicates a pH of 6.6 to 6.8 for the normal cytoplasmic matrix (Table 14.1). Delayed segregation of the indicators into vacuoles is observed in all cases. This is evident even when the centrifuged amebae are observed within 5 to 10 minutes after microinjection. Chlor phenol red becomes segregated in the upper vacuolar zone, coloring the vacuoles red-purple. This occurs in such a way that no coloration is ever observed in the hyaline zone. With brom cresol purple, segregation in the upper vacuolar zone giving this a blue color occurs far more slowly. The most interesting results are obtained with phenol red. The hyaline zone is colored a distinct yellow indicating a pH more acid than 7.0, while the vacuolar zone assumes the pink coloration of the more alkaline range of this dye. The gastric vacuoles in the heavy zone show tints varying from lemon yellow to deep red, due

[4] An identical value was obtained by Wiercinski (1944), who microinjected indicators into the hyaline zone of centrifuged amebae.

presumably to variations in pH according to the different digestive states of their contents. After an ameba microinjected with phenol red is removed from the capillary tube, the separated components become redistributed rapidly. This redistribution imparts a yellow-orange tint to the ameba as a whole, representing, to the observer's eye, an amalgamation of the yellow color of phenol red in the matrix and the pink color of the phenol red within the distributed minute vacuoles. The color imparted to the ameba by injections, therefore, varies according to the number of vacuoles in the cell interior and the degree to which the injected indicator segregates into them.

The injection of the pH color indicators, as sodium salts, causes a temporary cessation of streaming. However, a few minutes afterward, the ameba recovers and, provided the injection was not excessive, remains active for hours.

The existence of an acid of injury is not easy to demonstrate in the ameba. When an ameba, already injected with indicator is vigorously torn with a microneedle so that the injured area is ultimately discarded, acid production occasionally can be demonstrated. When death occurs in the presence of the indicators in the appropriate range, a transitory acidity as low as pH 5.5 may be observed. Often, however, acid production following mechanical injury of an ameba is not observed. In this respect the ameba differs from most cells, which release acid following mechanical injury. The frequent absence of an acid of injury in amebae may be due to a counteracting liberation of alkaline fluid from vacuoles which disrupt as the protoplasm undergoes disintegration.

In studies on the action of various electrolytes injected into the ameba, calcium chloride and magnesium chloride are found to differ from other salts by causing the production of acid when microinjected in sufficient quantity to induce solidification (reversible) or coagulation (irreversible).

If calcium chloride in a concentration of 0.005 M is injected into an ameba colored with phenol red, a flash of lemon color occurs at the site of injection (Reznikoff and Pollack, 1928). This is accompanied by a localized solidification of the region injected. At this dilution of the injected salt the clotting cytoplasm frequently reverts to its former fluid state, and the color returns rapidly to that of the normal pH. If the color does not revert within a few seconds, the condition becomes irreversible, and the affected area is pinched off. By injecting a series of indicators, it was found that the fall in pH is as low as 4.0 to 4.6 in the irreversibly coagulated portion of the ameba. This eventually is pinched off, and the pH of the dead mass assumes the reaction of the environment.

Following injection of the calcium and magnesium salts, a greater intensity of acid is produced than after mechanical injury alone. This may be due to the formation of insoluble calcium or magnesium compounds with the liberation of free acid.

The cytoplasmic matrix of the living ameba shows considerable buffering capacity. When the injection of a salt, an alkali, or an acid causes pH changes which last more than a few seconds, the buffering capacity tends to be exceeded and death ensues.

Solutions of 0.01 N hydrochloric acid, when introduced into an ameba containing previously injected indicator, cause an immediate decrease in pH of the injected area. If the injected acid is in minimal quantity, the pH reverts within a few seconds to that of the normal cytoplasm. On the other hand, if the pH does not immediately return to normal, the condition is irreversible and the injured portion is pinched off as a dead mass (Reznikoff and Pollack, 1928). Considerable amounts of 0.005 M phosphate buffer solutions may be injected without causing injury, as long as the pH is only momentarily altered (Pollack, 1928b). Buffers in the pH range of 5.6 to 6.0, when injected into amebae colored blue with brom cresol purple, produce a temporary yellow flash followed by a reversion to the normal blue alkaline color. The injection of phosphate buffers more alkaline than pH 7.0 into amebae colored yellow with phenol red produces transitory red flashes.

ROOT HAIR CELLS

The root hair cells of the water plant *Limnobium spongia* are very suitable material for microinjection work (Figure 13.4). The cell wall is thin, the osmotic pressure in the sap is low, the protoplasm is devoid of plastids and other inclusions, and streaming occurs constantly. Each root hair is a single cell and, according to age, of varying lengths, and about 40 to 50 micra in diameter. The nucleus always lies at the base of the cell, which is embedded in the epidermis of the root. Cyclosis, or protoplasmic streaming, is normally active and rapid. The protoplasm forms a thin layer which flows along the wall of the cell and continually heaps up at the tip. Here the direction of flow is reversed, and the protoplasm streams toward the base in the form of a rope-like central strand which traverses the sap vacuole.

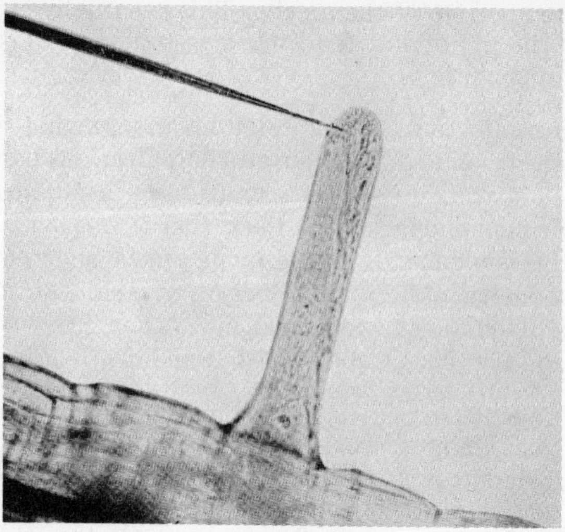

FIGURE 13.4. *Limnobium* root hair cell, 45 × 250 mm., with micropipette inserted into protoplast through cellulose wall at tip of cell. Cell nucleus lies at base of cell hidden by epidermis of root. Injection of monovalent salt solutions need not disturb cyclosis, or protoplasmic streaming, except by temporarily slowing streaming. This induces heaping of cytoplasm at tip of root hair and facilitates cytoplasmic microinjections of, for example, pH color indicators. (From Chambers and Kerr, 1932.)

Chambers and Kerr (1932) microinjected the sulfonphthalein indicators into the cytoplasm by puncturing the cell at its tip, where the protoplasm, during its streaming, accumulates, especially after the injection of sodium salts, to form a relatively deep layer (Figure 10.4b). Following an injection, streaming sometimes ceases temporarily, but shortly afterward the normal flow is resumed. When injected into the cytoplasm, the indicators produce a diffuse coloration, and there is no evidence of specific staining of granules or cell inclusions. The colors imparted to the cytoplasm are consistent with a pH of 6.8 ± 0.2 (Figure 13.1). The sulfonphthalein indicators slowly pass into the sap vacuole until the cytoplasm is colorless. In no case is there outward passage of indicator to the external medium, as could be verified by looking for color in the space between the cell wall and the external surface of a plasmolyzed protoplast. When injected into the sap vacuole, the sulfonphthalein indicators are restricted to the sap, where they remain without fading for over 24 hours. The living, uninjured cytoplasm never takes up the indicators from the colored sap vacuole. This unidirectional passage of the indicators from cytoplasm to vacuole suggests secretory activity (Chapter 16).

The pH of the sap vacuole was determined at 5.2 ± 0.2. This value is always the same in healthy cells of all ages. Neither diurnal nor seasonal changes are observed. Although the normal environment is pond water, which possesses an alkaline reaction, the pH of the sap vacuole is not altered by growing the root hairs in acid (pH 5.0) solutions.

Mechanical injury, as is true for the cytoplasm of most cells studied, produces an acid of injury, the pH of the injured region falling to a value of 5.2 ± 0.2. Interestingly, when a cell exhibits cytolysis, although the cytoplasm exhibits an acid of injury, the pH of the sap vacuole rises to a value of about 6.0. The vacuole still shows this temporary alkaline shift even when the root hairs are immersed and undergo cytolysis in an acid solution of pH 5.0.

The cytoplasm of the root hair cell possesses considerable buffering capacity. This has already been noted for the echinoderm egg and the ameba. When injected cells are exposed to carbon dioxide or to ammonia, the diffuse color of the indicators in the cytoplasm remains unchanged as long as the cell is alive. On the other hand, the sap in the vacuole is only weakly buffered, if at all, since ammonia alkalizes, and carbon dioxide acidifies, the vacuolar contents.

14

Evaluation of Intraprotoplasmic pH Determinations

In this chapter are incorporated an evaluation and a summary of the pH measurements which we have previously described in some detail. Numerous difficulties confront the investigator with regard to the meaning of the pH values he finds. In dealing with living tissues he is working with a heterogeneous system in which there are many possibilities to lead him astray. Since the pH determinations must be of aqueous solutions, many researchers attempted to secure aqueous extracts of cells, even though it was realized that extracts are, at best, mixtures of various components and that their use might produce abnormal acidic or basic reactions. Attempts to circumvent such errors included treating muscle tissue with anesthetics beforehand to avoid contracture, and also cooling or freezing tissue. Using such methods, all the experimenter can do is to hope that he is working in the right direction, but how far he can have no idea.

Many investigators have drawn conclusions about the pH of the cell interior by noting the colors of indicators which are naturally present or which appear within cells following their immersion in solutions of certain dyes. As previously noted, the results obtained concern only the color of indicators already segregated in the vacuoles, where the tints assumed are not directly related to the pH of the protoplasmic matrix. Furthermore, basic dyes had to be used, since these include the only indicators which penetrate cells in general and yet change color in the appropriate pH range. The basic dye indicators, however, are unreliable, in that they tend to form undissociated complexes and either dissolve as the free base in lipid components or rapidly segregate into cytoplasmic vacuoles within the cell.

The pH of the protoplasmic matrix can have meaning only when the measurements are of the hydrogen-ion concentration of its aqueous phase. The micromanipulative technique has shown that protoplasm, in general, possesses a continuous aqueous phase. Accordingly, from the theoretical point of view, inserting microelectrodes into the cell would be a good method for determining hydrogen-ion concentration. Although the technical difficulties have thus far made it impossible to obtain conclusive electrometric determinations on uninjured cells, Caldwell's (1954, 1956) glass microelectrode is promising, and every effort should be made to overcome the problems.

Another important contribution of the micromanipulative method has been the demonstration that the cytoplasmic proteins, dispersed throughout the continuous aqueous phase of the protoplasmic matrix, exist well to the alkaline side of their isoelectric points as proteinates and are peculiarly inert, especially with respect to their behavior at interfaces (Chapter 1). These characteristics of the cytoplasmic proteins in living cells permit the use of the Clark and Lubs series of overlapping indicators as a valid measure of intracellular pH.

Thus when the sodium or potassium salts of these fully dissociated sulfonphthalein indicators (anionic) are microinjected, they spread through the aqueous phase of the protoplasm, imparting to it a diffuse, even coloration. Combination of the injected dyes with cellular components does not occur unless the cell is damaged. The semifluid consistency of the cytoplasmic interior is unchanged, and the protoplasmic activities continue unimpaired. Indication of the inert state of the proteins, with respect to their possible interaction with the dye anions, is the absence of detectable protein error, since the colors imparted to the different indicators are all consistent for a given pH.

Since the sulfonphthalein indicators do not penetrate cells in general, they must be microinjected. This has proved the only reliable, generally applicable method for securing pH measurements of uninjured cells. Moreover, introduction of the indicators by microinjection provides the only means at our disposal for estimating the various pH values of the different intracellular components. The cytoplasmic pH, however, can be accurately measured, without resorting to microinjection, in one specialized type of cell, namely, the proximal convoluted tubule cell of the kidney, which takes up the sulfonphthaleins by secretory activity. During this process, the cytoplasm is colored diffusely, and segregation into vacuoles does not occur. The indicators are evidently in solution in the continuous aqueous phase of the cytoplasmic matrix, and their colors serve as a reliable measure of pH.

The only noncolorimetric method which has yielded significant results in uninjured cells is the method of carbon dioxide analysis. Since this provides an over-all, average value for the hydrogen-ion concentration of the cell interior, meaningful results can be obtained only on cells which possess a predominantly "homogeneous" cytoplasm, such as the muscle cell with its abundant sarcoplasm.

CONSTANCY OF THE pH OF CYTOPLASMIC MATRIX AND NUCELEOPLASM

By means of microinjecting indicators, Chambers and his co-workers have consistently found that the cytoplasmic matrix has a pH of 6.8 ± 0.2 and the nucleoplasm a pH of 7.6 ± 0.2 in a wide variety of cells. The most reliable pH measurements which we feel have so far been obtained are listed in Table 14.1. These include colorimetric determinations on the eggs of the starfish, the sea urchin, and the clam; on cells of the ciliated epithelium, the gastric and intestinal mucosa, the liver, the pancreas; on the unripe ova of the frog and *Necturus;* on ganglion cells of the fish *Lophius;* and on malignant cells of mammary carcinomas and sarcomas grown in tissue culture. The pH of the nucleoplasm was determined either by first introducing the indicators into the cytoplasm, following which diffusion of the indicators into the nucleus readily occurs, or by microinjecting the nucleus directly. The nuclear pH measurements refer only to the fluid nucleoplasm or nuclear sap. No determinations have been attempted on the "solid," or gelled, nuclei of amebae and of many plant cells.

The pH of the cytoplasmic matrix only was determined in *Amoeba,* eggs of the fish *Fundulus* and *Oryzias,* the skeletal muscle of the frog, and the root hair cells of *Limnobium.* In addition, the colors assumed by the sulfonphthalein indicators in their passage through the epithelial cells of the proximal convoluted tubules of the kidney and of the chorioid plexus, while being secreted, indicate a pH of 6.8 ± 0.2 for the cytoplasmic matrix.

A critical review of the determinations which have been made using the microinjection method[1] fails to reveal any evidence for

[1] Rapkine and Wurmser (1928) claim to have found the identical pH of 7.2 for both the nucleus and cytoplasm of salivary gland cells and echinoderm ova. Apparently this value was inferred from the observation of an orange tint imparted by the microinjection of phenol red. They failed, however, to confirm their finding with other indicators. Little is said about the methods used, and certainly the micrurgist's skill is taxed to the limit when he attempts the microinjection of nuclei.

TABLE 14.1. Results obtained by microinjecting colorimetric pH indicators into living cells. Abbreviations for indicators as in Table 13.1.

	Cytoplasmic matrix							Nucleoplasm			
	C.P.R.	B.C.P.	B.T.B.	P.R.	N.R.	C.R.	Inferred pH	P.R.	N.R.	C.R.	Inferred pH
Immature ovum (*Asterias*)[a]	>6.4	⋝6.6	⋝6.8	⋜6.8	<7.0	<7.2	6.7 ± 0.1	⋝7.6	7.6–7.8	⋜7.8	7.7 ± 0.1
Immature ovum (*Mactra*)[b]	>6.4	⋝6.6		⋜7.0		<7.2	6.8 ± 0.2	⋝7.6		⋜7.8	7.7 ± 0.1
Immature ovum (frog)[c]				⋜7.0				⋝7.6		⋜7.8	7.7 ± 0.1
Mature ovum (*Fundulus*)[d]	>6.4	>6.6	⋝6.8	<7.0		<7.2	6.7 ± 0.1				
Mature ovum (*Arbacia*)[e]	>6.4	⋝6.6	⋝6.8	⋜6.8			6.7 ± 0.1				
Mature ovum (*Lytechinus*)[f]	>6.4	⋝6.6	6.8	⋜6.8			6.7 ± 0.1				
Amoeba (centrifuged)[g]	>6.4	>6.6		<7.0	<7.0	<7.2	6.8 ± 0.1				
Root hair (*Limnobium*)[h]	>6.4	>6.6		⋜7.0		<7.2	6.8 ± 0.2				
Proximal tubule, chick kidney[i]	>6.4	>6.6	6.6–6.8	⋜6.8		<7.2	6.7 ± 0.1				
Ganglion (*Lophius*)[j]	>6.4			⋜6.8			6.7 ± 0.2	⋝7.6			⋝7.6
Gastric epithelium (frog)[j]	>6.4			<7.0			6.7 ± 0.2	⋝7.6			⋝7.6
Ciliated epithelium (frog)[c]	>6.4			<7.0		<7.2	6.7 ± 0.2				
Striated muscle (frog)[c]	>6.4			<7.0			6.7 ± 0.2				
Carcinoma (human)[k]		>6.6		<7.0			6.8 ± 0.1	⋝7.6			⋝7.6
Sarcoma (human)[k]		>6.6		<7.0			6.8 ± 0.1	⋝7.6			⋝7.6

[a] Chambers and Pollack (1927a); [b] Grand (1938); [c] Chambers, Pollack, and Hiller (1927); [d] Chambers (1932), for blastomere-developing *Oryzias* egg see Yamamoto (1936); [e] Pandit and Chambers (1932); [f] Kopac (1935); [g] R. Chambers (unpublished), Wiercinski (1944); [h] Chambers and Kerr (1932); [i] Chambers and Cameron (1932); [j] Chambers (1933c); [k] Chambers and Ludford (1932b).

the occurrence of pH values which differ appreciably from those obtained by Chambers and his co-workers (Table 14.1). The constancy of the pH of the cytoplasmic matrix in such a wide variety of cell types is particularly of interest, since the cells studied exist in environments whose pH values range from 8.2 to 8.4 for sea water, from 7.4 to 7.8 for the blood plasma of vertebrates, and from 6.0 to 8.0 for pond waters.

From the universally found constancy in the colorimetric pH of the cytoplasmic matrix there has come to our attention one exception. Injecting a series of Clark and Lubs pH indicators, Kopac (1935) found a consistent cytoplasmic pH of 6.8 $\pm$ 0.2 among a large series of mature, unfertilized echinoderm eggs, except in the ovum of one large sea urchin, *Tripneustes esculentus,* for which he recorded an unusually acid pH of 5.4 to 6.0 for the cytoplasmic matrix. A possible explanation for this discrepancy is that the cytoplasm of this egg contains a secretion product which either is exceptionally acid in nature or combines with the indicators.

An analogous situation is met with in the determination of the cytoplasmic pH of mucus-secreting cells of the gastric mucosal epithelium in the rabbit. Over two thirds of the epithelial cell is filled with mucus. The actual cytoplasmic pH of 6.8 was determined by microinjecting in the close vicinity of the cell nucleus at the base of the cell, while the intracellular mucus gave a decidedly alkaline reaction.

It should be emphasized that the microinjection method does not measure "intracellular" pH. This refers to an over-all average hydrogen-ion concentration, which undoubtedly varies under different metabolic or external conditions (Caldwell, 1956), especially in cells with abundant vacuoles or other fluid-containing inclusions. Constancy of pH value is claimed, within the limits of accuracy of the method (0.2 pH units), only for the hyaloplasmic matrix, in which inclusions of variable pH values are suspended.

ACID OF INJURY

Acid production generally occurs following injury to the cytoplasm. The increase in acidity due to injury may lower the pH two full units. If the injury has not been too severe, the pH rapidly reverts to the normal, while if the injury induces irreversible cytolysis, the acidity is short lived, and the pH rapidly changes to that characteristic of the surrounding medium. In all cases where the acid of injury has been observed the cells were at room temperature. Whether or not the acid of injury is suppressed by cold or by metabolic inhibitors has not been determined.

In certain cells, such as echinoderm eggs, the production of an acid condition may occur simply as a result of mechanical trauma, without involving irreversible injury. For example, a localized flash of color change indicating acid production is apparent in a previously injected starfish egg if the needle is introduced abruptly into the interior. In other cells, an acid production is not evident unless the mechanical trauma is vigorous enough to cause cytolysis of the disturbed part. The *Amoeba* is the only cell studied by the microinjection method in which, even after injury sufficient to cause death, an acid of injury is observed only inconstantly. As previously noted, this may be due to a variable number of alkaline vacuoles in the cytoplasm, which disrupt upon cytolysis.

The development of an acid of injury is generally a good criterion for establishing whether or not the region of a cell injured includes the protoplasmic matrix. The reaction may not be evident in cells which are composed largely of yolk. For example, in the freshly laid, unfertilized egg of *Fundulus* the cytoplasm occupies a very thin layer surrounding the voluminous yolk. When such an egg, immersed in an unbuffered alkaline (red) solution of phenol red, is punctured, only a flash of yellow at the site of puncture can be detected, indicating a momentary production of acid. If, on the other hand, the whole egg is crushed, no acid of injury can be detected in the debris because of the mass of inert yolk present. However, if the eggs are allowed to stand an hour or more in sea water, there develops a discoid cap of colorless and finely granular protoplasm. This cap is the only part of the egg which undergoes cleavage; the rest consists of yolk material. If the cap is microinjected with an indicator, there is the typical reaction of a cytoplasmic pH of 6.8 ± 0.2, and when this cap is torn, an acid of injury is evident (Chambers, 1932).

When calcium or magnesium salts are injected into the ameba, the coagulation or solidification which occurs is accompanied by a lowering of the pH. If the coagulation is irreversible, the pH falls to values considerably below those observed following simple mechanical injury. This may be due to the formation of undissociated calcium or magnesium compounds, with the liberation of hydrogen ions.

In contrast to the cytoplasmic reaction, injury to the nucleus does not induce an acid of injury.[2] Upon being irreversibly injured, the nuclear remnant maintains its alkaline reaction for a short period and then gradually assumes the pH of the environing medium.

[2] Chambers and Pollack (1927a); Chambers and Ludford (1932b).

BUFFERING CAPACITY OF CYTOPLASMIC MATRIX AND NUCLEOPLASM

Considerable amounts of acid and alkali may be injected into the cytoplasm of cells containing previously introduced indicators without causing more than momentary shifts from the normal pH value. If the injections are sufficient in amount to cause more than momentary changes in pH, death of the cells results. In addition, it is significant that the localized increase in intraprotoplasmic acidity caused by mechanical injury or by the injection of divalent salts is immediately neutralized if recovery occurs. The only observed naturally occurring instance of a pH change of the cytoplasmic matrix is a transient one, when the alkaline nuclear sap mixes with the cytoplasm in the maturating starfish ovum.[3]

When cells are immersed in neutral solutions of the basic dye, neutral red, the dye readily penetrates and is segregated in the vacuoles but does not visibly stain the cytoplasmic matrix. Exposure of the cells to an ammonia or a carbon dioxide atmosphere readily shifts the color of the indicator in the vacuoles to the alkaline or acid side. This suggested to the earlier investigators that the cytoplasmic pH can be changed easily by environmental conditions. If, however, cells previously stained with neutral red so as to color the vacuoles are injected with the sulfonphthalein indicators and exposed to carbon dioxide or ammonia, marked shifts in pH of the vacuoles occur, but no change can be observed in the colors of the indicators in the cytoplasmic matrix and nucleoplasm, as long as the cells remain healthy. If one attempts to speak of an over-all intracellular hydrogen-ion concentration, this most certainly has been altered. Evidently, however, cytoplasmic buffer systems are mobilized to maintain the pH of the cytoplasmic matrix at a constant level, at least within the range of detectable color change, that is, $\pm$ 0.1 pH units.

Wide changes in intracellular pH, calculated by the carbon dioxide method, have been reported to occur in muscle fibers when the carbon dioxide tension of the environment is changed.[4] Fenn and Cobb (1934) pointed out that these determinations refer rather to the average pH of muscle tissue. Furthermore, Wallace and Lowry (1942), also using the carbon dioxide method, were unable

[3] A burst of acid formation occurs very shortly after fertilization in the sea urchin egg (Runnström, 1933). It would be interesting to determine whether or not this acid formation is accompanied by a transient change in pH of the cytoplasmic matrix. For this purpose a transparent egg such as that of *Lytechinus*, or the hyaline zone of the centrifuged egg, would have to be microinjected.

[4] Stella (1929); Cowan (1933); see also Hill (1955).

to detect any significant shift of intracellular pH of muscle exposed to a medium in which the carbonic acid/bicarbonate ratio was varied. Recently, Caldwell (1956), using his glass microelectrode method, reported that the exposure of crab muscle and squid nerve to high carbon dioxide tensions causes the intracellular pH to fall a full unit. However, insertion of the necessarily large electrode undoubtedly damaged the sarcoplasm or axoplasm. Extrapolating the results obtained by the microinjection method, a plausible prediction would be that as long as the fibers remain fully normal and excitable little, if any, maintained alteration in pH of the sarcoplasmic substance should occur.

The hydrogen-ion concentration differences noted between the cytoplasmic matrix and the nucleoplasm, whose buffering systems are poised at different levels of pH, may be related to the presence in the nucleus of basic proteins, the histones, which in the isolated state have been shown to possess isoelectric points considerably to the alkaline side of neutrality. No studies relative to the state of the proteins in the living nucleus, however, have been made, while micromanipulative experiments have shown that the isoelectric points of the proteins of the cytoplasmic matrix lie well to the acid side of neutrality.

RELATION OF HYDROGEN-ION CONCENTRATION TO METABOLISM

As is well known, cells impose pH changes on their immediate environment as a result of their metabolic activities. For example, alkalization and acidification of muscle accompanying activity (Dubuisson, 1950) reflect changes of pH in the interstitial fluids, but not necessarily changes in pH of the sarcoplasm.

If the pH of any of the cellular components is altered during varying conditions of metabolic activity,[5] one would expect to find such changes most evident in the unbuffered or weakly buffered contents of the vacuoles. Rohde (1917) immersed filaments of *Spirogyra* in methyl red and observed that during daytime, in sunlight, the vacuolar contents were red (acid), while at night the color was yellow (alkaline).

Diurnal variations,[6] but in the opposite direction, have been

[5] Many pH measurements have been made of juices expressed from plant and animal tissues which had been, while in the living state, exposed to varying environmental conditions (Small, 1929, 1955). In view of the errors inherent in such methods, it is impossible to evaluate the results obtained.

[6] On the other hand, Hoagland and Davis (1923), who measured electrometrically the pH of the vacuolar fluid sucked out of *Nitella* cells, did not observe any diurnal or seasonal changes in the pH of the vacuolar contents. The same results were obtained by Chambers and Kerr (1932) for the non-chlorophyll-containing root hair cells.

observed in the sap vacuoles of *Bryophyllum* and have been correlated with changes in the content of malic and citric acids. In the Protozoa a cycle of change in the hydrogen-ion concentration of the fluid in the food vacuoles during digestion of their contents has been described by Shapiro (1927) and Howland (1928). While initially the pH of the gastric vacuolar fluid is equal to or more alkaline than that of the external medium, the pH falls to 4.0 to 4.5 during active digestion and finally rises to 6.8 to 7.0 prior to excretion. In a similar category are the acid-secreting cells of the stomach (Bradford and Davies, 1950). When these cells are stimulated to secrete, the colorimetrically determined pH in the intracellular canaliculi falls from about neutrality to less than 1.4.

In general, the observed acidification or alkalization of tissues or cells accompanying activity involves changes of pH in interstitial fluids or in the fluid contents of vacuoles or other cytoplasmic inclusions. All the valid evidence points toward the fact that no significant pH shifts occur in the cytoplasmic matrix during different functional states, at least within the accuracy of the intracellular colorimetric methods used (± 0.1 pH units).

The most remarkable feature concerning the relationship between metabolism and pH is not that correlated changes are detectable, but rather that metabolic activity is undoubtedly involved in maintaining the pH of the cytoplasm at a constant level. Even in measurements which purport to provide a value for "intracellular" pH, the striking feature is that the observed pH shifts fall far short of those that would be expected on a purely physical basis, such as Donnan's equilibrium (Hill, 1955; Caldwell, 1956). Metabolic activity undoubtedly underlies the extraordinarily constant pH values determined for the cytoplasmic matrix in a great variety of cells, maintained even when the cells are exposed to widely different environmental and experimental conditions.

VI

Penetration of Dyes into Cells and Vital Staining

INTRODUCTION TO PART VI. Vital dyes may be grouped into two main categories. In the first are included dyes which penetrate cells by a process involving only simple diffusion. In the second category are dyes whose passage into cells involves metabolic activity on the part of the cells; these dyes penetrate only when the cells are properly oxygenated and at temperatures compatible with normal functional activity.

In the first category are the water-soluble, acid and basic, dyestuffs. These are largely weak electrolytes, but include a few relatively strong electrolytes such as methylene blue. Many exist in solution as ion aggregates, or micellae. The second category includes dyes belonging to two different groups. In one group are the dyes which exist in aqueous suspension as particles of colloidal or larger dimensions. These dyes are ingested by cells which possess the property of phagocytosis, either at the microscopic or at the submicroscopic level. In the other group are the highly dissociated sulfonated acid dyes. These do not enter cells generally, but do penetrate specialized secretory cells, such as liver cells and the proximal tubule cells of the kidney, by a process of active transfer. There is no indication that vacuole formation is involved in the entry and transfer of these highly dissociated dyestuffs.

15

Penetration of Dyes by Diffusion

The feature which dyes of this group have in common is that their penetration into cells does not depend upon any specific vital activity of the cells.

Many of the dyes in this category are only partially dissociated in the physiological pH range (weak electrolytes). However, some (like methylene blue) are fully ionized and yet readily diffuse into cells. Many salts of weak acids, which are fully dissociated in aqueous solution, readily penetrate cells, although not as rapidly as the corresponding undissociated acids (Beck and Chambers, 1937). Furthermore, many of the dyes in this group exist in solution as aggregates of molecules or ions; their penetration by physical diffusion depends upon the sizes of the particles in aqueous suspensions.

The quantity of dye which enters a cell, and the rate at which it enters, may depend in large measure on whether or not a chemical change occurs after penetration. Basic dyes tend to form undissociated complexes with cytoplasmic proteins and, therefore, accumulate in large amounts in cells even when the concentration in the external medium is extremely low. Many basic dyes are preferentially oil soluble (for example, neutral red) and these accumulate as the free base in lipid deposits. Furthermore, dyes whose dissociation changes significantly within the physiological pH range move predominantly in the direction of regions where the prevailing pH favors maximal dissociation.

We have selected for detailed discussion two weak electrolytes: the basic dye neutral red and the acid dye methyl red. Both undergo changes in dissociation within the physiological pH range. A study of the penetration of these dyes into cells demonstrates clearly that the direction of diffusion of weakly dissociated substances depends on pH differences between the cell interior and the environment. In addition, we will discuss the penetration into cells of a series of selected acid dyes, ionized to a constant degree throughout the physiological pH range. This series demonstrates the relationship between the diffusion of dyes across the cell surface and the size of the dye particles in solution. The striking difference between the protoplasmic surface film and the nuclear membrane in regard to their permeability to dyestuffs is also discussed.

NEUTRAL RED AND METHYL RED

The significant feature of these two dyes is that they undergo marked changes in dissociation within the physiological pH range, the optimal pH values for their associated states lying to opposite sides of pH 6.8, the pH value for the aqueous phase of the cytoplasmic matrix (Figure 15.1).

Neutral red becomes increasingly associated and goes out of solution, as the undissociated base, when its solution is rendered increasingly alkaline. It is virtually insoluble at pH 8.2 and above. In acid solutions the solubility of neutral red increases in accordance with its increasing dissociation as the cation of a colored weak base, with equal numbers of associated molecules and cations in aqueous solution at a pH of 6.8.

In the case of methyl red the molecules are principally in the associated state at pH values from 4.0 to 4.4. When the pH of the medium is raised, the dye molecules increasingly dissociate as the colored anions, with equal numbers of associated molecules and anions at a pH of 5.1.

The penetrability of both dyes, neutral red and methyl red, is optimal when the pH conditions are such that the maximum difference exists in the degree of dissociation of the dye between the exterior and interior of the cell, the associated state being greater outside the cell. In the case of methyl red this occurs in solutions which are more acid than pH 6.0. For neutral red it occurs in solutions more alkaline than pH 7.0. Neutral red becomes increasingly dissociated as the acidity of the medium increases, while the converse is true for methyl red.

The penetrability of these dyes in relation to pH is well illustrated in the following experiment using sand dollar eggs (Chambers, 1930a).

> These eggs are highly translucent and contain no pigment, the cytoplasm being filled with minute, colorless vacuoles. By adding an aqueous solution of neutral red to a suspen-

sion of eggs in sea water at a pH of 8.2 (at which value neutral red is almost entirely in the associated molecular state), a sufficient concentration of neutral red is obtained to induce vital staining. The stained eggs are then transferred to sea water acidified with hydrochloric acid to a pH of about 6.0. All color in the eggs rapidly disappears. The experiment can be repeated several times by transferring the eggs from the acidified sea water to the more alkaline sea water and back again. With each transfer to the alkaline sea water, the neutral red reaccumulates in the eggs, to pass out again whenever the eggs are transferred to the acidified sea water. With methyl red the reverse occurs. The eggs are readily stained when immersed in sea water acidified to pH 5.0 to 5.4. Upon transfer of the eggs back to the alkaline sea water, the methyl red in the eggs passes out, to return again when the eggs are retransferred to the acid sea water.

The pH of the aqueous media in which cells generally exist is alkaline (pH 7.4 to 8.2) relative to that of the protoplasmic interior (pH 6.8). Within the limits of cell viability, the pH of the external aqueous medium may be shifted so as to be appreciably more acid or more alkaline than that of the protoplasmic interior. The constancy of the latter's pH (Chapter 14) and the shifts of the pH of the external medium determine not only the penetrability of the dyes but also the direction in which the dyes diffuse, that is, into or out of the cell (Figure 15.1).

The pH influences the direction of movement of the dyes not only across the protoplasmic surface film but also across the bounding surfaces of acid and alkaline vacuoles lying within the protoplasmic matrix. For example, in adult plant cells with acid sap vacuoles neutral red diffuses rapidly from the normal, rela-

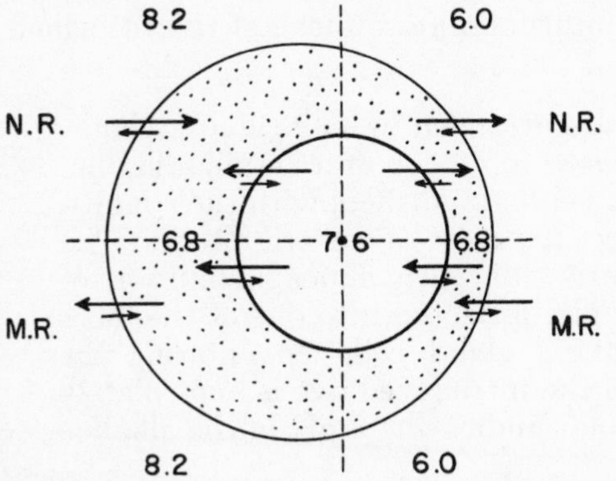

FIGURE 15.1. Diagram of cell, with nucleus at constant pH 7.6 and cytoplasmic matrix at constant pH 6.8. Arrows show directions of movement of neutral red (N.R., upper half) and of methyl red (M.R., lower half) into and out of cell, according to whether medium is pH 8.2 (left) or pH 6.0 (right).

tively alkaline environment into the more acid cytoplasm (pH 6.8) and from there into the still more acid vacuoles (pH 5.4). On the other hand, methyl red does not penetrate from the normal, relatively alkaline exterior; but, if injected into an acid vacuole, it diffuses rapidly into the cytoplasm and from there to the more alkaline exterior.

Similar directional movements depending on pH occur across the nuclear membrane (Figure 15.1). Methyl red, once in the cytoplasm, diffuses rapidly into the more alkaline cell nucleus, as Morita and Chambers (1929) demonstrated using the ameba. Vital staining of the nucleus by methyl red was first noted by Schaede (1924) when he immersed a strip of the epithelial tissue of an onion in an aqueous solution of the dye. He found that the nuclei were vitally stained with the yellow color of the extreme alkaline range of methyl red. The explanation of this staining is that in preparing the strips of epithelium he was tearing many of the epithelial cells and thereby releasing the acid contents of their sap vacuoles. This acidification of the medium promoted the associated state of the dye. The dissociation of the dye molecules being greater in the relatively more alkaline cytoplasm, the dye moved into the cell, and from there it accumulated in the nucleus, as would be expected, since the nucleus is still more alkaline than the cytoplasm (Chapter 14).

On the other hand, neutral red does not diffuse from the cytoplasm into the more alkaline nucleus (Figure 15.1), since the behavior of this basic dye is exactly the reverse of methyl red. Thus, Florence Sabin (1923) stressed long ago that a reliable criterion for the normal, living, uninjured state of a cell is the lack of staining of the nucleus by neutral red. Furthermore, if neutral red is microinjected into the nucleus, the dye rapidly diffuses out to segregate in the cytoplasm. This is strikingly illustrated in microinjection experiments on the large, multilobulated nuclei of the silk gland cells of the silk worm.[1]

> The silk gland cells are arranged in long tubular glands which can be readily teased in a drop of caterpillar serum. The individual cells can be distinguished, with their multilobulated nuclei which occupy most of the cell interior. The interior of the nucleus is fluid. An aqueous solution of neutral red chloride is microinjected into one of the lobes of the nucleus of a living gland cell. On entering the nucleus, the red color of the introduced dye is immediately converted to a yellow color, indicating a pH in the alkaline

[1] Hibbard and Chambers (1935); R. Chambers (1949a).

range of the dye. The color spreads rapidly through the various nuclear lobules giving to the entire transparent nucleus a yellow color sharply distinguished from the surrounding colorless cytoplasm. Within a few seconds after the injection there occurs a shower of fine, acicular brown crystals in the nuclear fluid, which progressively loses its diffuse color. This indicates an alkalinity of the nuclear fluid sufficient to cause precipitation of the neutral red. The crystals quickly become ranged along the boundary of the nuclear membrane and go into solution in the bordering, more acid cytoplasm, which becomes red in color. A few seconds later the cytoplasm loses its color, while the numerous vacuoles with which the cytoplasm is filled take on the vivid red of the acid, dissociated state of the neutral red. The final result is an unstained, translucent, multilobulated nucleus in a cytoplasm filled with red colored vacuoles.

The diffusion of neutral red from the nucleus into the cytoplasm occurs also following microinjection of the dye into the germinal vesicle of the immature starfish egg (Chambers and Pollack, 1927a).

This dye's characteristic property of diffusing through a membrane from an alkaline to a more acid medium explains the fact that neutral red is eliminated in the urine of a frog whose kidney has been perfused with Ringer's solution containing the dye (Figure 15.2).

When the frog kidney is perfused by way of the aorta with neutral red dissolved in Ringer's solution at the normal pH of 7.4 to 7.6 (Chambers and Kempton, 1937), the elimination of the dye in the urine is scanty, in spite of an acid reaction of the urine recovered from the ureter. With an aortic perfusion the dye filters through the blood vessels in the glomerular tuft into the lumina of the proximal tubules. The unfiltered balance of the neutral red passes on by way of the efferent glomerular vessels, which, in the frog, come into relation with the proximal tubules but supply only small parts of the distal tubules. Since the proximal tubular fluid is alkaline (about pH 7.4 to 7.6) and the cytoplasm of the tubular cells relatively acid (pH 6.8), the neutral red vitally stains the cells and thus tends to be removed as it courses through the lumina (Figure 15.2, left). In this way only a comparatively small amount of the dye reaches the lower end of the lumina of the tubules.

When the frog kidney is perfused by way of the renal portal vessels, neutral red in copious amounts is eliminated as long as the urine is acid in reaction. Perfusion via the

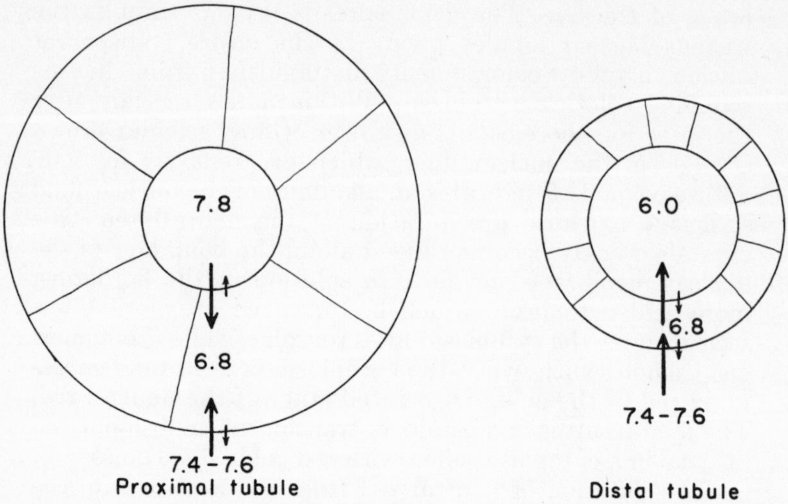

FIGURE 15.2. Diagram showing diffusion of neutral red in relation to pH of kidney tubules. Proximal tubule (cross section) with alkaline lumen, pH 7.8, and distal tubule (cross section) with acid lumen, pH 6.0. External surface of tubule cells bathed with perfusion fluid at pH 7.4 to 7.6. Cytoplasm of tubule cells at constant pH of 6.8.

With aortic perfusion, neutral red first enters lumen of proximal tubule, then the dye accumulates in cytoplasm of proximal tubule cells, resulting in removal of dye from lumen. Little or no dye passes on into lumen of distal tubule. With renal portal perfusion, neutral red reaches only external surface of distal tubule, via blood vessels. Dye passes through cytoplasm of tubule cells and accumulates in acid lumen of distal tubule. Copious amounts of neutral red appear in urine. (See Figure 16.1 for reconstruction of tubule and its arterial blood supply.)

portal system reaches not only the proximal but also the distal tubules and the collecting ducts. Because of the acid reaction of the fluid in the lumina of the distal tubules, neutral red diffuses from the alkaline perfusion fluid bathing the outer border of the distal tubule cells through the cytoplasm of these cells and thence into the acid lumina of the tubules (Figure 15.2, right).

In general, basic dyes at the normal pH of the cytoplasmic matrix have the property of combining with the protoplasmic proteins to form dye proteinates. The precipitating action of basic dyes can be observed by microinjecting a solution of neutral red into an ameba or into a starfish egg. A coagulum appears at the site of injection and either disperses a few seconds later, or, as in the ameba, is carried about as coagulated lumps in the protoplasmic currents (Chambers, 1920a). The dye gradually diffuses out of the coagulum and passes into the numerous minute, acid reacting vacuoles present in the cytoplasm. On the other hand, the salts of acid dyes (such as methyl red and eosin; for example, sodium eosinate) do not combine with the protoplasmic constituents. These, when microinjected, diffuse freely through the matrix of the protoplasm. Their lack of obvious reaction is explained on the grounds that the protoplasmic proteins are anionic, and the pH of the internal aqueous phase of the cytoplasm is such that the proteins are on the alkaline side of their isoelectric points.

The dependence on pH of the direction of diffusion of methyl red and neutral red has been found to be consistently the same in the many types of cells which have been investigated. The diffusion is always across cellular partitions from regions in which the dye is less soluble to those in which it is more soluble, that is, from regions of lower to regions of higher dissociation. This behavior of the dyes reveals that the protoplasmic surface film, or plasma membrane, appears to be, of itself, no obstacle to the passage of the dyes. The significant role of the surface film here is the maintenance of the protoplasm as an integrated body and the retention of constituents which maintain the internal pH constant. Diffuse coloration of the protoplasm never occurs as long as the cell is living, since the dyes are segregated according to the varying pH's of the different intracellular components.

The chemical properties of the basic dye neutral red ideally suit it for use as a vital stain, since the environment is generally more alkaline than the protoplasmic interior, and thus intracellular accumulation of the dye is promoted. Not only do the pH differences favor penetration, but accumulation within the cell is further encouraged by the tendency of the dye cations to form undissociated complexes with the cytoplasmic proteins.[2]

ACID DYES

We shall examine in this section only those acid dyes which are ionized to a constant degree throughout the physiological pH range. In considering the penetration of the acid dyes, the investigator must realize that the sizes of the particles in aqueous solution frequently bear no relationship to their molecular dimensions, since they exist as ion aggregates, or micellae. Furthermore, changes in salt concentration and in pH may profoundly affect particle sizes. Accordingly, Gordon and Chambers (1941), in their studies on the penetration of the salts of acid dyes into frog cells,

[2] Chambers, Cohen, and Pollack (1931); Chambers (1929b).

determined the size of the dye micellae in Ringer's solution at the prevailing environmental pH of 7.4 (Table 15.1, column 5). They also made measurements while the dyes diffused from this solution to Ringer's at pH 6.8, the pH characteristic of the cytoplasmic matrix (Table 15.1, column 4). After determining the particle sizes, they used the same solutions for ascertaining the penetration of the dyes into the frog cells (Table 15.1, columns 6, 7, 8). The results show that penetrability is not altered by change from aerobic to anaerobic conditions, nor is the ability of the dyes to enter related to their organo-solubility (Table 15.1, compare colums 2, 3 with columns 6, 7, 8). The size of the micellae appears to be the only factor affecting penetrability (Table 15.1, compare column 4 with columns 6, 7, 8). The dye micellae behave as if they were diffusing through pores of determinable size in the protoplasmic surface film. The experimental procedure is described more fully below.

> The rates of diffusion of the dye particles are determined by measuring the advance of a colored boundary from a chamber containing the dye dissolved in Ringer's solution into an adjacent chamber containing plain Ringer's. The two chambers are separated by a thin, removable partition. The solution containing the dye is at pH 7.4, while the colorless solution into which the diffusion takes place is kept at either pH 7.4 or pH 6.8. The value of pH 7.4 is selected as being that of the environing medium of frog cells and the value of 6.8 as being that of the cytoplasmic matrix. The experiment in which the dye solution at pH 7.4 diffuses into a colorless solution at pH 6.8 most nearly approaches the staining conditions encountered when using living frog cells in their normal environment.
>
> A comparison of the particle sizes (Table 15.1, columns 4, 5) shows that the pH of the medium affects appreciably the particle size of some dyes and not of others.
>
> The permeability of the cells to dyes is tested by noting the presence of color in ciliated cells and muscle fibers of the frog, and also by observing the passage of the dyes through bladder epithelium. These tissues are immersed in frog's Ringer's at pH 7.4 containing a given dye in the concentration used for the diffusion measurements. The results are shown in Table 15.1, columns 6, 7, 8.
>
> For the ciliated cells the upper limit of particle size of the dyes which penetrate lies between 5.5 ± 0.4 and 6.4 ± 0.5 Angström units. For the muscle fibers and bladder epithelium the limit is slightly higher: between 6.5 ± 0.4 and 7.3 ± 0.9 Angström units.

TABLE 15.1. Relation between penetrability and particle size*

Dye	Solubility		Particle radius in Angström units. Arithmetic means from diffusion rates in Ringer's solution at		Permeability of frog cells		
	Organo-soluble	Organo-insoluble	pH 7.4 to 6.8	pH 7.4 to 7.4	Ciliated cells	Muscle fibers	Bladder epithelium
Orange G	+	−	4.0 ± 0.3	4.0 ± 0.4	+	+	+
Patent blue V	−	+	4.7 ± 0.3	4.7 ± 0.3	+	+	+
Orange I	+	−	5.2 ± 0.3	5.0 ± 0.3	+	+	+
Erythrosin B	−	+	5.5 ± 0.4	5.5 ± 0.3	+	+	+
Light green SF	−	+	6.4 ± 0.5	6.6 ± 0.4	−	+	+
Eosin purple	−	+	6.5 ± 0.4	5.3 ± 0.5	−	+	+
Cyanol extra	−	+	6.6 ± 0.6	7.1 ± 0.5	−	−	−
Wool green BS	+	−	6.9 ± 0.5	6.6 ± 0.3	−	+	−
Amaranth	−	+	7.3 ± 0.9	6.1 ± 0.3	−	−	−
Azo rubin	−	+	7.5 ± 0.5	8.2 ± 0.4	−	−	−
Water blue	−	+	11.6 ± 0.5	14.2 ± 1.6	−	−	−
Congo red	−	+	12.3 ± 0.9	13.2 ± 0.8	−	−	−
Formyl violet 54B	+	−	14.0 ± 1.0	13.1 ± 0.7	−	−	−

* The results presented hold for both anaerobic and aerobic conditions. Of thirteen monodispersed dyes used (column 1), four are organo-soluble (column 2), and these cover as wide a range of particle sizes as the nine which are organo-insoluble (column 3). Values in column 4 are particle sizes of dyes in frog's Ringer's, calculated from experiments in which diffusion occurred from a medium at pH 7.4 into one at pH 6.8. Values in column 5 are particle sizes determined from experiments in which both media were at pH 7.4.

Columns 6, 7, 8 present data regarding the penetrability of dyes into cells. Plus signs indicate that cells became colored while still showing definite signs of activity or, in the case of bladder epithelium, permitted passage through of the given dye. Minus signs indicate either lack of any coloration or coloration only after death of cells. In the case of the bladder epithelium minus sign indicates that none of the dye passed through. (Gordon and Chambers, 1941.)

An interesting case is that of eosin, which exhibits a variation in particle size above and below the size that would penetrate the ciliated cells (Table 15.1, columns 4, 5). The radius of an eosin particle tested in Ringer's solution uniformly at pH 7.4 is found to be 5.3 ± 0.5 Angström units, which is smaller than the particle size of several other dyes which readily penetrate the ciliated epithelial cells of the frog. When the eosin, however, is tested while diffusing from Ringer's solution at pH 7.4 to one at pH 6.8, it shows an increase in particle radius to 6.5 ± 0.4 Angström units. This latter size is above the limit for penetration into the ciliated cells in question. It is, therefore, proposed that a ciliated cell is not stained by eosin because the dye particles, although of appropriate penetrating size in the bulk fluid, increase in size upon encountering a zone of increased acidity about the immediate periphery of the cell. A pH of 6.8 in this microscopic zone should be sufficient to cause the particles to enlarge beyond the penetrable size. The validity of such an explanation is strengthened by the findings of Danielli (1937) that the pH of a nonpolar oil–water interface is consistently lower than the pH of the bulk of the aqueous phase.

The highly dissociated sulfonated acid dyes do not penetrate cells in general (exceptions are certain cells possessing secretory activity, see Chapter 16). This has been well shown in extensive experiments on the permeability of starfish and sand dollar ova to the sulfonated oxidation–reduction and hydrogen-ion dye indicators.[2] Although the sulfonated dyes do not penetrate the protoplasmic surface film, it is a striking fact that when these dyes are microinjected directly into the cytoplasm, they readily diffuse through the nuclear membrane to color the nucleoplasm; or conversely, when injected into the nucleus, they diffuse out into the cytoplasm.[3] These data reveal the greater permeability of the nuclear membrane compared with the protoplasmic surface.[4]

The greater permeability of the nuclear membrane is also shown by injecting minute quantities of 0.0025 N hydrochloric acid into the cytoplasm in the vicinity of the nucleus of an ameba which has been vitally stained with methyl red. Prior to the injection both nucleus and cytoplasm are yellow, indicating the extreme alkaline range of the dye. The microinjected hydrochloric acid immediately

[3] Chambers and Ludford (1932b); Chambers and Pollack (1927a); Morita and Chambers (1929).

[4] The pores in the nuclear membrane described by electron microscopists (for example, Pappas, 1956, for the nucleus of *Amoeba proteus*) might account for the high permeability.

enters the nucleus, as evidenced by the fact that the yellow color of the nucleus changes to the red of the acid range. The cytoplasm in the vicinity of the microinjection also changes to red. The color change is momentary, and both nucleus and cytoplasm of the ameba recover completely. On the other hand, amebae stained with methyl red can be immersed in 0.0025 N hydrochloric acid without undergoing any color change. They remain viable for several minutes. A color change is observed only if irreversible injury has occurred. The injury associated with the change to a red color occurs in spots pinched off as dead remnants. In moving amebae the first parts injured are at the extreme ends of moving pseudopodia.

GENERAL FEATURES OF THE PENETRATION OF DYES BY DIFFUSION

The majority of the vital stains are included among the dyes that penetrate cells by simple diffusion. Their penetration does not depend directly upon the cells' metabolic activity.

The quantity of dye which penetrates a cell in a given length of time depends not only on the ease with which the protoplasmic surface is traversed by the dye molecules, but also on the concentration of the dye which will ultimately be attained within the cell when a state of equilibrium has been established between the cell and its environment. If the chemical nature of a dye is altered after its entry into the cell, by reaction or combination with an intracellular constituent, the diffusion gradient across the cell membrane is overwhelmingly in the direction of into the cell. Under such circumstances the dye continues to penetrate at the maximal rate almost indefinitely. Entry would cease only after the dye had been entirely removed from the external medium or after the available intracellular reactive groups had combined with the dye molecules. (Generally, the cell would die before such a saturation of reactive groups was achieved.) Basic dyes, whether they are weak or strong electrolytes, exhibit this behavior. Enormous amounts of these dyes will enter cells from extremely dilute solutions. An example is neutral red, which combines with proteins within the cell to form undissociated neutral red proteinates. Another example is methylene blue. This dye undergoes reduction within the cell to the colorless, leuco form; in addition, the dye forms undissociated complexes with proteins.

If a dye penetrates a cell without interacting with intracellular components, when an equilibrium is eventually attained, the internal concentration may be greater than, less than, or the same as the external concentration. If the dye is a non-electrolyte the concentrations inside and out should ultimately be identical. If it is a weak electrolyte, and the pH conditions favor greater dissociation inside the cell, the final intracellular concentration may be many times greater than that in the external medium, as we have seen. The reverse holds true if a greater degree of dissociation occurs in the external medium. If the dye is a strong electrolyte, the quantity of dye which ultimately enters the cell depends on the availability of intracellular ions for exchange. For dye anions of highly dissociated acid dyes the final concentration in the cell would be expected to be considerably less than in the external medium due to the preponderance of nondiffusible negative charges within the cell.

It is generally believed that the more dissociated a dye is in aqueous solution, the greater is the difficulty of penetration. However, methylene blue, which is completely dissociated in the physiological pH range, readily enters cells. The many experiments which have been carried out using weak electrolytes, have well established that the greater the degree of dissociation in the external medium, the less the tendency to penetrate the cell. These experiments have been interpreted as indicating that the dissociated compound does not penetrate the cell membrane. An alternative explanation is that the lesser tendency of the dissociated form of weak electrolytes to penetrate cells is a consequence of the pH differences between the cell interior and the external medium. Any measurement of the rate of penetration of a substance by diffusion into the cell is a measure of the rate of approach to an equilibrium state. In a system consisting of a weak electrolyte dissolved in two solutions at different pH's separated by a permeable membrane, equilibrium conditions require that the weak electrolyte diffuse into and accumulate in that solution which favors the greatest degree of dissociation. Thus, the "driving force" of the system favors the nonpenetration of the salts of weak electrolytes, quite apart from any question of the ease with which the dissociated forms can traverse the cell surface.

The fact that the highly dissociated sulfonphthalein indicators do not penetrate cells in general may be related to their molecular configuration or to micelle formation rather than to their high degree of dissociation.

The hydrogen-ion concentration of the external medium may affect penetration of dyes in a number of different ways. The state of aggregation of dyes which exist in aqueous solutions as micellae frequently is pH dependent. For example, a pH in the external medium which promotes the smaller particle size promotes entry of the dye into the cell. The distribution of weak electrolytes between the intracellular and extracellular phases is also determined by pH differences and the laws of simple diffusion. In this connection it is essential to realize that the various intracellular components have different pH values. Although the pH of the cytoplasmic matrix and of the nucleoplasm cannot be appreciably altered as long as the cell is living, the pH of the intracellular vacuoles can be readily changed by penetrating acids or bases. Accordingly, dyes which are weak electrolytes enter or leave the vacuoles, depending on the direction of the pH change.

A feature which should be emphasized is that changes in the hydrogen-ion concentration can affect penetrability of the dyes not only by changing the nature of the dye molecule but also by altering the structural characteristics of the protoplasmic surface film.

16

Penetration of Dyes by Metabolic Activity

UPTAKE BY INGESTION

Dyes of colloidal dimensions are taken into cells by ingestion. The better known dyes in this group are Congo red, trypan blue, trypan red, scarlet red, and vital red. These are all sulfonic acid dyes and are, accordingly, fully ionized. The ions, however, aggregate into particles of colloidal dimensions in aqueous suspensions. These cannot enter cells by simple diffusion, since the upper limit of size for penetration by diffusion, at least for the cells of frog tissues, is in the order of 6.5 to 7.0 Angström units.

In general, uptake by ingestion involves the invagination of the protoplasmic surface and the pinching off of this invagination as a vacuole, which moves into the protoplasmic interior. This process is referred to, perhaps rather arbitrarily, as pinocytosis[1] when the environmental fluid only is taken in, and as phagocytosis when, in addition, particulate material is incorporated in the forming vacuoles. Obviously, during these processes the cell ingests substances which are dissolved or suspended in the surrounding fluid. In addition, substances which have adsorbed to the surface of the cell[2] may be taken into the protoplasm because surface material is incorporated as a constituent of the inner wall of the vacuoles as these form.

For decades students of living cells have inferred that phagocytosis or pinocytosis must occur at a submicroscopic level, since certain types of cells can absorb minute particles even in the absence of any other visible sign of ingestive activity. Recent

[1] Lewis (1931).
[2] Odor (1956); Brandt (1958); Schumaker (1958).

studies with the electron microscope have provided morphological substantiation for the occurrence of such activity.[3]

In the discussion here the phrase "phagocytic activity" refers to an ingestive process, of both microscopic (light) and submicroscopic dimensions, whereby colloidal dyes are taken into the protoplasmic interior. This process requires the maintenance of metabolic activity by the cells and therefore is inhibited by exposure of the cells to anaerobiosis, to oxidation–reduction inhibitors, or to low temperatures.

Colloidal dyes are taken up in large amounts by cells of the reticulo-endothelial system, such as the Kupffer cells of the liver and the histiocytes which are dispersed throughout the connective tissues of the body. This uptake by phagocytosis also occurs in cells of columnar epithelium, such as the ciliated cells of the tracheal epithelium and the proximal tubule cells of the kidney.

An interesting feature is the functional polarity exhibited by the kidney tubule cells: phagocytic activity is restricted to their luminal borders. Excellent material for the study of this phenomenon is provided by the explants of proximal kidney tubules grown in tissue culture. When colloidal dyes are microinjected into the lumina of persisting segments, the particles are phagocytosed along the luminal border of the tubule. On the other hand, similar particles introduced into the external medium are not ingested, indicating the absence of phagocytic activity along the external surface of the tubule. When the epithelial cells are no longer organized in the form of tubules, they nonetheless retain their phagocytic property. This is seen in older cultures, where the cells at the cut ends of the tubules proliferate and form expanding sheets of epithelium. When colloidal dyes are added to the medium, the cells in the sheets pick up the particles.

UPTAKE BY SECRETORY ACTIVITY

Dyes taken up by secretory activity include the Clark and Lubs series of sulfonphthalein pH indicators. These are sulfonated acid dyes and thus are lipid insoluble and completely dissociated in aqueous solution throughout the physiological pH range. The dimensions of the dye particles are below the colloidal range.

Cells, in general, are impermeable to these dyes, and cells with phagocytic properties do not accumulate them.[4] The only cells

[3] Palade (1953); Odor (1956).

[4] It is conceivable that a cell in which pinocytic activity can be stimulated, such as an ameba suspended in a solution containing protein (Chapman-Andresen and Holter, 1955), might be made to accumulate the dissolved dye.

known to take these dyes up from the external medium are three types of secretory epithelia: the proximal tubules of the kidney,[5] the ependymal epithelium of the chorioid plexus,[6] and liver cells.

The transport of sulfonphthalein indicators has been intensively studied in tissue culture preparations of fragments of proximal renal tubules, the cut ends of which had sealed off to form closed sacs.[5] A typical experiment is as follows.

> A functioning mesonephros is excised from a chick embryo (Figure 16.1), teased, and the fragments planted in tissue culture. The segments of the proximal tubules regenerate at their cut ends, converting the fragmented tubules into closed sacs. A solution of phenol red (or of any of the other sulfonphthalein indicators of the Clark and Lubs series) is added to the culture medium, and after a few hours' incubation, the indicator is picked up by the cells lying in their normal position in the walls of the tubules. As the compound passes through the cells of the wall, it gives to the cells a diffuse color consistent with a cytoplasmic pH of approxi-

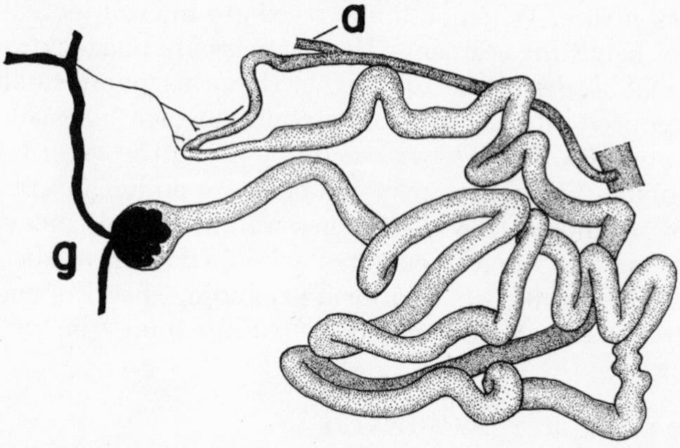

FIGURE 16.1. Reconstruction of representative tubule and Bowman's capsule of chick mesonephros from macerated material. Beginning at glomerulus (g), the tubule, at first relatively straight, becomes highly convoluted, then makes a complete turn, and returns to region of its glomerulus, where it suddenly narrows to become the distal tubule. This passes into mesonephric duct. Distal portion of tubule has no convolutions and is smaller in diameter along its entire extent. Occasionally, one of these tubules receives another as a side branch (a). Arterial blood supply shown in solid black. (From Chambers and Kempton, 1933.)

[5] Chambers and Cameron (1932); Chambers and Kempton (1933); Chambers (1935); Cameron and Chambers (1938).

[6] Cameron (1953).

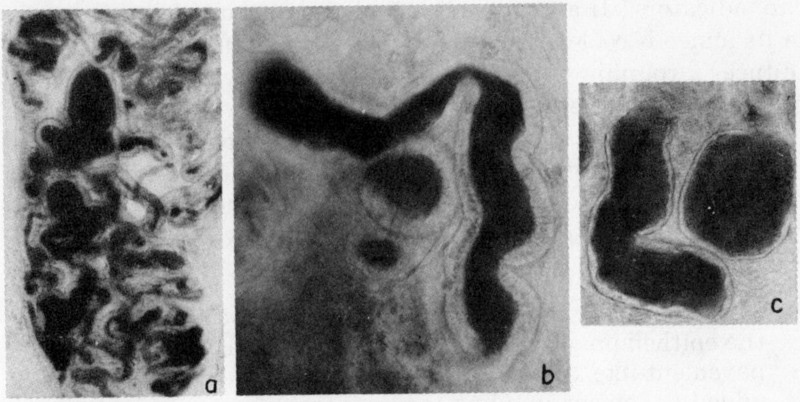

FIGURE 16.2. (a) Teased fragment of 9-day chick mesonephros in tissue culture incubated 24 hours with phenol red in medium. A mass of convoluted proximal tubules, lumina of which are colored red with phenol red (black in photograph). Walls are yellow. (b) Closed segment of proximal tubule from same culture. Closed-over ends shown at lower right and upper left corners. Columnar state of epithelium still evident. Lumina red with phenol red (black in photograph). Walls yellow. (c) Same culture after 72 hours' incubation. Two highly distended, closed proximal segments, now cyst-like through further secretion. Epithelial walls yellow, much flattened. Phenol red in lumina (black in photograph).

mately 6.8. No other cells in the culture are colored. After 24 hours' incubation, the healed tubular sacs contain colored fluid (Figure 16.2a, b), and 72 hours after incubation the sacs are distended (Figure 16.2c). All color has disappeared from the external medium and has passed through the cells and accumulated in the lumina of the sacs.

The proximal tubule cells take up the sulfonphthalein dyes exclusively from their external surfaces. The transfer continues until all the dye has been removed from the surrounding medium and accumulated in the lumina of the tubules. A significant feature is the diffuse coloration the indicators impart to the cells during the transport process, the color assumed by the indicator being in accordance with the cytoplasmic pH. This indicates that the dyes are transferred as particles dissolved in the continuous aqueous phase of the cytoplasm. Apparently the secretory process does not involve the formation of vacuoles and their passage across the cell.

When the tubule segments are cooled or exposed to a nitrogen atmosphere or to metabolic poisons, the unidirectional transport process is reversibly blocked, and the cells become impermeable to the indicators.[7] If a sac which has already accumulated indicator in its lumen is cooled, the color is lost. This occurs because the cold induces a rounding up of the cells, and the dye then passes out, not through, but between the tubule cells.

A point which should be stressed is that the renal epithelial cells pick up the indicators only as long as they are arranged in their natural positions composing the walls of tubules. If the tubular organization is altered, the cells become impermeable (Chapter 5).

> Occasionally the end of a tubule does not heal over, and the epithelium grows out into the medium as a flat sheet of pavement-like cells. When a sulfonphthalein indicator is added to the surrounding medium, the sheets of renal epithelial cells never show any trace of color. They have lost their specific property of secretion but not their potentiality for secretion. When, for example, they become reoriented as tubular cysts in the sheet, their secretory function is resumed.

The transport of the sulfonphthalein indicators is in striking contrast to the passage across the renal tubule cells of dyes which penetrate by simple diffusion (Chapter 15). Dyes in this latter group penetrate readily regardless of the cells' metabolic status and organizational pattern (in sheets or tubules); these dyes are not accumulated in the lumina of the tubules.

In general, we can distinguish three characteristics of the renal secretory cell: (1) it takes up the sulfonphthalein pH indicators only from the external surface of cells organized as tubules, and only when cellular metabolism is retained intact; (2) it ingests colloidal dyes only when cellular metabolism is normal, irrespective of the organizational pattern of the cells, but only from the internal surface of cells organized as tubules; (3) it stains with dyes which penetrate by simple diffusion, quite irrespective of the cells' metabolic status or organizational pattern.

As we have already stated, the sulfonated acid indicators do not penetrate the protoplasmic surfaces of the great majority of plant and animal cells. However, these dyes can be introduced into cells by microinjection. When this is done, the dyes readily spread through the cytoplasm and color the matrix diffusely. In animal cells the indicators eventually become segregated in cytoplasmic vacuoles, showing either a basic or an acidic reaction; the dye persists in the vacuoles for long periods of time. When injected into the cytoplasm of plant cells (for example, the root hair cells of

[7] Chambers, Beck, and Belkin (1935); Beck and Chambers (1935).

Limnobium), the dyes accumulate in vacuoles, notably the central, or tonoplastic, vacuoles, until the cytoplasm becomes colorless (Chambers and Kerr, 1932). If the dyes are microinjected into the tonoplastic vacuoles, they remain there without a visible trace diffusing into the cytoplasm. Although the vacuolar sap of the root hair cell is considerably more acid (pH 5.2) than the aqueous phase of the cytoplasmic matrix, differences of pH cannot account for the unidirectional passage, since the sulfonated dyes are fully dissociated in the physiological pH range. These experiments demonstrate that the external surface of the protoplast has permeability properties differing from those of the internal, vacuolar membranes. While the external surface is impermeable to the sulfonated dyes, these can pass unidirectionally across the vacuolar membranes to accumulate in the vacuoles. Evidently the sulfonated dyes are being secreted from cytoplasm into the vacuoles in a manner resembling the secretion of these dyes into the lumina of the kidney tubule.

VII

The Aster, the Spindle and Cell Division

INTRODUCTION TO PART VII. Suitable material for experimental study, especially study of cell division, is provided by the eggs of some of the more common marine echinoderms. In these, the formation and development of the aster play a prominent role in the fertilization process and in the subsequent cleavages of the ovum. The eggs of the sand dollar *Echinarachnius,* common along the New England coast, and of the sea urchin *Lytechinus,* common along the southwestern and southeastern coasts, are so translucent in the living state that many internal cytological features are visible to the eye. These echinoderm eggs, when mature and ready to be fertilized, measure about 120 micra in diameter and possess a fluid cytoplasmic matrix filled with minute granules and spherical vacuoles which average 1 to 3 micra in diameter. The eggs are hololecithal, and the cell nucleus, when it undergoes mitosis, lies in the center and controls the division of the entire egg.

The physical nature of the protoplasmic interior and of various differentiated regions within the echinoderm egg and within several other selected cell types is readily tested with microneedles. Micromanipulation is successfully performed on fertilized ova after denuding the eggs of their tough fertilization membranes. This is easily accomplished immediately after insemination, when the membranes have not yet hardened, by sucking the eggs through the narrow mouth of a capillary tube or by vigorously shaking the eggs, suspended in a small amount of sea water, in a test tube. A fine microneedle readily penetrates the denuded egg without causing injury. The extremely delicate movements of the microtips, even under the highest magnifications of the light microscope, permit the investigator to determine the consistency of the finest structures without causing their alteration or inducing gel–sol reversals. The micromanipulative technique has played a particularly valuable role in the elucidation of the physical structure of the aster and the spindle in the living cell.

17

The Aster

A radial arrangement of cytoplasmic granules collected around one or more centers constitutes the aster. It is a prominent feature as the single sperm aster (monaster) in the fertilization process of ova in general and as the polar asters of the amphiaster (Figure 17.1) in dividing embryonic animal cells, or in maturing ova forming polar bodies.

The characteristic alignment of the granules suggested the term aster. Already recognized in the early 1840's, the aster was first associated with cell division by Hermann Fol in 1873. Many

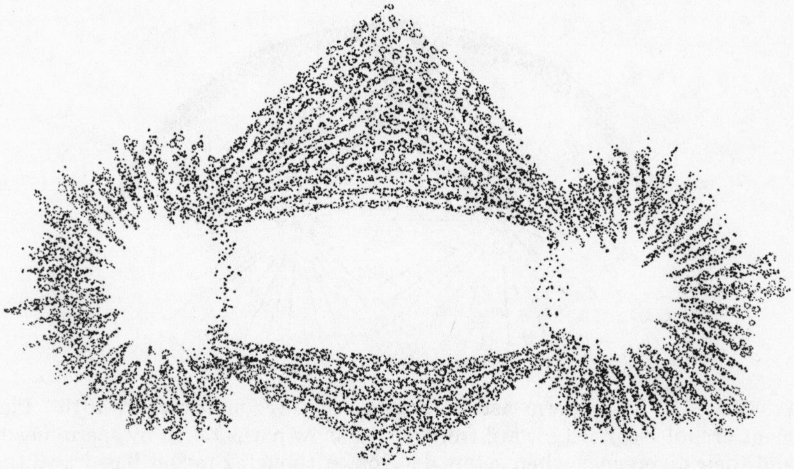

FIGURE 17.1. Fully formed amphiaster, in anaphase, in living sand dollar egg. Only central portion of amphiaster shown. Hyaline region extending between the two asters represents anaphase spindle. Fine granules mark off end region of spindle from astral lakes at poles of amphiaster. Chromosomes invisible. (From Chambers, 1917b.)

of the earlier investigators, but especially Fol,[1] had suggested that the radial figure is due to centripetal diffusion currents which arrange the cytoplasmic granules in rows radiating from a center. To quote Fol (1879): "... les lignes claires de l'aster ne sont en réalité que de courants de sarcode qui viendraient confluer en un amas central." This view, largely based upon the fact that the central hyaline area, or lake, of the aster increases in volume as the aster develops in prominence and extent, has been upheld by the microdissection experiments of Chambers (1917b).

An additional feature stressed by Fol (1879) and subsequently confirmed by other investigators is that when several asters appear in the cytoplasm, which occurs when several spermatozoa enter an egg, these asters always occupy positions indicating a mutual repulsion as they extend in size. A striking demonstration of this mutual repulsion between growing asters is the observation of Albert Brachet (1910). He was able to observe in a frog egg the paths taken by the heads of two spermatozoa which penetrated the egg at a short distance from one another. The paths of the sperm heads within the egg could be traced by the pigment granules which each sperm carried with it from the pigmented cortex of the egg. Figure 17.2 shows the paths of the two contiguous sperm heads. The sperm heads maintained a parallel course for a short period; then an aster developed about the head of each

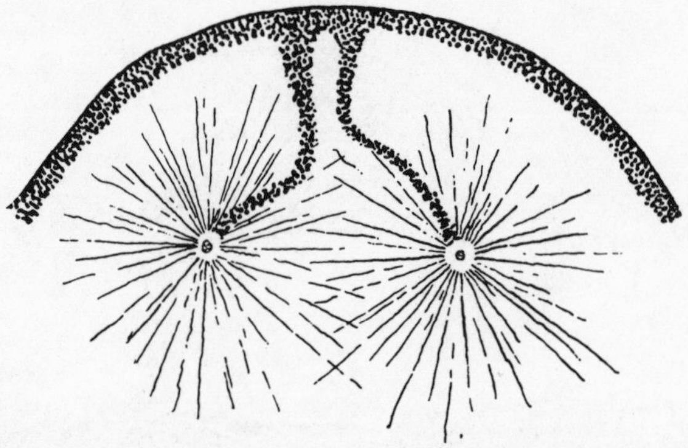

FIGURE 17.2. Two sperm asters in frog's egg. (From Brachet, 1910.) Pigment granules carried inward from cortex show paths taken by sperm heads and their divergence when asters develop. Although Brachet has drawn the astral rays as though these interlock, we believe the appearance of overlap is due to optical illusion resulting from location of asters in slightly different focal planes.

[1] Fol (1873, 1879), Bütschli (1894), and later Rhumbler (1898), Morgan (1900), Wilson (1901a), and Mathews (1907).

sperm, whereupon the paths diverged, and the extent of the divergence increased as the asters grew in size. From these observations Brachet concluded that the growing aster is a region of an expanding solidification, a concept[2] which was later confirmed by Chambers (1917b).

Normally, the aster is always associated with the nucleus. However, under experimental conditions of artificially induced activation of the mature ovum, as in artificial parthenogenesis, one or several non-nucleated asters may appear in different regions of the cytoplasm. These are known as cytasters, and their occurrence is associated with abnormalities in further development of the activated egg. Under optimal conditions of parthenogenetic activation, resulting in normal development of the ovum, cytasters are not formed (Chambers, 1921a).

DEVELOPMENT OF THE SAND DOLLAR EGG TO FIRST CLEAVAGE

In order that the reader may the more readily follow the subsequent description of microdissection studies of the physical nature of the aster and spindle, there follows a résumé of the development of the echinoderm egg from fertilization to first cleavage, using the sand dollar egg as an example (at room temperature).

Several minutes after the sperm has entered, a diminutive sperm aster appears at the periphery of the egg, around the neckpiece of the recently fertilizing spermatozoon. As the astral radiations extend and the hyaline astral lake enlarges, the aster as a whole moves toward the center of the egg. At the height of its development (10 to 15 minutes after fertilization) the rays of the monaster extend throughout the interior of the egg. Meanwhile, the egg pronucleus has moved into the astral lake, where fusion with the sperm nucleus soon occurs. By 15 to 20 minutes after fertilization, the monastral radiations begin to fade. The disappearance of the radiations commences in what will be the equator of the dividing egg and gradually extends to the two poles. Although the monastral radiations disappear completely, the central lake of the monaster persists as a distinctly visible hyaline region. This lake, now containing the prophase nucleus, elongates in the direction of the future long axis of the dividing egg, and an amphiastral configuration of the first cleavage evolves with the appearance de novo of radiations about each of the two poles (35 to 40 minutes after fertilization).

[2] This concept had already been tentatively advanced by Teichman (1903), Fischer and Ostwald (1905), and Delage (1907).

Each of the two polar asters grows to a size equal to approximately one fourth the diameter of the egg and then maintains this size through prophase and metaphase, during and for some time after the chromosomes of the fusion nucleus become arranged in the equator of the mitotic spindle. At the beginning of anaphase, while the daughter chromosomes are separating, the asters start growing rapidly in size, and concomitantly the spindle elongates (without increasing in width). During anaphase there can be seen the interzonal portion of the spindle which lies between the oppositely moving groups of chromosomes and a polar zone which extends from the chromosomes to the astral lake of each of the asters of the amphiaster. After the chromosomes have reached the two opposite poles, the entire spindle may be considered as the interzonal body.

As the asters and spindle grow in dimensions, the astral lakes increase in volume. By late anaphase or early telophase the radiations of both asters extend throughout the cytoplasm, reaching close to the surface of the egg. As the asters and the spindle attain their maximum dimensions, the egg undergoes a general elongation along the long axis of the amphiaster (early telophase), preparatory to the sinking in of the furrow.

THE FULLY FORMED ASTER

The fully formed sperm aster or a polar aster of the amphiaster, as seen in the conspicuously translucent egg of the sand dollar (Figure 17.1), is a radially arranged alignment of cytoplasmic granules surrounding a central, hyaline, more or less spherical region known as the lake (or centrosphere of E. B. Wilson). From the lake radiate innumerable delicate rays of the same hyaline substance as the central lake. The granular cytoplasm surrounds the hyaline rays and projects between them into the central lake in the form of conical processes (Figures 17.1, 17.3a).

The astral lake consists of a clear, nonviscous liquid. The microneedle can be inserted into the lake and moved in it without meeting any resistance. The nucleus, consisting of either the closely associated male and female pronuclei or the fusion nucleus, lies within the lake of the monaster and may be pushed about with ease in this liquid region. The hyaline rays which extend radially from the periphery of the lake are also liquid. On the other hand, the narrow, cone-like projections of granular cytoplasm, which extend between the hyaline rays and whose apices outline the border of the lake, are solid. These granular projections, when tested at the border of the lake, may be bent (Figure 17.3b) and

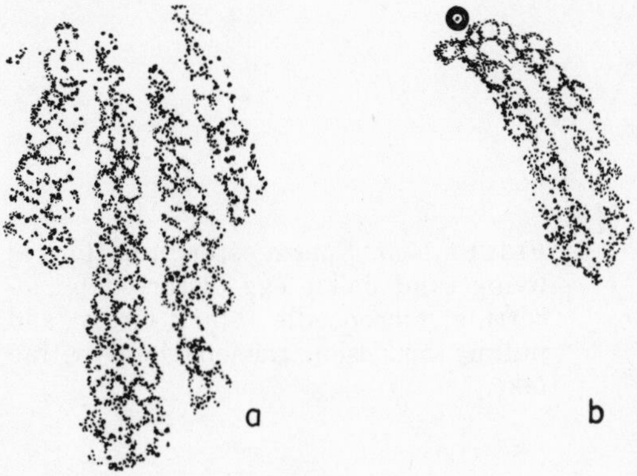

FIGURE 17.3. Highly magnified drawings of aster at border of central lake (above) in living sand dollar egg. (a) Central ends of radial strands of gelled granulocytoplasm of aster projecting into periphery of lake (above). Hyaline rays lie between gelled strands. (b) Same, but projecting tips of gelled conical processes bent with microneedle (black circle). (From Chambers, 1917b.)

pushed from side to side. They act like comparatively rigid gel structures with the cytoplasmic granules imbedded in them. With the tip of the microneedle the radiations can be distorted in a localized region (Figure 17.4) or the whole aster can be twisted into spiral and other distorted shapes (Figure 17.5). The rigid state of the gel is most pronounced at the border of the lake and diminishes as the radiations approach the periphery of the aster. This gel state of the granular cytoplasm in the aster is in marked contrast to the sol state of the granular cytoplasm beyond the peripheral border of the aster and of the cytoplasm of the resting egg. Even when the aster is fully formed and occupies almost the entire mass of the egg, the radial configuration falls short of the gelled cortex of the egg. When the egg is held with a supporting needle, the aster can be rotated by means of a second needle without any indication that the periphery of the aster is mechanically attached to the cortex of the egg.

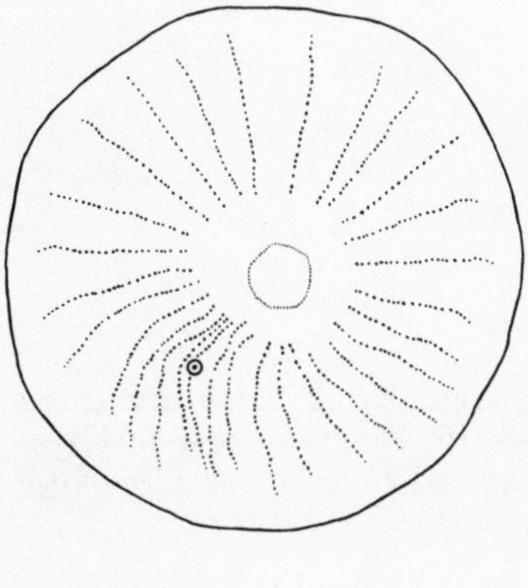

FIGURE 17.4. Local bending of sperm aster radiations by inserting microneedle (black circle) in living sand dollar egg.

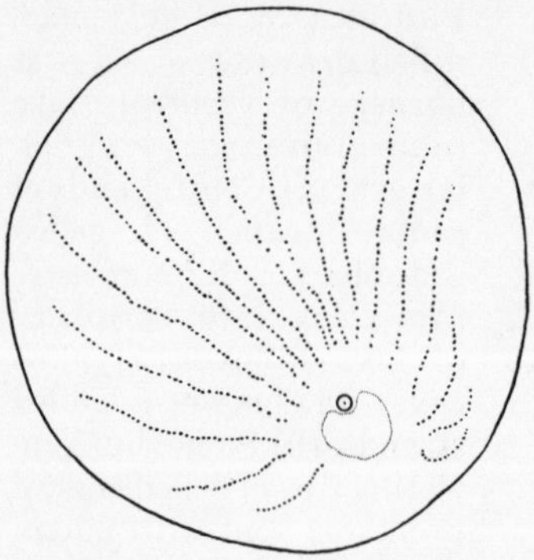

FIGURE 17.5. Sperm aster radiations in living sand dollar egg distorted by inserting microneedle (black circle) and pulling on fusion nucleus held within lake.

While the astral radiations of the monaster and the polar asters of the amphiaster in the dividing egg are not fastened to the cortex of the egg, the polar body aster in the amphiastral maturation figure of the *Cerebratulus* egg is anchored to the egg cortex (Figure 17.6).

In short, microdissection shows that the fully formed aster is a gelled body. The rays of the aster and its central lake are fluid.

Mechanical agitation with the tip of a microneedle inserted into an aster will cause disappearance of the radiations. The agitation can be so localized as to cause a disappearance of only a few rays, indicating that the reversal of the gel to the sol state can be limited to local regions. When the needle is withdrawn, the solated region returns to its gel state and thereupon reacquires its original radial appearance. More extensive agitation of the aster causes a complete disappearance of the ray-like structure, all the cytoplasm

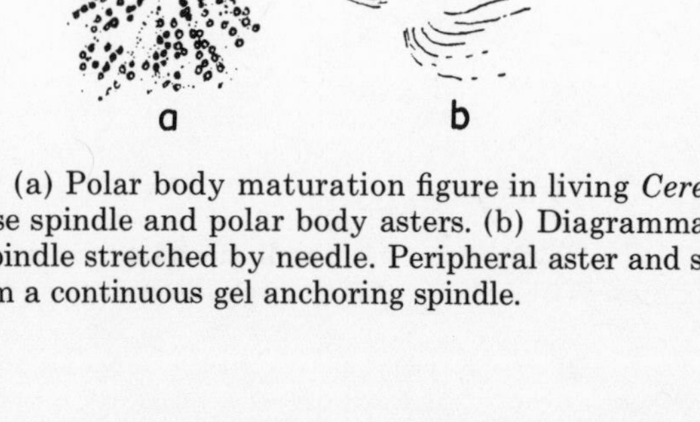

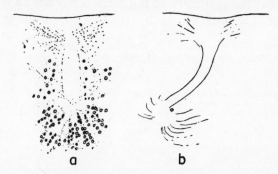

FIGURE 17.6. (a) Polar body maturation figure in living *Cerebratulus* egg with anaphase spindle and polar body asters. (b) Diagrammatic sketch of same, with spindle stretched by needle. Peripheral aster and surface of egg evidently form a continuous gel anchoring spindle.

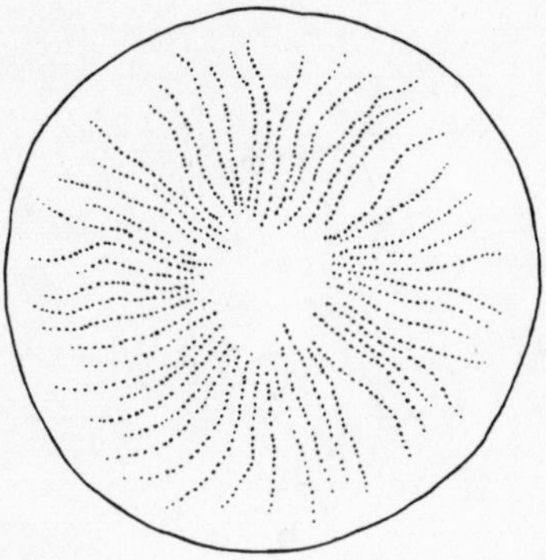

FIGURE 17.7. Twisted radiations of sperm aster which has reappeared after churning, in living sand dollar egg.

of the egg then reverting to the sol state. If the mechanical agitation has been not too severe, the central hyaline lake still persists and the rays reappear after a short interval but are twisted in appearance (Figure 17.7). The monaster, when mechanically agitated with the needle, is found always to be more stable than the asters of the amphiaster, indicating that the monaster is a more rigid structure than the amphiaster.

THE SPERM MONASTER

Within a minute or two after the sperm has entered a sea urchin or sand dollar egg, the microneedle may be passed through any region of the cytoplasm, and the cytoplasmic granules flow by the moving needle. Two to 3 minutes after sperm entry the diminutive sperm aster appears at the periphery, and that the granular cytoplasm within the early aster is already solid is shown by manipulation with the microneedle (Figure 17.8). The sperm nucleus is held within the gelated border of the lake. From the beginning, the aster contains a liquid central region, the expanding lake.

The formation of the sperm aster depends upon an interaction between the entering spermatozoon and the cortex of the mature egg. This is indicated by the irresponsiveness to an entering sperm of an endoplasmic exovate or fragment which lacks cortical material. The sperm remains quiescent in the cytoplasm; the sperm aster fails to develop.[3] Allen (1954) has shown that the sperm aster does not form when a sperm penetrates a region of egg surface which is injured so that it lacks cortical granules.

The sperm aster, at first diminutive in size, grows as it moves toward the center of the egg. Along with its growth, the central lake enlarges by a centripetal flow of the hyaline liquid in the con-

[3] See Chapter 1; Chambers (1921e); Runnström and Kriszat (1952).

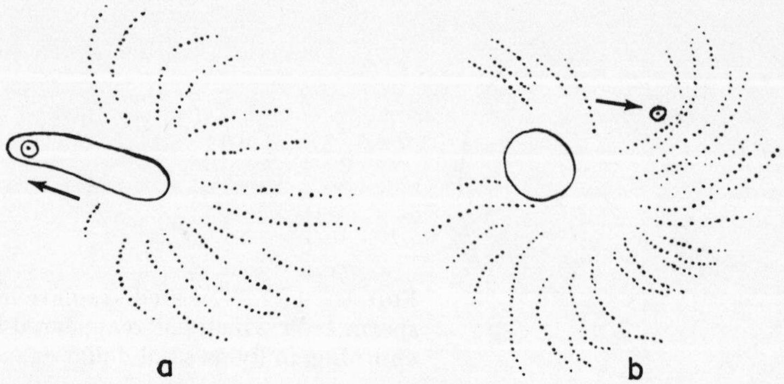

FIGURE 17.8. Two sketches of early sperm aster being manipulated with tip of microneedle, in living sand dollar egg. (a) Microneedle inserted into sperm pronucleus, which is pulled into a ribbon shape, showing that nucleus is held in a gel. (b) Early sperm aster spirally twisted by moving inserted microneedle.

verging rays and also by the incorporation within it of the egg pronucleus. When the astral rays extend throughout the cytoplasm, the lake in the central region of the egg has attained its maximum size. At this stage the cytoplasmic granules throughout the egg are held in a gel.

The path followed by the growing aster as it moves to the center of the egg is related to the egg's contour. As long as the egg is spherical, the path described by the center of the astral lake, as the aster moves from the point where it first appeared, is always in a straight line at right angles to the surface of the egg. In irregularly shaped eggs, the path of the growing aster is at right angles to the surface of the egg only during the initial phases of the growth of the aster (Figure 17.9, A to B). Later, as the aster continues to grow in size, its gelated boundary approaches the egg cortex, and this alters the direction of the path (Figure 17.9, B to C). The path followed by the center of an expanding gelated mass, the growing aster, is analogous to that of an eccentrically located balloon as it is inflated within a hollow container.

Relation of the sperm aster to movements of the pronuclei

The sperm pronucleus, which lies in the astral lake, moves toward the center of the egg because it is surrounded by a growing, expanding, gelated body, the aster (E. L. Chambers, 1939). The egg pronucleus is at first eccentrically located; it lies near the periphery of the egg, where it had given off the polar bodies during maturation. Since the fertilizing spermatozoon may enter the egg at any point on the periphery of the egg, the sperm aster forms at a variable distance from the position of the egg pronucleus. It is significant that the movements of the sperm pronucleus are not influenced by the egg pronucleus. In the spherical egg the sperm pronucleus, surrounded by the growing monaster, moves in a straight line to the center of the egg. Only irregularities in contour of the egg cause deviations from this pathway.

Interesting observations on pronuclear movements were carried out by Allen (1954) on fertilized sea urchin eggs made cylindrical by being confined in narrow capillaries. He observed that the inward migration of the sperm nucleus coincides with the growth of its aster and that the depth to which the sperm nucleus penetrates in the cylindrical egg is approximately equal to the radius of the capillary. This behavior is what we should expect if the sperm nucleus is moved inward as the result of the growth about it of a gelated body.

The movements of the egg pronucleus as it migrates to the sperm nucleus are determined by the sperm aster.[4] The egg pro-

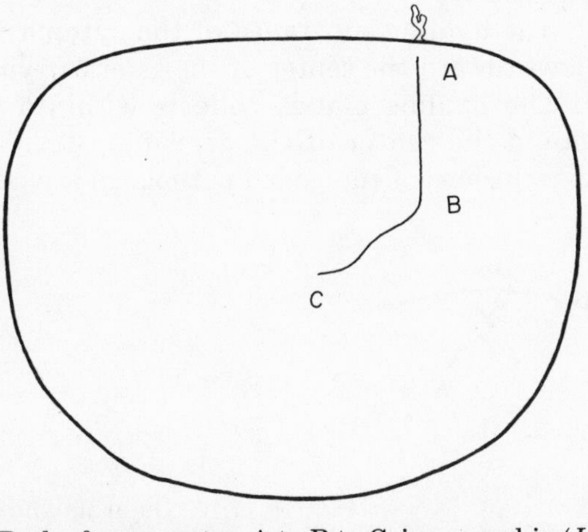

FIGURE 17.9. Path of sperm aster, A to B to C, in sea urchin (*Lytechinus*) egg rendered non-spherical. Flame-like process extending above flattened surface is exudation cone at site of sperm entry. A to B, path of sperm aster; being off center, it later changes in direction and becomes directed toward center of egg along path B to C. (E. L. Chambers, 1939.)

[4] We do not intend to claim that the only movements of which the egg pronucleus is capable are related to the sperm aster. For example, in an unfertilized sea urchin egg which has been allowed to stand in sea water, the egg nucleus gradually moves from its initially eccentric position to the center of the egg. Allen (1954) has noted that following fertilization of elongated sea urchin eggs sucked into a glass capillary, the egg nucleus first moves to the axis of the capillary before it starts to migrate toward the sperm nucleus.

nucleus, after the sperm aster has appeared, moves toward and finally into the central lake of the monaster. Since the monaster itself is moving, the path taken by the egg pronucleus is curved according to where the egg nucleus happens to lie with regard to the position of the growing monaster. The path taken by the egg pronucleus is straight only if it lies directly in line with the path of the moving sperm aster. These features are illustrated in Figure 17.10.

Analysis of the path followed by the egg nucleus during its migration to the sperm nucleus in the sea urchin egg was carried out by Kuhl and Kuhl (1949) using time-lapse photography. Their observations are in agreement with those here described.

The path of the egg pronucleus indicates that it is carried by a centripetal flow directed to the center of the monaster. When the egg pronucleus is within the confines of the growing monaster, its movements are conditioned by the fact that it is being held in a gelled aster, which itself is moving to the egg center (E. L. Chambers, 1939). The path traced by the egg pronucleus while it is moving within the aster is therefore the resultant of two forces acting simultaneously.

Evidently the hyaline substance of the cytoplasm within the astral rays flows toward the center of the aster carrying with it the egg nucleus. The hyaline matrix collects within a progressively enlarging lake in the center of the growing aster. The granular and vacuolar inclusions of the gelled cytoplasm are not involved in

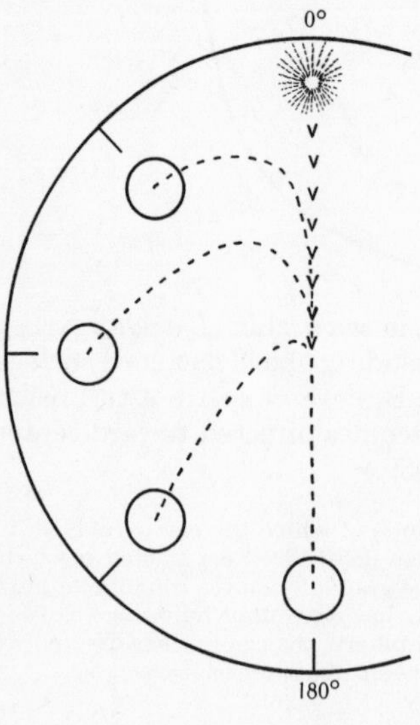

FIGURE 17.10. Diagram showing four possible paths of egg pronucleus (interrupted lines) and path of center of sperm aster (arrow heads) in sea urchin egg. Site of sperm entry at 0°. Sperm aster, as it enlarges, moves straight to center of egg along path indicated by arrow heads. Four possible initial positions of egg pronucleus (circles) and corresponding paths (interrupted lines) of egg pronucleus as it moves toward center of growing and moving sperm aster. (E. L. Chambers, 1939.)

this flow, since they are held in the gelled substance of the growing aster.[5]

Striking evidence for the centripetal flow is the finding that when a minute oil drop, which happened to be present in the cytoplasm, was pushed by the microneedle into the periphery of the aster, it moved slowly toward the lake (Chambers, 1917b).

Boveri (1918) made a remarkable observation on a batch of sea urchin eggs in which each egg nucleus accidentally happened to consist of many partial nuclei lying at random in the cytoplasm. He inseminated these eggs and found that as the sperm aster moved to the egg center the partial nuclei migrated into the astral lake, where they all united with the sperm nucleus to form a single fusion nucleus. This movement of the partial nuclei from various regions in the egg toward and into the astral lake is well explained by the streaming of the cytoplasm toward the center of the aster.

As the egg nucleus moves from its initial position in the cytoplasm, where it is spherical, it changes to an ovoid shape while moving through the aster, and becomes pear shaped when it reaches the border of the central lake. These changes in shape are in conformity with the radially aligned gel structure of the aster. When the egg nucleus enters the fluid astral lake, it again becomes spherical in shape.

A striking demonstration of the fact that the growing aster is an expanding gelated body, toward the center of which a flow is directed, is shown in observations on fertilized eggs with artificially produced exovates. The reaction of the egg varies according to whether the egg nucleus or the sperm nucleus lies in the exovate (Figures 17.11, 17.12).

The first case concerned a recently fertilized *Lytechinus* egg. After the fertilization membrane had elevated, pressure was exerted on the egg, causing rupture of the membrane. This occasioned the extrusion of an exovate through the rupture. The exovate contained the egg nucleus, while the sperm nucleus with its developing monaster lay in the larger portion of the egg which remained within the fertilization membrane (Figure 17.11a). During the further progress of events the centripetal currents in the sperm monaster carried the egg pronucleus into the central lake of the monaster. This resulted in the withdrawal of the entire exovate

[5] Observations on the fertilized *Lytechinus* egg reveal that the egg pronucleus is accompanied by the neighboring cytoplasmic granules while it migrates from its initial position to the periphery of the sperm aster. Once within the aster, no such simultaneous movement of the granules is observed. Wilson (1901b) described the failure of the granules surrounding the egg pronucleus to move with it. If Wilson's observation referred to movements within the aster (this is not stated), the behavior described for the granules is what we should expect. Further investigation of this point is desirable.

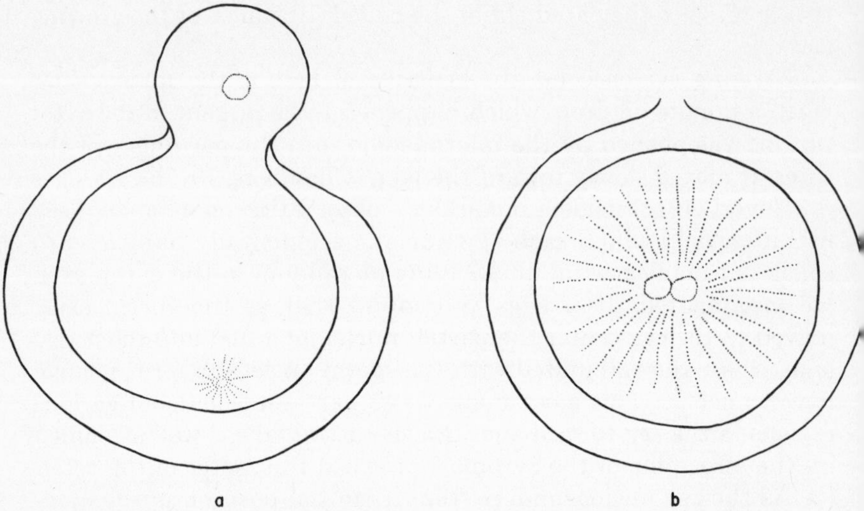

FIGURE 17.11. Recently fertilized living *Lytechinus* egg with exovate. Sperm nucleus with developing sperm aster lies within fertilization membrane. (a) Diminutive sperm aster develops in main body of egg while egg nucleus lies in exovate outside fertilization membrane. (b) Fully developed sperm aster; exovate containing the egg pronucleus has been drawn back into body of egg within its fertilization membrane.

into the egg within the fertilization membrane (Figure 17.11b). Further development occurred normally, with the fusion nucleus in its usual position in the egg lying completely within its fertilization membrane.

The second case also concerned a *Lytechinus* egg recently fertilized. In this case it was the exovate which contained the sperm nucleus surrounded by the developing monaster (Figure 17.12a). As the monaster developed, the exovate progressively increased in volume at the expense of that part of the egg which lay within the fertilization membrane. The currents streaming to the central lake of the growing sperm monaster involved more and more of the egg, until finally all of the egg became incorporated in the exovate (Figure 17.12b).[6] This egg continued its development outside its fertilization membrane.

[6] Chambers (1938c). Moore (1937) made the following observation on a sand dollar egg deformed by centrifuging to a dumbbell shape. The lighter portion of the egg contained the egg nucleus and was joined to the heavier and larger portion by a constricted stalk of cytoplasm. A sperm entered the lighter end, fused with the egg nucleus, and the fusion nucleus moved through the narrow connecting stalk into the larger portion. Moore believed that the fusion nucleus is pulled across the stalk and cited this as evidence that a contractile process is involved. Further investigation of this experiment with reference to the relation of the sperm monaster to the movement of the fusion nucleus is indicated. In addition, the possibility that centrifugal separation of the cytoplasmic components may influence the behavior of the pronuclei requires consideration.

The significance of these observations on eggs with differently located exovates lies in the centripetal currents occasioned during the growth of the monaster. The center of the monaster becomes the center of the entire egg, whether the egg lies within the fertilization membrane or whether it becomes incorporated in the exovate outside the fertilization membrane.

THE AMPHIASTER

The disappearance of the monaster is accompanied by a reversal of the gel to the sol state. This is shown by the fact that as soon as the monastral radiations are no longer visible the microneedle moves the cytoplasmic granules as if they lay in a liquid. Soon the astral radiations blaze out again, but this time around the two poles of the hyaline region (the persistent but now elongated monastral lake) in the center of the egg, producing the amphiaster. When tested with the microneedle, the spindle and the two small polar asters of the metaphase amphiaster behave like relatively rigid bodies. The cytoplasm beyond is fluid.

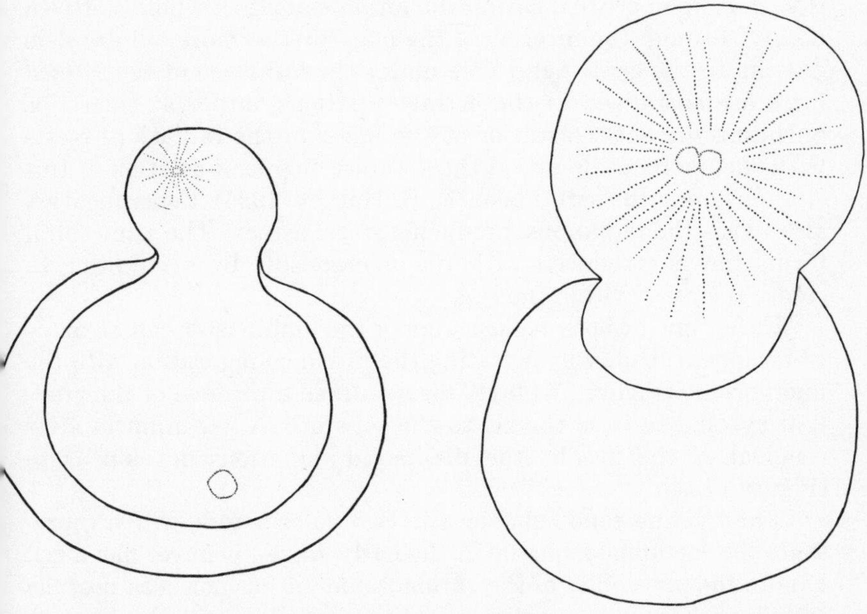

a b

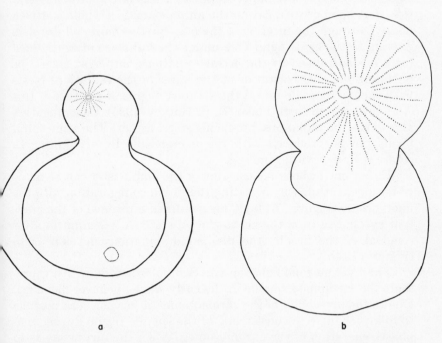

FIGURE 17.12. Recently fertilized living *Lytechinus* egg with exovate. In this case, sperm aster lies in exovate outside fertilization membrane. (a) Diminutive sperm aster develops in exovate while egg pronucleus lies in main body of egg within fertilization membrane. (b) Fully developed sperm aster; all of egg has been incorporated in exovate outside fertilization membrane.

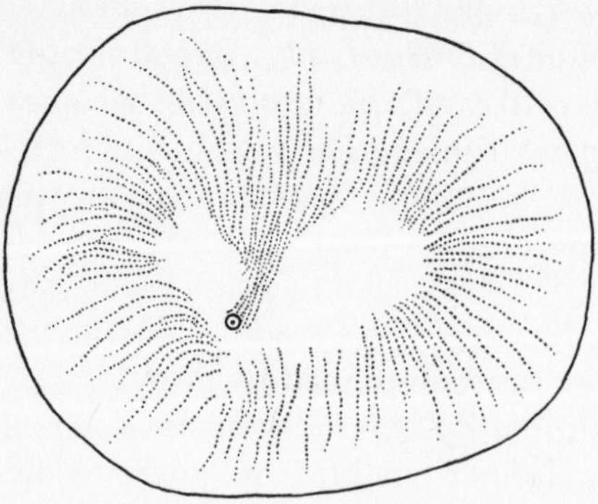

FIGURE 17.13. Amphiaster in living sand dollar egg with tip of microneedle inserted into aster at border of lake. Needle is moved across lake, dragging with it gelled granulocytoplasm.

When the asters have reached full development (Figure 17.1, late anaphase or early telophase) and the radiations extend almost to the periphery of the egg, it is possible to show with the microneedle that the granular cytoplasm contained within the asters is a gel and that the rays can be distorted (Figure 17.13). Additional evidence that the developing amphiaster is gelated is the observation that upon centrifugation the amphiaster as a whole is driven bodily through the interior of the egg. Furthermore, oil droplets moving to the centripetal pole under the influence of centrifugal force are entrapped in the astral radiations, either at their tips outlining the entire aster or at the bases of the radiations (tests with the microneedle reveal the greatest degree of gelation at this site) outlining the astral lake (E. B. Harvey, 1935). Centrifugation also twists the radiations, producing spiral asters.[7] The same spiral forms can be produced with the microneedle by attempting to rotate the aster within the egg.

Either one or both polar asters of the amphiaster can be made to disappear readily by agitating the radial configuration with the microneedle (Figure 17.14a). This results in a reversal of the granular cytoplasm from the gel to the sol state. A few minutes after removal of the needle, the dissipated radiations develop again (Figure 17.14b).

The hyaline zone between the two polar asters, which represents the metaphase spindle in the early stages, behaves like a gel. Due to the invisibility of the chromosomes, it has not been possible to follow changes in consistency of the spindle region during anaphase separation of the chromosomes. When the furrow starts to sink in, however, the hyaline region, now representing the interzonal region of the spindle, or spindle rest, has reverted to a sol

[7] Morgan (1910); Spooner (1911); Conklin (1917).

state. This liquefaction shortly precedes the karyokinetic lengthening of the egg and the sinking in of the furrow. The liquefaction of the spindle would account for the invasion of granules into the region of the spindle rest described, for example, by Conklin (1902) for the *Chaetopterus* egg.

We have discussed the physical state of the amphiaster with reference to the polar asters, the metaphase spindle, and the spindle rest. In addition, the granular cytoplasm in the equatorial region of the egg, where the cleavage furrow will ultimately sink in, requires careful consideration (Chambers, 1917b). When the amphiaster first reaches full development (Figure 17.1), radiations from the two polar asters of the amphiaster appear to be continuous strands in the granular cytoplasm of the equatorial region bordering the hyaline anaphase spindle. In the periphery of the equatorial zone the radiations from the polar asters meet each other from opposite directions, but the rays are never seen to cross each other. Testing with the microneedle reveals that the cytoplasmic granules in this region are embedded in a gel. Subsequently, when karyokinetic lengthening of the egg occurs, just prior to the sinking in of the cleavage furrow, the granular cytoplasm in the equatorial region of the egg loses its streaked appearance, and testing with a microneedle indicates that the granules are no longer held in a gel (Figure 17.15). Shortly after the disappearance of the ray-like structure in this region, the cleavage furrow sinks in. These observations reveal that both the interzonal region of the amphiaster (spindle rest, or interzonal body) and that part of the equa-

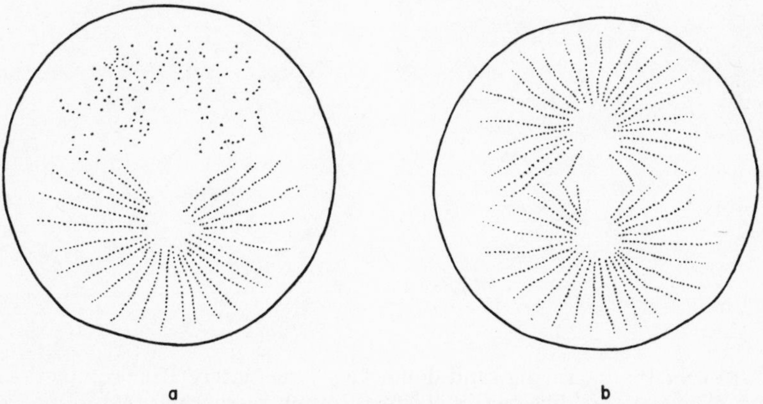

FIGURE 17.14. (a) Amphiaster in living sand dollar egg with one polar aster destroyed by local gentle mechanical agitation with microneedle. (b) Same egg a few minutes later. Polar aster of amphiaster has reappeared.

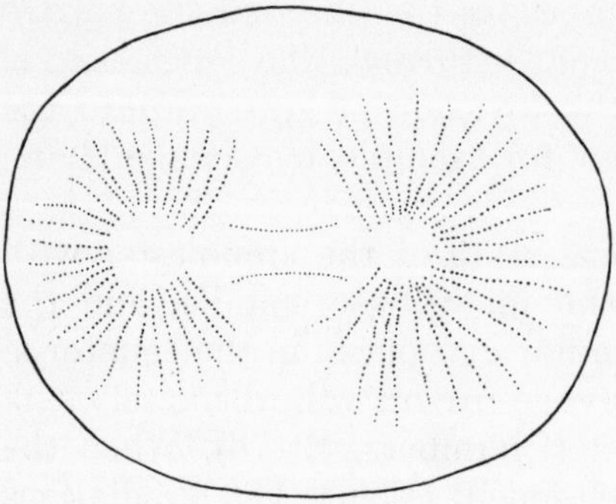

FIGURE 17.15. Amphiaster in living sand dollar egg just prior to sinking in of cleavage furrow. Interzonal hyaline region is dumbbell shaped and liquid. Astral radiations in equatorial region have disappeared.

torial region which contains astral radiations undergo liquefaction immediately prior to cleavage of the egg.

When cleavage has been completed, the two polar asters, one in each daughter cell, are fully maintained for a time. In the absence of pressure, the gel state of the two asters tends to keep the blastomeres spherical. This is best seen in an egg whose investing fertilization membrane has been removed just prior to cleavage (Figure 17.16a). The removal of the fertilization membrane does not free the blastomeres, for they still remain surrounded by a delicate pellicle, the so-called hyaline layer, which is not strong

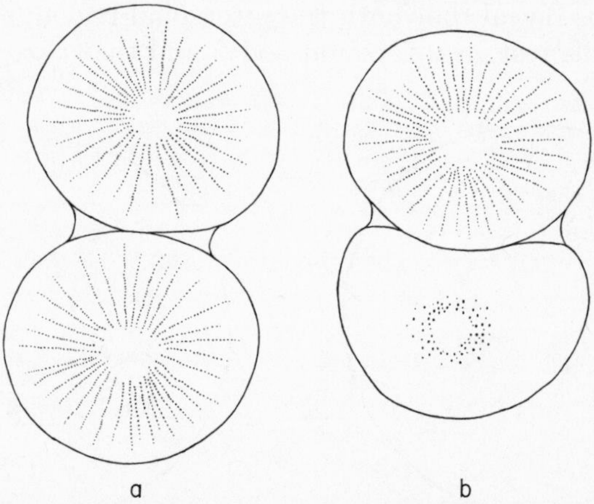

FIGURE 17.16. (a) Living sand dollar egg immediately after completion of first cleavage and divested of its fertilization membrane. Asters present. Both blastomeres spherical and held together with hyaline layer material. (b) Same, with aster in one blastomere dissipated by microneedle. This has caused it to flatten against spherical neighbor.

enough to distort the blastomeres as long as the asters persist. With a needle, the aster in one of the blastomeres was destroyed by puncturing, whereupon the blastomere lacking the aster flattened against its neighbor (Figure 17.16b). The aster reappeared, and the blastomere rounded again. When left undisturbed, both asters eventually faded and both blastomeres flattened against one another preparatory to further development.

Attempts have been made to compare and correlate the microdissectionist's results, obtained by testing the rigidity of various structures in the cell with the microneedle, with "viscosity" measurements of the interior protoplasm, using as criterion the rate or extent of movement of granules, vacuoles, or oil droplets in a centrifugal field. The comparisons are hardly valid. The microdissectionist is keenly aware of the changing, "multistructured" nature of the interior of the dividing cell. Gel and sol exist side by side, with infinite gradations in rigidity or fluidity. He can only wonder what significance is to be attached to attempts at describing the over-all "viscosity" of such a protoplasmic interior.

Heilbrunn (1952, 1956) stresses that "when a cell begins to divide the appearance of the spindle is preceded by a sharp increase in viscosity of the main mass of protoplasm." This statement is based on "viscosity" measurements by the method of centrifugation carried out on marine ova during the period from fertilization to first cleavage. In every case the development of the mitotic spindle is preceded by the growth and disappearance of the sperm aster. We cannot see why the viscosity rise and fall should be related more to the mitotic spindle than to the developmental phases of the sperm aster. We know from microdissection studies that the sperm aster is a gelated body even more rigid than the mitotic asters. Heilbrunn continues with the statement that the bulk of the protoplasm is fluid *as* the mitotic spindle appears. This is not in disagreement with the microdissectionist's results, since even through metaphase the amphiaster (spindle with polar asters) remains relatively small and surrounded by a broad region of fluid cytoplasm. A large proportion of the protoplasmic inclusions would then be free to move in a centrifugal field.

However, at the height of development of the amphiaster (just prior to cleavage) or of the earlier formed sperm aster, a large proportion of the inclusions are held in the regions of gelation. At corresponding stages the interior of the egg is stratified with difficulty; these are, according to Heilbrunn, periods of high "viscosity."

THE ASTER IN ARTIFICIAL PARTHENOGENESIS

Artificial parthenogenesis is brought about by exposing eggs to certain chemical agents. These procedures activate the eggs to undergo successive mitoses and further development. The physical nature of the asters and the sequence of astral formations are essentially the same in parthenogenetically activated and sperm-fertilized eggs. If chemical activation results in normal development, a monaster corresponding to the sperm aster of the sperm-fertilized egg precedes the amphiaster of the first cleavage figure. However, the early development of the monaster in the artificially activated egg differs in that circumscribed hyaline regions first appear in the center of the egg and then coalesce. The monastral radiations subsequently develop around the central hyaline zone. In chemically activated eggs irregularities in formation and fusion of the central hyaline regions, at this early stage, are associated with abnormalities in the later development of the ova.

Unfertilized sand dollar eggs are immersed in butyric acid in sea water and then exposed to hypertonic sea water for less than half an hour (Chambers, 1921a). Under this treatment a large proportion of the eggs undergo normal cleavage. After the lifting of the artificially induced fertilization membrane, the first sign of further activation is the appearance of one or several small, circumscribed, hyaline regions in the vicinity of the center of the egg. Within a few minutes, the hyaline regions coalesce to form a larger, central, clear region of about one tenth the diameter of the egg. The egg nucleus lies close to or within this region. Gradually, rays appear, extending in all directions from the hyaline region. The extended rays gradually become more numerous, more pronounced, and spread through the entire egg, which finally is occupied by a large monaster. This corresponds to the fully developed monaster of the sperm-fertilized egg. During this time the hyaline central region increases in volume, and probing with microdissection needles indicates that the center is a liquid lake of hyaloplasm. From now on further development is closely analogous to that of a sperm-fertilized egg. When the monaster disappears, the hyaline fluid occupying the centrally located lake persists, and the major part of the lake accumulates at the two poles of the nucleus into two polar areas. A gelation process now sets in with these two regions as centers, and there results a typical amphiaster preparatory to the first cleavage of the egg.

In the mode of aster formation the major difference between the sperm-fertilized and the parthenogenetically activated egg is the manner in which a liquid separates out of the cytoplasm during the formation of the monaster. In the sperm-fertilized egg radiations appear quickly about the sperm head, and the accumulation of the hyaline lake is from the very start through the agency of the ray-like channels of the growing monaster. In the parthenogenetically activated egg the process of gelation is much slower, and the separating out of an hyaline phase takes place slowly and before the granular cytoplasm has become sufficiently gelated to exhibit the radial astral configuration.

Abnormalities met with in parthenogenetically activated eggs are connected with irregularities in the merging of the several hyaline regions which sometimes appear in the cytoplasm. In eggs undertreated with a parthenogenetic agent, the monaster develops as usual, but upon the disappearance of the monaster, the central lake, instead of flowing to the two polar regions of the egg nucleus, remains a single body. With the return of the gelation period, a single aster again forms, and more fluid accumulates in the lake, which continues to increase in size. This process repeats itself several times, and segmentation of the egg never occurs.

Eggs overtreated with a parthenogenetic agent exhibit several circumscribed hyaline regions in the cytoplasm. Instead of collecting in the center of the egg, they increase in number throughout the cytoplasm. Radiations appear about these hyaline regions with the result that the egg becomes filled with many small cytasters. Most if not all of these asters develop independently of one another. Some of the asters which lie close to the periphery of the egg exhibit furrows forming from the surface of the egg and partially enclosing the cytasters, as shown in Figure 17.17. Occasionally these aberrant cleavage furrows actually complete their course to

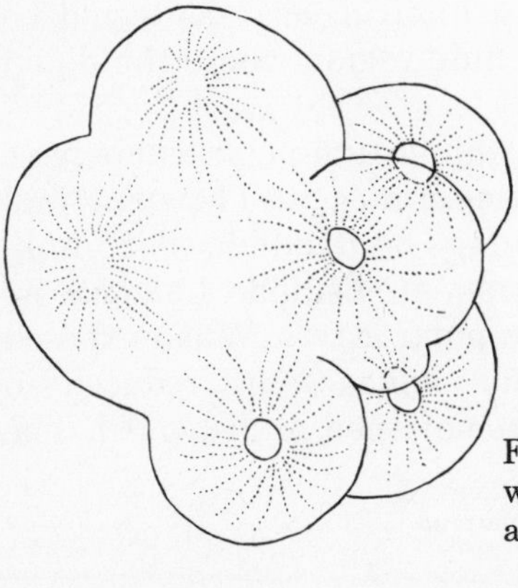

FIGURE 17.17. Nuclear asters and cytasters with partial furrows in parthenogenetically activated living *Lytechinus* egg.

the extent of enclosing a cytaster, resulting in a spurious kind of segmentation.

NATURE OF THE ASTER

Results obtained largely by the method of microdissection reveal that the growing aster, whether this be the sperm monaster or one of the polar asters of the amphiaster, is a centrifugally expanding, gelated body of granulocytoplasm. Interspersed throughout the gelated granulocytoplasm are numerous radially directed fluid channels of hyaloplasm, flowing toward, and collecting in, the enlarging fluid central lake.

Confirming the gelated nature of the aster is the observation that the aster in the living egg is eliminated by any agent which solates gelated structures, such as mechanical agitation or exposure to high hydrostatic pressures or to solutions of colchicine. Furthermore, that the aster is the manifestation of converging streams of hyaloplasm is supported by the fact that it is suppressed by agents which inhibit streaming movements (for example, cold, metabolic poisons such as cyanide, anesthetics).

Microincineration of sections of frozen-dried echinoderm eggs in the amphiaster stage reveal a radial orientation of ash. This result is in accord with a differential distribution of salts between the hyaloplasmic rays and the gelated inter-ray regions in which numerous minute vacuoles and granules are imbedded.

The extreme delicacy of the astral radiations, in both living and fixed preparations, has long suggested a structural, radial orientation of submicroscopic particles. Such an orientation is indicated by the birefringency of the aster when observed under polarized light. (Might this be flow birefringency due to orientation of submicroscopic fibrillar elements in the hyaloplasm as it flows centripetally through the innumerable very narrow radial channels?) The central astral lake is lacking in birefringency, as would be expected on the basis of its being a fluid region where the liquid hyaloplasm collects.

Mitotic figures consisting of the spindle and polar asters have been successfully isolated from echinoderm eggs.[8] The structural integrity of the isolated figures depends on disulfide (S–S) linkages.[9] Enhancement of the formation of disulfide linkages by oxidants stiffens the spindle and the polar asters. When exposed to sulfhydryl reagents, the disulfide linkages are reduced to sulfhydryl groups and the figures are softened or dispersed. The

[8] Mazia and Dan (1952); Dan and Nakajima (1956).

[9] Mazia (1955); Dan (1956); Mazia and Zimmerman (1958).

asters and the spindle in the living egg are also disorganized by the penetrating sulfhydryl reagent, mercaptoethanol.

ROLE OF THE ASTER

The sperm monaster plays a major role in bringing about the union of the male and female pronuclei. Following insemination of the ovum, the growing sperm monaster, an expanding, gelated body which develops around the neckpiece of the entering spermatozoon, carries the sperm pronucleus from its initial position just under the egg surface to the center of the egg. Meanwhile, the centripetally flowing hyaloplasmic channels in the growing sperm aster carry the egg pronucleus to the central lake of the sperm aster, where it joins with the sperm pronucleus to constitute the amphinucleus of the first cleavage figure.

Asters are a prominent feature of cell division in cells which possess an abundant amount of cytoplasm and cleave rapidly, such as embryonic animal cells, especially developing echinoderm eggs. When the amphiaster is fully developed, the major part of the cytoplasm is occupied by the asters of the amphiaster, of which the spindle is small. In these cells changes in shape associated with cleavage occur according to a geometrical pattern.

In many animal tissue cells the asters are very small or absent, and the spindle occupies the major part of the cell. The irregular form of cleavage in a cell such as the dividing fibroblast has been correlated with the small size of its asters (Hughes and Swann, 1948). Asters have not been observed in dividing cells of the higher plants, in which the major part of the cytoplasm becomes included in the spindle.

Experimental studies on the role of the aster have been restricted almost exclusively to marine ova, mainly the ova of echinoderms.

The feature with which the asters are most closely correlated is rapidity of the cleavage process. In general, this occurs far more rapidly in cells with prominent asters than in those with very small or absent asters. In experimental studies the asters have been suppressed by anesthetics, and mitosis and cell division still occur, but are greatly delayed.

The currents of hyaloplasm flowing into the central lakes of the polar asters of the amphiaster undoubtedly achieve a redistribution of the hyaloplasm between the incipient blastomeres. From the astral lakes the hyaloplasm may pass into the spindle to contribute material for its enlargement and elongation. Suggestive evidence for this is the observation that the rapid phase of growth of the asters and the anaphase elongation of the spindle occur coincidentally (Swann, 1951b).

Let us restrict our attention now to the role of the amphiaster in the actual furrowing process (exclusive of the already completed mitotic process). Originally, Chambers (1919a) suggested that the karyokinetic elongation of the egg and the ingrowth of the furrow at right angles to the long axis of the amphiaster could be accounted for by the growth in size of two semisolid spheres within the egg, at the expense of all but a small peripheral part of the egg substance. When the combined diameters of the two asters grow to the extent of exceeding the diameter of the spherical egg, elongation would necessarily occur. With the incorporation of still more fluid cytoplasm (largely that remaining in the equatorial region), a furrow would have to sink in between the two asters.

A special feature of the aster is that as it grows in size its gelated periphery can exert a pushing action. This may be related to the fact that the aster represents more than a radially extending zone of gelation, there being a simultaneous, radially directed centripetal flow of nongranular liquid. The pushing action of the asters is such that when two grow side by side they mutually repel one another. As we have seen in the case of the sperm monaster, when an aster is located eccentrically, during the process of growth it pushes against the nearest surface of the egg. This causes the aster to move toward the egg center. If, however, the growing aster meets an obstruction, such as another growing aster, a pushing action is exerted against the egg surface, causing an outward bulge and the appearance of a furrow between the two asters. Such behavior is seen, for example, when cytasters are caused to form by parthenogenetic activation of non-nucleated fragments. Furrows form between the asters, whether or not the asters are joined by spindles. Occasionally the furrows sink through. This, however, is exceptional unless chromosomes are present on the spindles.

A number of years later Chambers (1938b) sought to establish with greater certainty the role of the asters in cell division by observing the results of causing their disappearance at different stages of development. This was readily accomplished by mechanical agitation of the asters with a microneedle. If the asters are suppressed well before the cleavage furrow starts to sink in, the cytoplasm reverts to a fluid state, and cell division is completely inhibited. However, suppression of the asters immediately before or after the furrow has started to sink through does not prevent the continued cleavage of the ovum. (This feature is discussed more thoroughly in Chapter 19.)

While it is conclusively established, therefore, that the furrowing process itself does not require the presence of the amphiaster, it is nonetheless an interesting fact that two asters expanding side by side are quite capable of causing furrow formation. It is as though, in the normal process of cleavage, the egg were provided with a dual mechanism for achieving furrow formation. Mitchison (1953) called attention to the irregularity of the cleavages which occur when the asters are destroyed during cell division, and he suggested that one of the roles of the asters in cell division is to assure a regular cleavage pattern.

Although the late amphiaster is not essential, the early amphiaster is evidently a determining factor in the cleavage process.

18

The Spindle

CHARACTERISTICS OF THE SPINDLE

The brief description at the beginning of the previous chapter of the spindle and chromosome movements during mitosis in the sand dollar egg holds true for cells in general. Variations in the mitotic figure in different cell types occur principally with respect to the size of the spindle in relation to size of the cell as a whole, and the presence or absence of asters. Elongation of the spindle during anaphase is a phenomenon of widespread occurrence.

The metaphase spindle probably originates from the nuclear gel[1] in which the prophase chromosomes within the intact nucleus are imbedded (Duryee and Doherty, 1954). From studies of mitosis in the endosperm cells of plants, Bajer (1957, 1958) has concluded that the mitotic spindle is composed of both a granule-free cytoplasmic constituent and the nuclear sap.

A considerable amount of evidence indicates that the metaphase spindle, in both plant and animal cells, has the properties of a gelated body, frequently of considerable rigidity. Microdissection studies on echinoderm eggs (Chapter 17), on pollen mother cells, on insect spermatocytes (Chambers, 1924a, b), and on grasshopper neuroblasts (Carlson, 1952) indicate that the metaphase spindle as a whole has the consistency of a gel, to which the metaphase chromosomes are firmly attached.[2] Furthermore, the metaphase spindle, together with the attached chromosomes, can be moved

[1] Chromosomes have been demonstrated to be imbedded in a jelly-like matrix not only in the nuclei of immature frog ova (Duryee, 1950) but also in the interphase nuclei of salivary gland cells (Chapter 1).

[2] This had already been indicated in 1905 by Foot and Strobell, who found that the metaphase spindle remained intact with the chromosomes in situ when an earthworm egg containing the spindle was squashed and the contents allowed to flow out. Evidence of the rigidity of the metaphase spindle was also obtained by Bĕlăr (1929), who observed that when insect spermatocytes were compressed between slide and coverglass, the long axes were mostly oriented parallel to the slide.

bodily through both plant and animal cells by the application of centrifugal force. Oil globules, in their passage to the centripetal pole, evidently unable to penetrate the stiff metaphase spindle, outline its centrifugal borders (Shimamura, 1940; Figure 18.1a, a′).

A thorough investigation of the physical state of the anaphase spindle is that of Carlson (1952), who tested with the microneedle the various regions of the spindle in grasshopper neuroblasts. He found that, although tough interzonal fibers connect the two sets of separating chromosomes, the spindle matrix of the interzonal region (between the chromosomes) undergoes liquefaction. The liquefaction begins in the equatorial region at the start of anaphase and spreads toward both poles with the advance of the chromosomes.

A weakening in the interzonal region of the anaphase spindle is also indicated by the centrifugation studies of Shimamura (1940) on pollen mother cells (Figure 18.1b, b′). Centrifugation causes the mid-anaphase spindle to break in the mid-region, forming two widely separated, tilted spindle halves. The gelated state of the polar halves is indicated by the maintenance of their integrity and by their being outlined by centripetally moving oil droplets, which, however, readily pass through the interzonal region.

In the sand dollar egg the physical state of the interzonal region has not been determined for the various stages of anaphase, but in early telophase at the time of karyokinetic lengthening just prior to cleavage, the interzonal region of the spindle (or spindle rest) has the consistency of a fluid (Chapter 17). Whether the region between the separating chromosomes is fluid during anaphase in cells generally, as Carlson (1952) indicated it is for the neuroblast, has yet to be determined.

Longitudinal orientation of the structural elements of the metaphase spindle and of the polar regions of the anaphase spindle is well established, especially on the basis of their birefringency in polarized light.[3] In the interzonal region of the anaphase spindle birefringency is minimal. The loss of birefringency starts in early anaphase at the equatorial region and spreads toward the poles as the chromosomes move apart. This pattern fits admirably with Carlson's description of solation in the interzonal region between the separating chromosomes. Liquefaction of the gelled spindle would be expected to result in disorientation of its submicroscopic components. Inoué (1953) has described longitudinal birefringent fibrillar elements in the polar regions of the metaphase and anaphase spindle. He also found less prominent birefringent fibrillar

[3] Schmidt (1939); Hughes and Swann (1948); Inoué and Dan (1952); Inoué (1953).

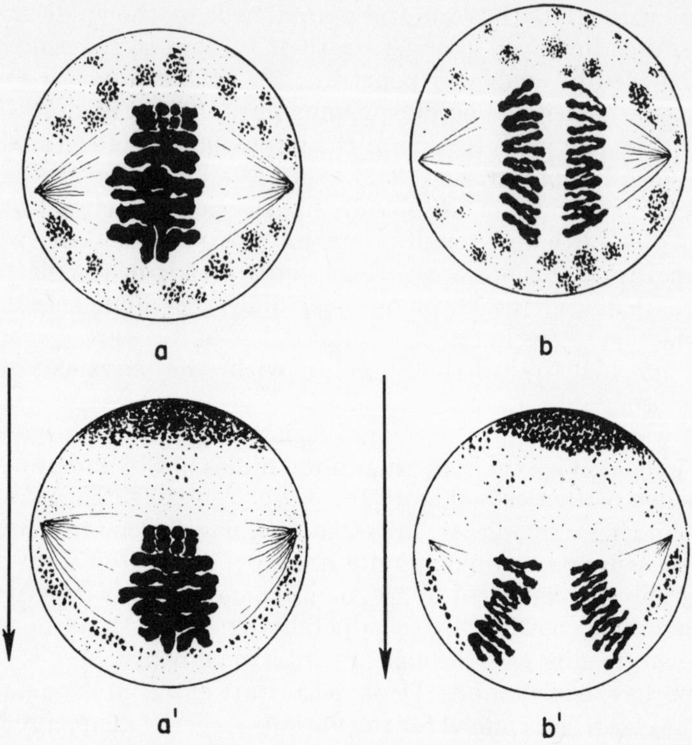

FIGURE 18.1. Diagrams showing effect of centrifuging pollen mother cells of *Lilium* during mitosis. (a), (b) Controls. (a′), (b′) After centrifugation. Arrow indicates direction of centrifugal acceleration. (a) Control, at metaphase. (a′) Same after centrifugation. Entire spindle with chromosomes displaced centrifugally. Oil droplets outline whole centrifugal border of spindle; all other oil droplets driven to centripetal pole forming oil cap. (b) Control at anaphase. (b′) Same after centrifugation. Polar halves of spindle tilted and pulled apart at interzonal region. Oil droplets outline centrifugal border of spindle only in polar regions; all other oil droplets driven to centripetal pole to form oil cap. (From Shimamura, 1940.)

elements in the interzonal region of the anaphase spindle. These may represent the tough interzonal fibers between the separating chromosomes which Carlson (1952) demonstrated by the method of microdissection.

ROLE OF THE SPINDLE

From time to time the anaphase elongation of the spindle has been implicated as a cause of cell division in eggs with prominent polar asters, such as echinoderm eggs. This postulate is based on the belief that the two polar asters are pushed apart by the lengthening spindle. Such an explanation seems unlikely, since in the sand dollar egg, at the time the cleavage furrow sinks in, the spindle rest (or interzonal body) is fluid.

Although the spindle in cells in which this occupies a large part of the cell interior may be involved in the redistribution of hyaloplasmic materials between the daughter cells, its essential role is concerned with the movement of the chromosomes. An extraordinary feature concerning the anaphase movements of the chromosomes, noted for the grasshopper spermatocyte, is that once separation has started, the movement of the chromosomes cannot be hindered by mechanical trauma short of destroying the cell (Chambers, 1924b). The separation of the chromosomes at anaphase occurs both by a movement poleward along the spindle and by an elongation of the spindle itself. While gel structure in the spindle is essential for these anaphase movements (Pease, 1946), Ris (1949) has presented evidence that the two types of movements occur by different mechanisms.

Substantial evidence indicates that the elongation of the spindle during anaphase occurs by an active, intrinsic elongation of the interzonal region. This was indicated in 1929 by Bělăr, who observed a pushing apart of the chromosomes by an extensive elongation of the interzonal region of the spindle of plant cells immersed in hypertonic sucrose solutions. As evidence of intrinsic elongation Berkeley (1948) cited the buckling of anaphase spindles in their interzonal regions when two elongating spindles met end to end during mitosis in multinucleated amebae. The same feature was indicated by Carlson (1952) in his description of the ability of the elongating anaphase spindle to exert pressure against the microneedle or to press against and deform the cell surface.[4]

The actual mechanisms whereby the anaphase movements occur remain in the realm of speculation. Bělăr suggested that, if the interzonal region is a gel, the pushing apart of the chromosomes could occur by an extension in length of this gelated column. Such a process would involve solation immediately in advance of the chromosomes, with reversal to the gel state in the interzonal region (Chambers, 1951). The pushing action of the column would resemble, for example, that of the expanding monaster. While supporting the concept that the chromosomes are

[4] The spindle can be set free of all attachments, yet separation of the chromosomes and elongation of the spindle occur normally, indicating the intrinsic nature of the process (Carlson, 1952) and the absence of stretching, which might be exerted if the poles of the spindle (or the polar asters) were fastened to the cortex of the cell. Evidence that elongation of the spindle is an intrinsic process has been reviewed by Dan (1943).

pushed (rather than pulled) apart, Carlson (1952) suggested that the pushing force may result from the lengthening of interzonal connections between the separating chromosome groups, since he found the interzonal region to be fluid but traversed by tough interzonal fibers. We have yet to determine to what extent liquefaction of the spindle substance occurs during anaphase in cells other than the grasshopper neuroblast, but the lengthening of interzonal fibers or of a gelated column by accretion at their ends could well represent essentially similar processes.

While a pushing apart of the chromosomes almost certainly occurs, at least in association with elongation of the spindle, we have to consider the possibility that the chromosomes are also pulled to the spindle poles. This could be accomplished by a shortening of longitudinally oriented fibrillar elements in the gelated polar portions of the spindle, as has been repeatedly proposed. The fact that in certain cell types, such as the grasshopper spermatocyte, the microneedle can be passed transversely through the polar region without disturbing the metaphase chromosomes (Chambers, 1924b)[5] does not vitiate such a possibility, but does cast doubt upon the existence of morphologically distinct traction fibers. The absence of thickening in the polar region as the chromosomes move to the poles does not necessarily speak against a contractile mechanism, since new contractile material could be brought into play as previously contracted material undergoes solation.

Since a polar contractile mechanism manifestly cannot explain that phase of chromosome separation associated with elongation of the spindle, a reasonable proposal is that the chromosomes are being pulled, as well as pushed, apart. The chromosomes may be pulled to the poles by contractile fibrillar elements in a structure which itself, as a whole, is expanding longitudinally. Duryee and Doherty (1954) have suggested a mechanism along these lines. Or the chromosomes, while being pulled to the poles, may be pushed by a lengthening interzonal body or lengthening interzonal fibrillar elements.

A final possibility to consider is that the spindle may have a structure like that of the aster, with fluid channels running longitudinally (in the gelated body of the spindle) rather than radially (as in the aster). Granules being absent, such a structure would be homogeneous in appearance when

[5] Carlson (1952) attempted the same experiment in the grasshopper neuroblast, but owing to the pronounced rigidity of the metaphase spindle, the effect of moving the needle sideways within the spindle was only to cause its dislocation. Evidently the metaphase spindle of the grasshopper spermatocyte is less rigid, and the spindle tends to revert to the sol state when the tip of the microneedle is thrust into it.

viewed through the light microscope. According to this concept the birefringency in polarized light and the fibrillar structure noticeable after fixation would have a common basis in both spindle and asters. The movement of chromosomal vesicles or of individual chromosomes in the rays of an aster resemble the anaphase movements of chromosomes (Boveri, 1918; Bělăr, 1933). Presumably, the chromosomal vesicles are moved through the aster in the same manner as the intact egg pronucleus. The factors responsible for movement of the egg pronucleus are (1) streaming movements of the hyaloplasm, (2) solation of gel in advance of the nucleus, and (3) a movement imparted to the egg nucleus as the periphery of the expanding gelated aster pushes against the surface of the egg. By analogy, we may conceive of streaming movements in channels in the gelated spindle carrying the chromosomes toward the spindle poles (instead of contractile fibers), solation in advance of the moving chromosomes, and expansion of the spindle as a whole, further separating the chromosomes. A final decision concerning the mechanism of separation of the chromosomes awaits further information about the physical state of the anaphase spindle in different cell types.

19

Cell Division in Echinoderm Eggs

In this chapter we shall devote our attention to cell division in eggs of the sea urchin, starfish, and sand dollar. The eggs of all these echinoderms contain a relatively sparse amount of yolk, at least as compared with squid or amphibian eggs.

SURFACE LAYERS OF THE EGG

We must begin our discussion of cell division by examining the surface layers of the egg. These include coatings applied to the external surface and, immediately under the protoplasmic surface film, the cortex of the egg (Figures 2.1, 19.1). Closely applied to the external surface of the unfertilized egg is the vitelline membrane. After fertilization the vitelline membrane lifts, and the surface film becomes coated with the relatively thick hyaline layer. Both these external coats can be removed without detriment to cell division or to the life of the cell. Immediately beneath the protoplasmic surface film of both the unfertilized and the fertilized egg is the gelated cortex of several micra in thickness. Extending from the protoplasmic surface of the fertilized egg are fine fibrils, presumably cytoplasmic in nature and extensions of the cortex (Gray, 1931). These are imbedded in the hyaline layer (Dan and Ono, 1952). In the case of denuded eggs in calcium-free sea water the fibrils constitute the radially structured surface halo.

Critical study of changes in the cortical protoplasm during cell division frequently requires the prior removal of the external coats. The eggs can then be maintained in a denuded state by allowing them to develop in calcium-free media, which promotes the dissipation of newly secreted hyaline layer material. Since the daughter blastomeres separate from one another as the naked egg cleaves, the investigator is enabled to observe clearly the walls of the advancing furrow. Furthermore, removal of the hyaline layer permits the experimentalist to differentiate properties of the protoplasmic cortical region from those of the extraneous coats. In view of the evidence that the cortex is fastened to the hyaline layer, cortical movements during cleavage may be considerably more restrained in the egg with an intact hyaline layer than in the denuded egg.

During the interkinetic periods the cortex encloses the relatively fluid endoplasm. During the growth of the sperm aster or of the mitotic figure the cell interior becomes increasingly occupied by the gelated asters and spindle until these structures occupy almost the entire interior. Even at the height of the monastral or amphiastral stage, a thin layer of fluid subcortical cytoplasm intervenes between the gelated cortical layer and the periphery of the asters. Katsuma Dan, however, has insisted for many years that the peripheral astral radiations of the fully formed amphiaster are fastened to the cortex. His conclusion is based on the deformation, or stretching, of the rays when the aster is displaced either by the microneedle or by centrifugal force. In any event, in describing cyclical changes in the physical properties of the cortical layer of a denuded egg the investigator frequently is faced with the difficult task of deciding whether the change ascribed to the cortex might not, in fact, result from an alteration in the physical state of the endoplasm.

Cortex of the egg

Imbedded in the cortex of the unfertilized egg are the cortical granules (or vacuoles) and, frequently, pigment granules (or vacuoles). Shortly after fertilization the cortical granules are extruded leaving, in certain species of sea urchins, a peripheral hyaline, or extragranular, zone (Figure 19.1, e.g.z.). This outer hyaline zone of the cortex persists through the early cleavage cycles. The pigment granules, when present, are located beneath this zone (Figure 19.1).[1]

Although the inner part of the cortex (Figure 19.1, i.p.l.) of the echinoderm egg, in which the pigment granules are imbedded when these are present, is indistinguishable to the eye from the underlying cytoplasm, its presence can be detected with a microneedle[2] thrust directly through the interior of an *Arbacia* egg until it

[1] An interesting phenomenon is the migration of pigment granules into the cortex of the *Arbacia punctulata* egg after fertilization (Harvey, 1911; Allen and Rowe, 1958). In the unfertilized egg only a few pigment granules are located in the cortex, the remainder being distributed at random in the endoplasm. In certain other species, however, all or most of the pigment granules are already imbedded in the cortical gel of the unfertilized egg and remain in this location after fertilization.

[2] Chambers (1917a, 1921b, 1938a).

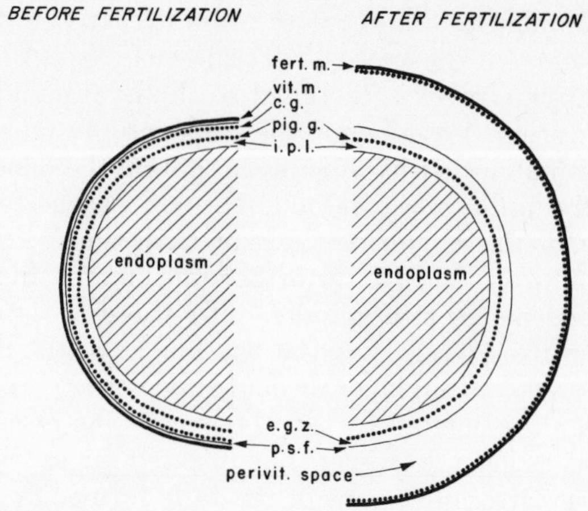

FIGURE 19.1. Diagram showing structural components of cortex of sea urchin egg. Left half before fertilization; right half after fertilization. Extragranular zone (e. g. z.); inner protoplasmic layer, or inner border of cortex (i. p. l.); cortical granules (c. g.); pigment granules (pig. g.); protoplasmic surface film (p. s. f.); vitelline membrane (vit. m.); fertilization membrane (fert. m.) consisting of altered vitelline membrane and material of cortical granules; perivitelline space (perivit. space). Hyaline layer not shown; it is probable that some of the material from cortical granules contributes to hyaline layer. (Adapted from Motomura, 1941. Motomura, 1957, describes two types of cortical granules and shows, by diagram, their contribution to formation of fertilization membrane in greater detail than the above.)

comes into contact with the opposite side of the egg. Upon withdrawal of the needle, strands of adhering gelled material can be seen being pulled into the more fluid interior of the egg. The fact that the cortical gel is of appreciable thickness can be demonstrated by injecting a minute drop of paraffin oil into the fluid interior of the egg. When the oil drop is pushed, at the tip of the micropipette, toward the surface from inside the egg, the boundary of the oil always remains an appreciable distance of several micra from the protoplasmic surface.

Recently Hiramoto (1957) used an adaptation of the microneedle technique to demonstrate the cortex and determine its thickness in eggs of the sea urchin *Hemicentrotus* (formerly *Strongylocentrotus*) *pulcherrimus*. He inserted a microneedle with a flat tip 2 micra in diameter through the interior of a denuded egg in calcium-free sea water and observed that the contour of the egg did not change as the tip was moved through the fluid endoplasm (Figure 19.2a). When, however, the tip reached a position about 2 micra from the surface, further insertion caused an outward bulge (Figure 19.2b, c). This indicated that the tip had come into contact with the gelated cortical layer.

The existence of a relatively firm cortex beneath the surface of the egg is also indicated by the following experiment. A drop of paraffin oil exuding from the tip of a pipette close to the margin of a denuded sea urchin egg was caused to coalesce with the egg in such a manner that the drop remained connected with the oil within the lumen of the pipette during the procedure. The lowered tension within the egg caused oil to flow from the pipette, thus increasing the size of the drop within the endoplasm of the egg. The pipette was then removed. Where the oil drop had entered the egg, a cytoplasmic excrescence developed, and the oil drop was carried out of the egg through the gap in the cortex caused when coalescence occurred. As the oil drop passed out through the cortex, it assumed the shape of a dumbbell. Since the oil drop had increased in volume within the egg, it accommodated itself to the outward passageway by narrowing as it emerged.

The presence and thickness of a gelated cortex at the advancing tip of the furrow in the dividing sea urchin egg, and the force exerted by the cortex as the furrow advances, can be determined by introducing a drop of paraffin oil into the equator of a dividing sea urchin (*Lytechinus*) egg (Chambers and Kopac, 1937b). It was

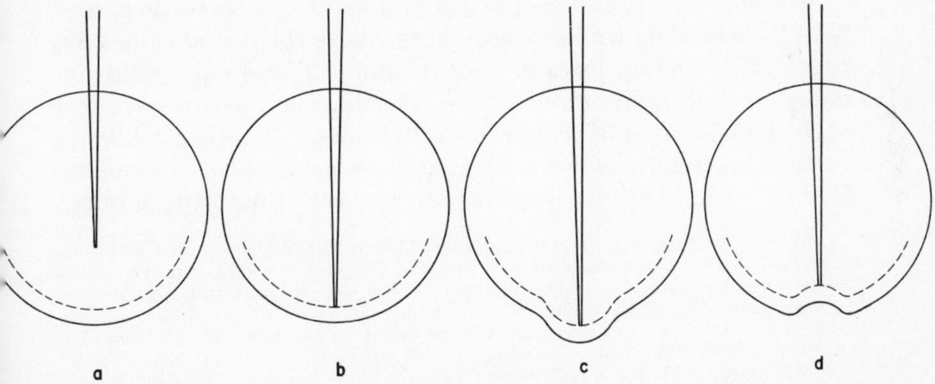

FIGURE 19.2. Diagram showing methods of detecting cortex in living sea urchin egg. When tip of microneedle inserted into egg is within endoplasm, contour of egg scarcely changes with movement of needle (a). After tip reaches cortex (b), surface of egg bulges out upon further insertion of needle (c) and is indented upon withdrawal of needle (d). Width of cortex exaggerated. (After Hiramoto, 1957.)

found that the oil drop, even if initially located a little to one side of the equatorial center, tends to move to the center. The oil drop, at first spherical, remains so until the external surface of the advancing furrow on each side is 4 to 5 micra from the surface of the drop. This distance of 5 micra indicates the thickness of the cortex in the region of the furrow. The oil drop then becomes increasingly deformed into the shape of an hourglass as the opposing walls of the furrow approach one another. Finally, the drop becomes so constricted that it separates in two except for a narrow thread of oil connecting the two droplets. The pinching of the oil drop represents work being done by the advancing cleavage furrow. By the time of final separation the two drops lie far apart in the central axis of the telophase spindle of the daughter cells. Each oil drop is again divided in two during the succeeding cleavage.

The rigidity of the cortex in the region of the cleavage furrow accounts for the fact that when an egg starting to undergo cleavage is cut in two, parallel or diagonal to the long axis of the spindle, the furrow follows its original course, and four fragments result (Figures 19.6, 19.7, 19.8; for further description see page 243).

In addition to the evidence obtained in microdissection experiments, the gelated state of the cortex has been inferred from the resistance the imbedded pigment granules exhibit to displacement in a centrifugal field.[3] The pigment granules are the densest component of the egg interior, and a far greater centrifugal force is required to drag them out of the cortex than to drive them and the granular inclusions of the cell interior through the endoplasm.

Besides having the properties of a solid, the cortex is elastic. This is clearly shown in experiments where the denuded cleaving eggs (of *Arbacia punctulata*) developing in isosmotic potassium chloride or in sodium and potassium chloride mixtures are torn with the microneedle. The interior cytoplasm flows out, while the cortical layer shrinks down without showing any sign of wrinkling (Figure 19.3).[4] This experiment indicates that the cortex normally

[3] Brown (1934) commented on the visibility of this cortical gel surrounding the hyaline zone of *Arbacia* eggs centrifuged after fertilization. The visibility was due to the pigment granules which remain imbedded in the cortical gel in spite of centrifugation sufficient to induce stratification of the egg interior. Motomura (1935) described the fixation of pigment granules in the cortex of the unfertilized egg of the sea urchin *Hemicentrotus pulcherrimus* in spite of the application of high centrifugal force.

[4] Mitchison (1953) was unable to duplicate the above described "deflating balloon" effect in eggs of different species of sea urchins. He claims that the media used in Chambers' experiments were too abnormal to permit drawing conclusions as to conditions in the normal egg. Actually, the procedures used by Mitchison may not have been adequate to ensure outpouring of the egg contents. Chambers' "deflating balloon" experiments were carried out in media from which all traces of calcium were removed, and the unoperated eggs gave every evidence of undergoing normal cleavage (page 44; Chambers and Grand, 1933).

exerts elastic tension on the egg interior. The magnitude of this tension in isosmotic media, however, is relatively small (Sichel and Burton, 1936). The elastic behavior of the cortex is strikingly shown when a cleaving, dumbbell-shaped, denuded *Arbacia* egg is transferred from an isosmotic to an hypotonic sodium and potassium chloride mixture. The interior cytoplasm is shot out through a rupture in the wall, and the cortical layer shrinks down smoothly and rapidly until there remains only a tiny, deep red, dumbbell-shaped replica of the original egg. This is the shrunken cortex with the imbedded pigment granules.

CHANGES IN SHAPE OF THE DIVIDING EGG

Both for eggs developing in sea water with their full complement of extraneous coats and for denuded eggs developing in calcium-free media, the initial change in shape from the spherical is flattening of the equatorial region and elongation of the cell. Shortly thereafter the furrow starts as a shallow, flat pit at the mid-equatorial region.

In denuded eggs a greater elongation occurs prior to the sinking in of the furrow, due to the lack of the restraining extraneous coats. Furthermore, the furrow maintains a broad contour throughout cleavage, since the daughter blastomeres are free to separate. However, in eggs developing in sea water with intact extraneous coats and well-defined hyaline layers, the blastomeres are prevented from rounding away from each other, and cortical movements are restricted by the hyaline layer (Dan, 1954a; page 241). As the furrow advances the walls rapidly become steeper, until in the deeper parts of the furrow the walls are almost vertical and the surfaces closely pressed against one another.

Furrow formation can be converted to the knife-edge-like type in eggs cleaving in sea water by compressing the egg within its fertilization membrane between slide and coverslip. This prevents any lengthening of the egg, so that the walls of the deepening furrow are in intimate contact almost from the very start (Figure 5.1; for further description see page 75). The knife-edge-like type of furrow frequently occurs in nature, as in ctenophore eggs (Ziegler, 1898) and in amphibian eggs with intact vitelline membranes. The fact that this type of furrow can occur in echinoderm eggs suggests a similar mechanism of cleavage in these diverse types.

An extraordinary feature of the kinetics of fission is the persistence of a narrow, connecting stalk after the termination of the advance of the cleavage furrow. In the echinoderm egg the stalks may persist through several cleavages (Herbst, 1900). Eventually, however, the connections break as the blastomeres slip over one another during their successive divisions. When the denuded egg is made to develop in an isosmotic mixture of sodium chloride and potassium chloride, the persisting stalks are very evident. This is because the dividing blastomeres round away from one another, since there are no restraining external coats to keep them closely appressed. Evidently the forces which induce the formation of the cleavage furrow do not carry the cell division to completion. There remains a narrow cylinder of cytoplasm sheathed by hyaline extraneous material.

CHANGES IN PHYSICAL PROPERTIES OF THE SURFACE LAYERS DURING THE CLEAVAGE CYCLE

Changes in thickness of the cortex

Measuring the width of the cortex requires the use of mechanical methods, such as testing with an oil drop or a microneedle, to locate the border between fluid endoplasm and cortical gel. Using his blunt microneedle technique, Hiramoto (1957) estimated the thickness of the cortical layer of naked eggs in calcium-free sea water at different stages of development and in various regions of the dividing sea urchin egg. Following fertilization the cortex increases in thickness, the width being uniform over the entire circumference as long as the egg remains spherical. As furrowing begins, the cortical layer continues to thicken in the equatorial region while simultaneously becoming thinner at the poles.[5] During the succeeding interkinetic period, the cortex at the furrow region becomes thinner while the polar cortex thickens.

These changes in thickness of the cortex are corroborated by the observations of Dan and Dan (1940) on the same species of eggs in calcium-free media. As the furrow appears and deepens, these investigators observed that the outer hyaline, or extragranular, zone of the cortex becomes noticeably thinner at the polar regions. The reverse changes occur after cleavage is completed.

> Mitchison (1956), however, has concluded that no appreciable changes occur in thickness of the cortical layer during the cleavage cycle of sea urchin eggs in sea water or in calcium-free media, and that no differences in thickness exist between polar and equatorial regions of the dividing sea urchin egg. He bases his conclusions on measurements of the thickness of a granular layer at the periphery of the

[5] The terms "equatorial" and "polar" as used in this book are defined as follows: The "equatorial region" girdles the egg at right angles to and at the mid-region of the spindle. The sinking in of the surface of the egg at the equator constitues the furrow. In contradistinction, the "polar regions" of the egg cap the opposite poles of the spindle.

hyaline zone in centrifuged eggs, both living and sectioned. With regard to these contradictory observations, we believe that the greater reliance should be placed on measurements made using the microneedle method of actually testing where sol ends and gel begins. The border of the cortical gel need not correspond to a demarcation visible to the eye.

Changes in rigidity or stiffness of the surface layers

Shortly following fertilization of the *Arbacia* egg an increase occurs in the amount of centrifugal force required to dislodge pigment granules from the cortex (Zimmerman, Landau, and Marsland, 1957). The amount of force required continues increasing, with several oscillations, to reach a maximum a considerable time before cleavage. This maximum is maintained through the early stages of furrowing, but as cleavage is completed the centrifugal force needed to dislodge the granules falls off abruptly. These results indicate that an over-all increase in stiffness, or thickness, of the cortical gel occurs prior to and during the early stages of cleavage. The *over-all* nature of these results should be stressed, since Marsland (1939) has shown that the pigment granules are more firmly held in the cortex at the equatorial region of the dividing egg than at the poles.

On the other hand, concerning the eggs of another species, those of the worm *Chaetopterus,* Wilson (1951) reports that shortly prior to and during cell division a decrease occurs in the amount of centrifugal force required to disrupt the continuous string of granules in the cortex. Wilson interpreted this result as indicating a decrease in cortical rigidity, but Marsland and his co-workers have pointed out that the criterion used is inadequate. For example, by Wilson's criterion a localized weakness of the cortical gel at the polar regions, for which there is evidence in the eggs of several different species, would have been interpreted as a decrease in rigidity of the cortex as a whole. Furthermore, the results obtained might easily have been anomalous owing to the weak gelational state of the cortex in the eggs of this particular species and their great susceptibility to the application of centrifugal force.

The assumption that Marsland's measurements of cortical rigidity are independent of the physical state of the cytoplasmic interior is probably valid. Far greater centrifugal force is required to dislodge the pigment granules from the cortical region than to drive them through the cytoplasmic interior, even when this is filled with the gelated asters. There is no indication that the changes in centrifugal force required could be due to alterations in density or size of the pigment granules. However, variation in the depth to which the pigment granules are imbedded in the cortical gel could markedly alter the centrifugal force necessary for their dislodgement, quite apart from changes in rigidity of the gel.

Estimates of changes in stiffness of the surface layers of fertilized sea urchin eggs have been carried out by Mitchison and Swann (1955). The mouth of a pipette was applied to the surface of a denuded egg. Their measure of "stiffness" was the amount of sucking force needed to cause a hemispherical bulge of the egg surface into the mouth of the pipette. They observed that the surface of the egg becomes far more resistant to deformation just prior to and at the beginning of cleavage. During the later stages of cleavage the resistance to deformation decreases. Differences could not be found in the stiffness of different regions of the egg at cleavage. The extent to which these "stiffness" determinations are related to the properties of the egg cortex is questionable. Nor is it certain that changes in the physical state of the cytoplasmic interior would not have affected the results.

The estimates of cortical rigidity by the centrifugal method, and of surface stiffness by the deformation method, both show high values prior to and during the early stages of cleavage in the sea urchin egg. It is tempting to correlate these observations with Hiramoto's finding of a thickening of the cortex prior to cleavage and thickening of the equatorial cortex during cleavage.

Differences in stability of equator and poles

Prominent differences in the surface reactions of the polar and of the equatorial regions of cleaving sea urchin eggs exposed to various cytolytic and toxic agents have been described recently by Kuno (1954, 1957). The differences were already evident prior to the sinking in of the furrow. Although the reations observed are undoubtedly related to the differing properties of the various regions of the cortical layer, these experiments do not reveal the nature of these differences.

The relative weakness of the poles, as compared with the equator, immediately prior to the sinking in of the furrow was noted by Just (1922) for echinoderm eggs. Following immersion in hypotonic sea water, the eggs burst at the poles rather than at the equator. Since Just's results could have been due to regional differences in strength of the hyaline layer, we repeated his experiments using denuded *Arbacia* eggs developing in a calcium-free medium. When the dividing eggs are immersed in an hypotonic sodium and potassium chloride mixture (9 to 1), the eggs burst anywhere in the polar regions, but not at the site of the furrow. Similar results are obtained by compressing denuded eggs in calcium-free media isosmotic with sea water.

The stability of the cortical region, particularly at the equator, during the division of the *Arbacia* egg is strikingly demonstrated by tearing the polar surface of a naked, dividing egg in a solution of potassium chloride isosmotic with sea water (page 44). The absence of the coagulating effect of divalent cations in the medium permits the continued outflow and dissipation of the granular interior of the egg an appreciable time before the disintegration of the cortex. In spite of the ensuing destruction of the torn incipient blastomere, the cleavage furrow continues its advance. In the case illustrated in Figure 19.3 the constriction had not sufficiently ad-

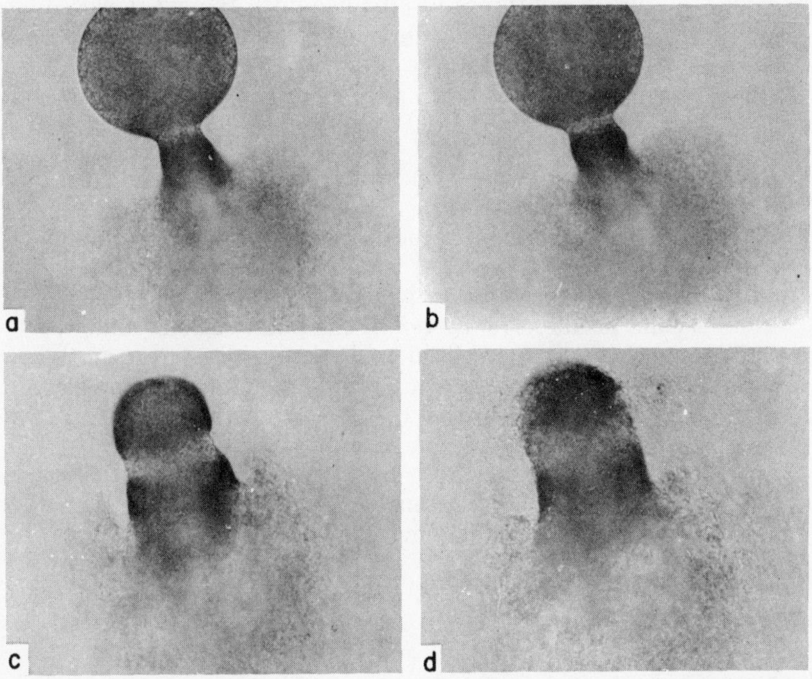

FIGURE 19.3. Polar end of one of two incipient blastomeres of naked living *Arbacia* egg torn in potassium chloride solution isosmotic with sea water. (a) Contents of torn blastomere dissipate while furrow persists. Outflow through open connection causes other blastomere to shrink. (b, c) Continued outflow and widening of constriction in floor of furrow. (d) Disintegration of cortex of shrunken blastomere and persisting wall at original floor of furrow. (From Chambers, 1938b.)

vanced to complete a separation between the two forming blastomeres. Consequently the contents of the untorn blastomere flowed out through the wide-open connecting stalk. The surface of the shrinking, untorn blastomere showed no evidence of wrinkling. The gelated cortex underlying the egg surface became increasingly condensed, and the brown pigment granules approached one another during the shrinkage of the blastomere, causing a visible darkening of the cortex. Finally, there occurred a disintegration of the entire surface of the egg, followed by a gradual breakdown and dissipation of the blastomeric remnant. It is significant that the most resistant part was that which constituted the original floor of the furrow, or the bottom of the annular trough which constricts the egg during cleavage (Figure 19.3d). Evidently the strongest part of the cortex is that which underlies the floor of the furrow.

If the daughter blastomere is torn fairly late during furrow formation, the furrow continues to advance until the gap is closed off, leaving a single intact blastomere with a superficial, temporary excrescence representing the remnant of the torn blastomere (Figure 19.4).

The greater stability of the walls of the furrow can be shown also by tearing the polar ends of both incipient blastomeres simultaneously. An *Arbacia* egg with a deep cleavage furrow is seized across the furrow with a "holding" needle while the two ends of the dumbbell-shaped cleaving egg are ruptured as simultaneously as possible with the tip of a second needle. The cortex in the polar regions collapses and, together with the fluid endoplasm, dissipates in the medium, while the equatorial zone persists for a time as a definitive, hollow, narrowing cylinder of cortical material.

If, on the other hand, the two opposite sides of the furrow are seized and simultaneously torn the entire egg disintegrates without the persistence of a cortical remnant. This is because the more stable part has been destroyed.

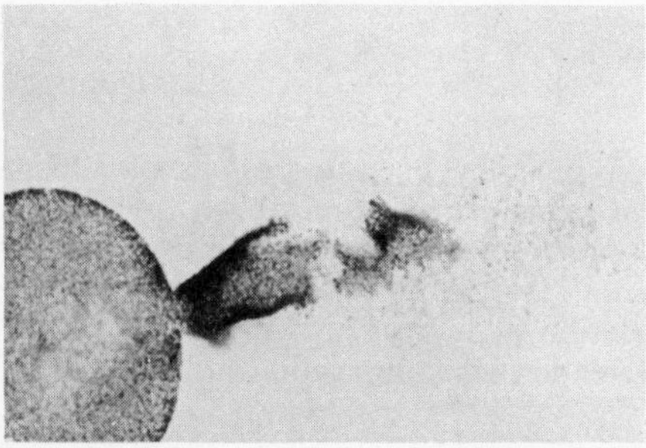

FIGURE 19.4. Same as Figure 19.3, with tearing done at later stage. Constricting stalk preserves one blastomere attached to remnant of other blastomere. (From Chambers, 1938b.)

Another interesting experiment showing regional differences in the properties of the cortex was performed by Marsland (1939). When a sea urchin (*Arbacia pustulosa*) egg is centrifuged as furrowing starts, the pigment granules are more easily dislodged from the cortical layer at the poles than at the equator. This leaves a fairly conspicuous pigment band girdling the equator of the centrifuged egg. The greater difficulty of dislodging pigment granules from the equator could be due either to greater thickness or to greater rigidity of the gel in this region.

As mentioned previously, Mitchison and Swann (1955) found no differences in surface stiffness at the polar and equatorial regions of dividing sea urchin eggs during early stages of cleavage. It should be noted that their micropipette method for testing surface deformation requires that a relatively large area be tested, amounting to about one quarter the total surface area of the ovum. Possibly, therefore, differentiation in the equatorial region was too localized to be detectable.

We have seen that many experiments on naked eggs indicate a greater stability of the cortex at the equator than at the poles. These results are consistent with Hiramoto's measurements, which show a thickening of the cortex at the equator during cleavage.

MOVEMENTS OF THE SURFACE LAYERS DURING CLEAVAGE

Movement of pigment granules imbedded in the cortex

In the fertilized, uncleaved egg the pigment granules are distributed uniformly at the periphery of the egg and imbedded in the cortical gel. As the naked egg cleaves in a calcium-free medium, the pigment granules tend to accumulate in the walls of the cleavage furrow, while becoming sparser at the poles.[6] Hiramoto (1957) has described a displacement of the accumulated pigment granules into the endoplasm at the edge of the advancing furrow, although in the walls of the furrow and elsewhere the granules remain in the cortical gel. After the furrow has cut completely through, the accumulated pigment granules disperse.[6]

The piling up of pigment granules in the walls of the advancing furrow of the denuded egg occurs coincidentally with a thickening and strengthening of the cortex in this region and a thinning and weakening at the poles. One possible explanation for these observations is that solated cortical material is transferred from the polar to the equatorial region, with gelation occurring at the latter site. Another possibility (which could occur together with the suggested sol–gel changes) is that the cortex thickens by contracting

[6] Dan and Dan (1940), Dan (1954a, b).

or shrinking circumferentially at the equator, while becoming stretched at the poles.

All the above observations on pigment granule movements, and the correlations described, refer to denuded eggs in calcium-free media. Motomura (1935) has pointed out that during cleavage in sea urchin eggs with intact and dense hyaline layers, in sea water, following an initial indentation of the cortex containing pigment granules, the further extension of the walls of the furrow is pigment-free. The same process is repeated in successive cleavages, the pigmented cortical cytoplasm remaining only at the surface in contact with the hyaline layer. According to Dan (1954a), the lack of pigment in the walls of the furrow of an egg with intact hyaline layer results from the fact that the cortical layer is held relatively immobile. After the initial contraction and bending in of the equatorial cortex, further extension of the furrow would be due to either (1) addition of new unpigmented gel in the walls of the furrow as the result of a sol–gel change, or (2) the stretching out of a narrow strip of the cortex which originally girdled the equator of the spherical egg.

Movement at the surface of the denuded egg

Katsuma Dan[7] has made a series of beautiful observations on the movement of minute particles stuck to the surface of denuded eggs developing in calcium-free sea water. From these movements he calculated the changes in surface area of different regions of the egg. From the beginning to the end of furrowing the surface at the equatorial region shrank in area, while the surface adjacent to the furrow as well as the polar region expanded in area. The expansion in surface area which must occur when the spherical egg divides into two spheres, therefore, is associated with an actual contraction in surface area of a band-shaped region encircling the equator.

The question arises to what extent the movements of particles adhering to the surface of denuded eggs in calcium-free sea water are indicative of movements in the cortical layer. A halo of delicate, radially directed fibrils surrounds the egg in calcium-free sea water. The particles are undoubtedly enmeshed in these strands, which are, presumably, protoplasmic extensions of the cortical

[7] The results given here are those of Dan and Ono (1954), based on computations of regional changes in surface area. Dan's work with kaolin particles starts with Dan, Yanagita, and Sugiyama (1937) and extends through a series of six papers, all of which are quoted in the 1954 article. In the earlier papers results are described in terms of changes in the linear distance between particles. Shrinkage at the equator *throughout* furrowing becomes evident only in the surface area calculations. After cleavage is completed, the particles are carried out of the furrow region, due to stretching of this surface (Dan, 1954a).

layer. The movement of the particles, therefore, should faithfully represent movements in the cortical layer.

The regional changes in surface area which Dan has calculated from the movement of adherent particles provide evidence that the cortex of the naked dividing egg shrinks circumferentially at the equatorial region, while stretching elsewhere. This picture of cortical behavior is consistent with the observed thickening of the equatorial cortex, its greater stability, and the accumulation of pigment granules at this site during cleavage of denuded eggs in a calcium-free medium.

Movement of the hyaline layer

When the dividing sea urchin egg, in sea water, elongates, and the furrow starts to sink in, the hyaline layer material tends to heap up and even wrinkle at the equator, while at the poles the layer becomes thinner (Gray, 1931). These changes suggest shrinkage of the protoplasmic surface at the equatorial region. Dan[7] also followed the movements of particles attached to the hyaline layer during cleavage. In the sea urchin egg with a well developed layer the movements were minimal and uniform over almost the entire surface, with some exceptions at the furrow region. Dan (1954a) explains the restricted movement of the particles (as compared with the extensive and regionally different movements at the surface of a denuded egg) as being due to the inelasticity of the hyaline layer. As we have noted previously, the hyaline layer is attached to the surface protoplasmic layers. It is easy to imagine that when the egg divides, the result of the attachment is to restrict cortical movement except in a narrow region girdling the equator. In the case of an egg with a thin hyaline layer, however, the movements observed in this extraneous coat, and of pigment granules imbedded in the cortex of the egg, are essentially similar.[7]

Movement in the subcortical cytoplasm

As stated previously, a thin layer of fluid, subcortical endoplasm intervenes between the gelated cortex peripherally and the gelated asters internally even at the height of development of the amphiaster. By observing the paths taken by individual granules as the egg elongates and cleaves, currents of flow have been detected in the subcortical cytoplasm extending from the two poles, around the gelated asters, and into the equator (Chambers, 1938b). This flow occurs in the same direction as the movements of pigment granules in the overlying gelated cortex and of particles attached to the external surface of denuded eggs.

The subcortical flow may be only secondary to movements occurring in the cortex, such as a contractile process in the equatorial region. On the other hand, the vortical currents may play an essential role in building up the wall of the advancing furrow by carrying into the equatorial region fluid cytoplasm which is then gelated. Furthermore, the cortical and surface shrinkage at the equator might be interpreted as due to a "dragging effect" induced by the subcortical currents.

STRUCTURAL AND KINETIC FEATURES OF THE DIVIDING EGG REVEALED BY MICRO–OPERATIONS

Cutting

The cutting process consists of pressing an horizontally placed fine shaft of a microneedle up against the egg, which lies in contact with the under surface of the coverslip (Figure 19.5). The operation must be done with extreme care, since any shearing movement will cause delicately poised gelated structures to revert to the sol state, upon which cleavage stops. Successful cutting experiments can be carried out only on selected types of eggs which have a high resistance to mechanical injury and in which the movements of the internal structures preparatory to cleavage are not easily upset. In this regard Yatsu (1908) was particularly fortunate in his selection of the egg of the nemertine worm *Cerebratulus*. Chambers (1919a) used the egg of the starfish *Asterias*. The eggs were cut in various directions just as the cleavage furrow was beginning.

One of the earliest and yet one of the most convincing demonstrations of the independence of the furrowing process itself from the mitotic figure is the following micro-operative experiment of Yatsu (1908). A *Cerebratulus* egg was cut at right angles to the cleavage plane and to one side of the center of the egg, just where

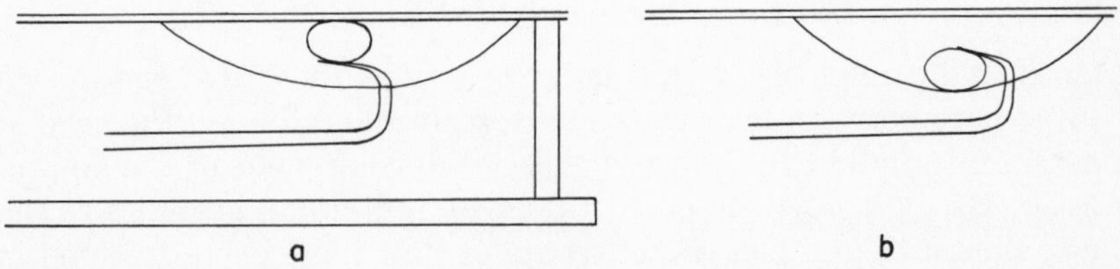

FIGURE 19.5. Methods for cutting an egg in two. (a) Side view of moist chamber magnified to show needle in position with its limb so placed as to compress egg between it and coverslip. Continued pressure of needle cuts egg in two. (b) Cutting egg by bringing end limb of needle down on egg so as to press egg against lower surface of hanging drop.

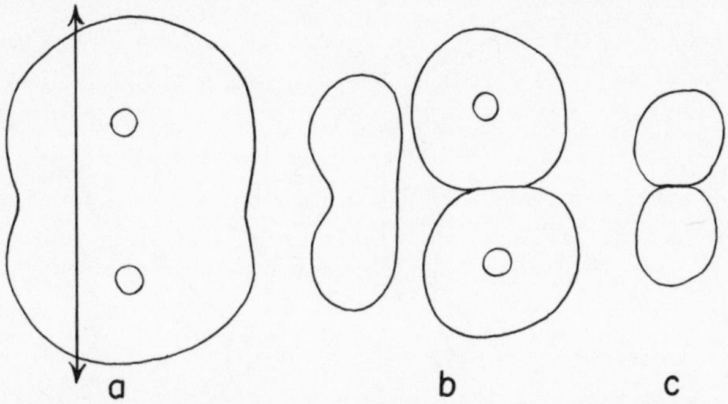

FIGURE 19.6. Segmenting *Cerebratulus* egg cut through to one side of long axis. (Asters not shown.) (a) Arrow shows direction of cut. (b) Original cleavage furrow completes its course in nucleated fragment (right) at same time that it persists in non-nucleated fragment (left). (c) Furrow finally cuts through non-nucleated fragment. (From Yatsu, 1908.)

the cleavage furrow had begun to sink in (Figure 19.6a). The original furrow persisted and completed its course in both fragments, even though no asters were present in the non-nucleated fragment (Figure 19.6b, c).

The independence of the furrowing process from the late mitotic figure can also be shown by cutting an *Asterias* egg, just beginning to cleave, in a plane diagonal to the cleavage furrow (Figures 19.7a, 19.8a). The diagonally transected furrow persists, partially segregating each fragment into a larger segment containing a prominent polar aster and a smaller segment in which the astral rays, cut away from their respective centers, fade from view (Figures 19.7b, 19.8b). The furrow continues to deepen along its original plane until each fragment becomes divided into a small non-nucleated and a larger nucleated fragment (Figures 19.7c, 19.8c). The nucleated fragments continue to cleave while the non-nucleated pieces lie as inert masses within the fertilization membrane (Figures 19.7d, 19.8d). The persistence of the furrow in the transected fragments in both these and Yatsu's preceding experiments is indicative of the gelated state of the equatorial cortex. The interior of the equatorial region is fluid at the time the furrow sinks through (page 212).

The polar ends of dividing eggs can be cut away without impairing the cleavage process. Yatsu (1908) made the interesting observation that if the polar region of the *Cerebratulus* egg is cut away at or prior to metaphase, the resulting blastomeres are of

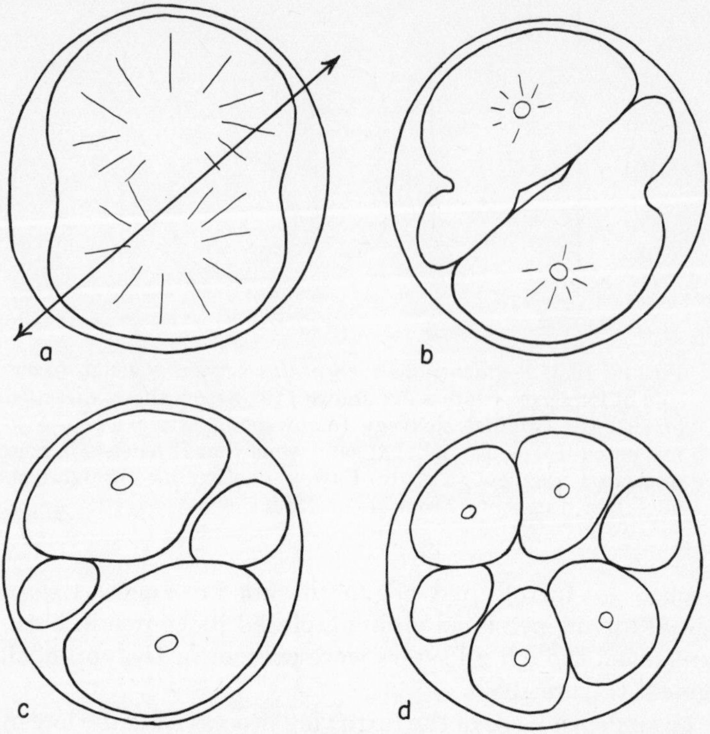

FIGURE 19.7. *Asterias* egg, beginning to segment, cut through diagonally. Cut has not disturbed physical state of egg. (a) Arrow shows direction of cut. (b) 5 minutes later. Cleavage furrow has persisted and is deepening in original plane. (c) 25 minutes after cutting. Non-nucleated fragments have pinched off. (d) 45 minutes after cutting. At time of second cleavage nucleated fragments cleaved. (From Chambers, 1919a.)

equal size. On the other hand, if the polar part is cut away immediately prior to or during cleavage, the original furrow persists and continues deepening to divide the egg into two blastomeres of *unequal* size (Figure 19.9). The same results were subsequently obtained by Chambers using *Asterias* eggs. An entire incipient blastomere can be destroyed by tearing with the microneedle, yet the furrow continues its advance (Figure 19.4; Chambers, 1938b).

Similarly, Mitchison (1953) has shown that nearly all of one incipient blastomere can be cut away from a dividing sea urchin egg without stopping the advance of the furrow at its original site.

Inducing solation of gelated structures

The rigid wall of the cleavage furrow can be solated and the furrow made to disappear when the furrow region is mechanically agitated

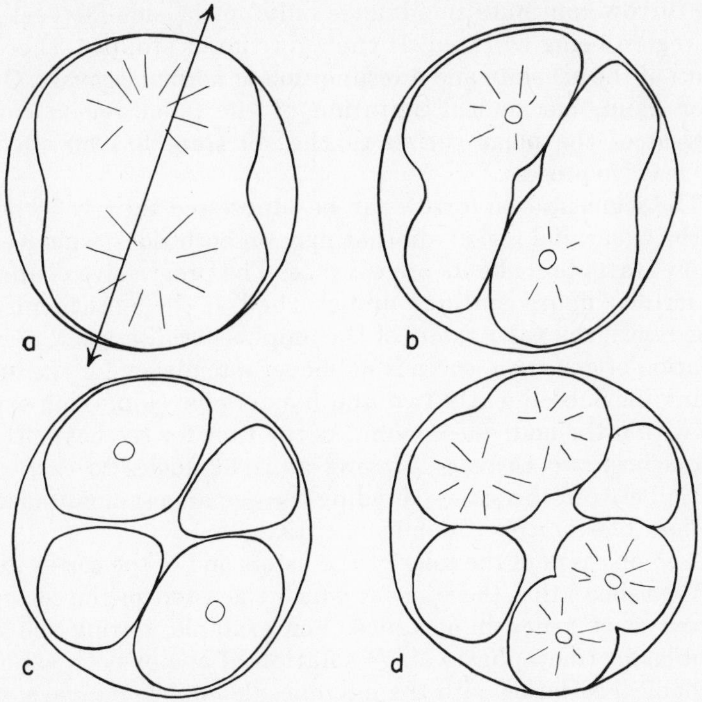

FIGURE 19.8. Same as Figure 19.7, except that diagonal cut is more nearly perpendicular to cleavage plane, with result that larger non-nucleated fragments are pinched off by furrow. (a) Arrow shows direction of cut. (b) Furrow deepens. (c) Non-nucleated fragments have pinched off. (d) Nucleated fragments continue to cleave. (From Chambers, 1919a.)

by puncturing it with a microneedle and moving the needle back and forth. If this operation is confined to one side of the dividing egg, only the agitated region of the furrow flattens out, and, if the agitation is continued, the cleavage of the egg becomes one-sided.

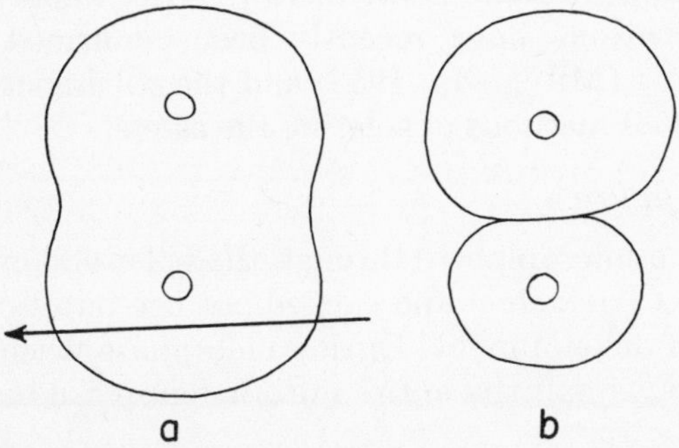

FIGURE 19.9. Segmenting *Cerebratulus* egg with one end cut off. (Asters not shown.) (a) Arrow shows direction of cut. (b) Original furrow has completed its course, resulting in two unequal blastomeres. Non-nucleated fragment not shown. (From Yatsu, 1908.)

The furrow continues to advance only on the side farthest from the region being agitated. If the agitation is stopped, there is a return of the gel state and a resumption of normal cleavage. On the other hand, mechanical agitation of the polar region causing reversal of the polar cortex to the sol state has no effect on the cleavage process.

The segmentation furrow can be suppressed entirely by agitating the equatorial region alternatingly on both sides, wherever the furrow starts to make its appearance. The progressive changes in the mitotic figure continue undisturbed. If the agitation is continued until the polar asters of the amphiaster disappear, then, on cessation of agitation, there is no longer a tendency for the furrow to sink in. Subsequently two amphiasters develop, lying side by side within the undivided ovum. As the time for the next cleavage approaches, two cleavage furrows at right angles to each other sink in between the asters, dividing the egg almost simultaneously into four blastomeres (Chambers, 1919a).

A comparison of the roles of the asters and of the cortex during cleavage shows that the stage at which the asters or the cortex are solated is of prime importance. For example, during the early amphiaster (metaphase) stage solation of one or both asters by mechanical agitation with the microneedle inhibits cleavage; while solation of the cortex alone, as by tearing at it and drawing out strands of protoplasm from it with a needle, has no effect. On the other hand, at the time the egg starts to elongate prior to cleavage, or as cleavage occurs, solation of the asters by mechanical agitation does not prevent completion of the furrow, while solation of the cortical layer in the equatorial region immediately suppresses cleavage (Chambers, 1949a).

Solation of the asters can also be induced by exposing the eggs to colchicine (Beams and Evans, 1940). As is to be expected from the experiments with the microneedle, colchicine treatment just prior to or during cleavage, resulting in disappearance of the asters, does not prevent the sinking through of the furrow. These important experimental observations have recently been confirmed, using both the microneedle (Mitchison, 1953) and the colchicine (Swann and Mitchison, 1953) methods of solating the asters.

Sucking out the mitotic figure

Hiramoto (1956) inserted a micropipette through the polar end of eggs (of the heart urchin *Clypeaster*) and sucked out the mitotic figure at various stages of development. During metaphase when the amphiaster is relatively small the entire mitotic figure can be removed. Although neither the spindle nor the asters re-form, nonetheless a cleavage furrow is initiated at its original presumptive site in some eggs. The furrowing process, however, is not carried through to completion.

When the operation is carried out just prior to or during cleavage, the spindle and central parts of the asters can be sucked out. Furrowing continues, although slowed, to completion and at its original site, cutting across the stretched rays remaining from one of the polar asters (Figure 19.10). Furthermore, it was found that a considerable portion of the protoplasm can be removed, so that the cell surface shrinks, without altering the position of the cleavage plane.

Displacing the mitotic figure by centrifugal force

When the mitotic figure is displaced by centrifugal force (Harvey, 1935) just prior to cleavage, the furrow cuts in at its original presumptive site, irrespective of the new position of the spindle. If the figure is displaced at earlier periods, up to early metaphase, two furrows may appear: one at the original presumptive site, and the other at right angles to the displaced spindle. The latter is evidently induced by the post-centrifugal position of the mitotic figure. Finally, if the mitotic figure is displaced very early, at prophase, the furrow always develops at right angles to the new position of the figure. These data indicate that the position of the

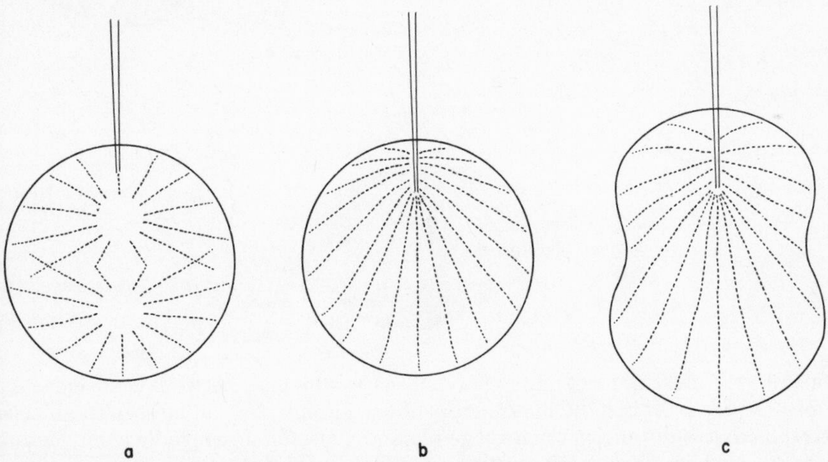

FIGURE 19.10. Cleavage of sea urchin egg when spindle is sucked out using micropipette. (a) Before removal of spindle. (b) After removal. (c) Furrow appears at predetermined position. (From Hiramoto, 1956.)

furrow is determined by the early mitotic figure, but that once this position has been established furrowing can take place independently of the mitotic figure.

Stretching the cleaving egg

Stretching a dividing *Arbacia* egg at right angles to the cleavage plane, by inserting microneedles through opposite sides of an egg and then separating the needles, does not impede cell division (Figure 19.11a). The stretched egg strives to resume its former ellipsoidal shape prior to continuing cleavage even if it has been stretched into a long ribbon.

Similarly, when the egg is stretched diagonally to the cleavage plane just before furrowing starts, the stretching does not alter the plane of the cleavage furrow (Figure 19.11b). However, when the needles are thrust through the walls of the incipient furrow and the egg is stretched parallel to the plane of the furrow, cleavage is completely stopped (Figure 19.11c).

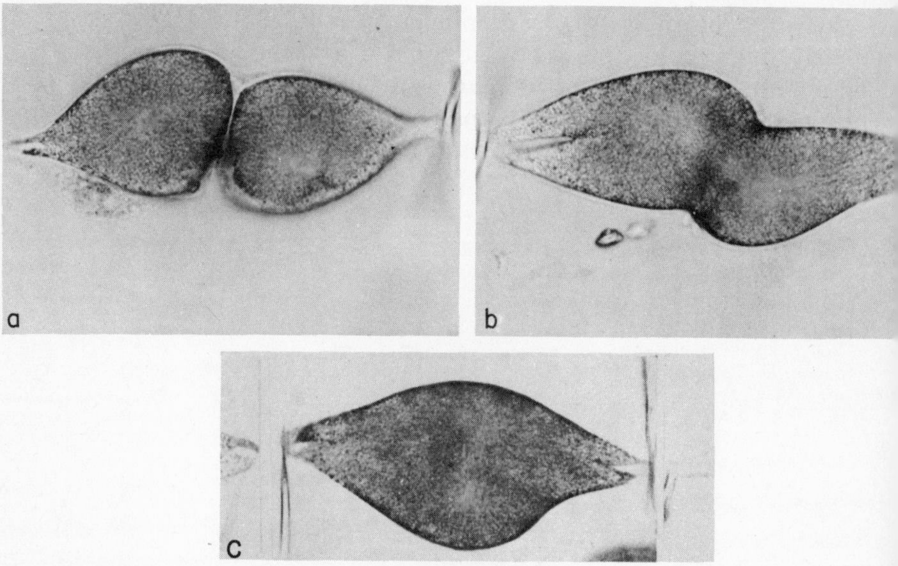

FIGURE 19.11. Effect of stretching cleaving sea urchin eggs between microneedles. clear zone at the center of the polar asters of the amphiaster can be seen in all cases. Egg stretched at right angles to cleavage plane (longitudinal to spindle axis). Protopl flows away from needles, with partial rounding-up of egg as cleavage continues. When stretched diagonally to cleavage plane, egg divides at right angles to mit spindle although diagonal to long axis of stretched egg. (c) When stretched paralle cleavage plane (right angle to spindle) division is prevented. (From Chambers, 193

Compressing the cleaving egg

Eggs with their hyaline layers intact, in normal sea water, divide readily even when considerably compressed, irrespective of whether they are compressed at right angles or parallel to their furrows. Evidently the more liquid inner part of the hyaline layer material, drawn into the advancing furrow, keeps the protoplasmic surface films of the opposing furrow surfaces from coming into direct contact.

> This feature is indicated in the knife-edge-like type of furrow formation seen in the sand dollar egg compressed between slide and coverslip (Figure 5.1). The furrow always starts as a pit with noncontiguous walls. With the deepening of the furrow, the walls, just behind the advancing tip, come into intimate contact and leave a tear-drop-shaped space at the extreme tip of the advancing furrow. The presence of this space ultimately results in the formation of a narrow bridge of appreciable length connecting the two blastomeres.
>
> Hyaline layer material is carried inward within the tear-drop-shaped space, coating the protoplasmic surfaces of the furrow walls and preventing their coalescence.

For thoroughly denuded eggs developing in calcium- and magnesium-free media, even moderate compression parallel to the furrow inhibits cleavage. For example, eggs lying in a groove transverse to the long axis of the mitotic figure will not divide if the groove is sufficiently narrow to prevent karyokinetic lengthening. The furrow attempts to sink in, but reverts back. The failure to achieve cell division evidently results from the coalescence of newly formed surfaces when these are closely appressed and lack a coating of extraneous material, rather than from any mechanical factor.

CONCLUSIONS[8]

Experimental studies have revealed that the actual sinking in and cutting through of the furrow occur independently of the late mitotic figure. This is clearly shown in the micro-operative experiments on dividing eggs by the continued advance of the furrow after the egg has been cut at right angles or diagonal to the cleavage plane, after the mitotic figure has been solated by mechanical agitation, and after the amphiaster has been displaced by centrif-

[8] Throughout this section the final results listed are experiments described in detail in the body of the chapter.

ugal force or removed by suction with a micropipette. The mitotic figure can even be displaced or eliminated shortly prior to cleavage without preventing furrowing or altering its presumptive location.

It is true that removal or displacement of the mitotic figure during or just prior to cleavage delays or renders irregular the furrowing process, but we cannot at this time state whether this is due to the micro-operative insult or to elimination of some influence exerted by the mitotic figure.

While the furrowing process occurs independently of the late mitotic figure, there is no question that the ultimate location and timing of cleavage depends on influences emanating from the early mitotic figure. This had already been indicated by Yatsu in his cutting experiments. If he cut away the polar end of the eggs prior to or at metaphase, the presumptive site of the cleavage furrow was altered so that two equal blastomeres resulted. If the same operation was carried out just prior to cleavage, however, the original site of the furrow remained unchanged, and two unequal blastomeres were formed. As is well known, the cleavage plane is located at right angles to the long axis of the developing spindle. Evidently the mitotic figure exerts the determinative influence responsible for this relationship comparatively early, so that by the time the later stages of the mitotic figure are attained, the differentiation in the cortical region responsible for the furrowing process has already been well established. Thus, solation or removal of the early mitotic figure inhibits cleavage, and displacement of the prophase mitotic figure by centrifugation displaces the original presumptive site of the cleavage furrow. Centrifugal displacement at later stages is without effect.

Another important contribution of the experimental studies on cell division is the demonstration that the sinking in of the furrow depends on factors located in the neighborhood of the furrow itself. This is clearly shown by the continuation of the furrowing process at its original site when fragments are cut from the equatorial region of the egg, when the polar portion of the dividing egg is cut away or destroyed, when the cortical region at the poles is liquefied by mechanical agitation, and when the egg is stretched longitudinally or diagonally between microneedles.

Since furrowing occurs at its original site after removal, displacement, or solation of the late mitotic figure, and since the active process is located at the equator, the gelated cortex in this region must play the essential role. Additional support for this conclusion is the observation that cleavage is inhibited when the cortical gel at the equator is solated by mechanical agitation, while solation at the polar region is without effect. Furthermore, even if the interior of the equatorial region is filled with oil, cleavage still occurs, dividing the droplet in two. In this latter experiment the only remaining "functional" part of the equatorial region is the cortex.

The essentiality of the gelated state for the furrowing process has been fully demonstrated by Marsland (1957) in experiments based on the solating effect of high pressures on protoplasmic gels. He has shown a proportional relationship between the amount of hydrostatic pressure needed, on the one hand to inhibit cleavage, and on the other to decrease the strength of the cortical gel as measured by the centrifugal force required to displace the cortical pigment granules. The prevention of furrowing by high pressures is undoubtedly due to solation of the cortical gel at the equator, in view of the abundant evidence that the cortex at this region plays a primary role in cleavage.

In view of the primary role of the equatorial cortex in the furrowing process, the observed increase in thickness and the greater stability of the cortex in this region during cleavage of naked eggs are of peculiar interest. As discussed previously, we believe the most reliable measurements of cortical thickness during the cleavage cycle have been obtained by Hiramoto (1957), using his blunt microneedle method. He describes a thickening of the cortex at the equatorial region as furrowing begins and a thinning of the cortex at the poles. These changes are corroborated by Dan's observations on the changes in thickness of the outer, or extragranular, zone of the cortex during cleavage. The greater stability, or the greater rigidity, of the equatorial cortex could be due partly if not entirely to its greater thickness as compared with the polar cortex.

A plausible explanation for the observed relationship between cortical changes and furrowing in denuded eggs is that the cortex at the equatorial region contracts circumferentially and consequently thickens at the expense of the polar cortex. In denuded eggs the accumulation of pigment granules in the cortex at the equator, and Dan's demonstration that the surface area of the equatorial region shrinks during furrowing, support this explanation. If furrowing were due to a contracting and thickening equatorial band, we might expect that the blastomeres would remain attached by a broad gelated stalk. Under normal circumstances, however, this does not occur, and the furrow advances until the blastomeres are connected only by a tenuous thread. This indicates that the cortical gel at the tip of the advancing furrow must ultimately undergo dispersal. Evidence that such a gel–sol reversal occurs is Hiramoto's (1957) observation that pigment granules are displaced from the inner part of the cortex into the endoplasm at the tip of the advancing furrow. Marsland (1957) explains the furrowing process as a contraction of the equatorial cortex followed by solation of the contracted material at the tip of the furrow.

Another possibility is that the advance of the cleavage furrow is effected by an inward advance of gelation from the cortical region at the equator, involving the reversal to the gel state of solated cortical cytoplasm carried to the equator by subcortical currents emanating from the poles (Chambers, 1938b, 1951). Solation of the cortical gel at the poles and addition of new cortical material at the equator could explain the regional differences in distribution of pigment granules and in thickness of the cortex observed in the naked egg. Conceivably the areal shrinkage of the equatorial surface of the denuded egg in calcium-free media could be accounted for by a "dragging effect" exerted by the vortical currents. In view, however, of the considerable rigidity of the gel, it is probable that a contraction of the equatorial cortex is largely responsible for its shrinkage in surface area.

In all likelihood both contraction of the gelated cortex and an inward growth of gel, the two processes localized at the equatorial region, are involved in furrow formation. Possibly in the denuded egg, where the floor of the furrow is broad throughout cleavage, contraction of the cortex at the equator is the principal mechanism. On the other hand, in an egg with a dense hyaline layer or in a compressed egg, in which the furrow is of the knife-edge-like type, ingrowth of gel at the equator may play a major role in furrow formation.

20

Cell Division in Cells other than Echinoderm Eggs

CELL DIVISION IN ANIMAL CELLS WITHOUT YOLK

Cleavage in animal cells containing little or no yolk, such as somatic cells in tissue culture or spermatocytes, closely resembles that in the echinoderm egg.

The rounded form of the cell about to divide is a characteristic feature of animal cells, and this change in form coincides with the conversion of its nucleus into the prophase stage. For example, the flattened blastomeres of the sea urchin blastula round up prior to division. In the fresh water ameba about to divide the extended pseudopodia are withdrawn, and the resulting rounded cell becomes studded with numerous short pseudopodia. An interesting example is the fibroblast: When division is about to occur, the extended processes of the fibroblast become released from their extraneous attachments and are withdrawn; the result is a spherical body of cytoplasmic material containing a centrally located cell nucleus.

Among the irregularly shaped epithelial cells of, for example, a sheet of intestinal mucosa, it is generally possible to detect a cell about to undergo cleavage by its rounding up and thereby deforming its more plastic, interkinetic neighbors. The rounding-up process suggests that the intercellular cement which binds the cells together becomes loosened prior to cell division. The cement is an extraneous, closely applied coat secreted by the cells and depends for its maintenance upon the presence of calcium in the medium. The cement can be weakened either by a deficiency of calcium in the medium or by a lowering of the pH (Chapter 4). The latter can occur by the liberation of large amounts of carbon dioxide in the immediate vicinity of cells preparing for cleavage.

The evidence indicates that in many different types of animal cells, as in the dividing echinoderm egg, the cortex at the poles of the cell is weaker, or less rigid, than the cortex in the equatorial region.

In the spermatocytes of the grasshopper (*Dissosteira*) the difference can be detected by manipulation with microdissecting needles. The cortex in the region of the furrow of the dividing cell offers considerable resistance to distortion. On the other hand, merely touching the surface at the poles tends to cause the formation of blebs which rise and fall, indicating a lack of rigidity at the polar surfaces. Possibly the polar expansions observed by Roberts (1955) in the neuroblasts of the grasshopper embryo at the beginning of cell division were the result of a weakening of the cortex at the poles.

A spontaneously appearing difference of surface conditions at the poles and the equator of the dividing cell can be detected in speeded-up motion pictures of tissue cultures showing cells undergoing division. When the cells become rounded, prior to division, it is difficult to determine where the well-known bubbling (Strangeways, 1924) first appears. Frequently, however, fibroblasts begin to undergo mitosis before the protoplasmic strands extending from their poles are completely retracted. The polar axis of the cell is then well defined and, moreover, the strands tend to anchor the forming daughter cells so that the cleavage furrow becomes longer than usual and, therefore, better visible (Figure 20.1). Blebs, or hyaline protrusions, appear first on the strands beyond the poles of the fibroblast (Figure 20.1a–c). Bleb formation then spreads to the polar regions and progressively over the surfaces of the two incipient daughter cells, but never develops in the floor of the cleavage furrow (Figure 20.1d–f). Both Strangeways (1924) and Burrows (1927) have called attention to the absence of blebs in the narrowing equatorial zone of the advancing furrow.

A remarkable demonstration of the semirigidity of the cleavage furrow was observed in tissue culture in which an epithelial cell was undergoing division on the margin of a sheet of epithelium. Figure 20.2 shows several successive stages of the dividing cell taken from a motion picture. A flow of material, with a reversal, occurred three times from one daughter cell into the other, the polar regions of each cell swelling and shrinking in correspondence with the direction of the flow. During the reciprocal flow the furrow was maintained and continued to advance until eventually division was completed at the time that the two daughter cells were approximately equal in size. After the division, the two cells

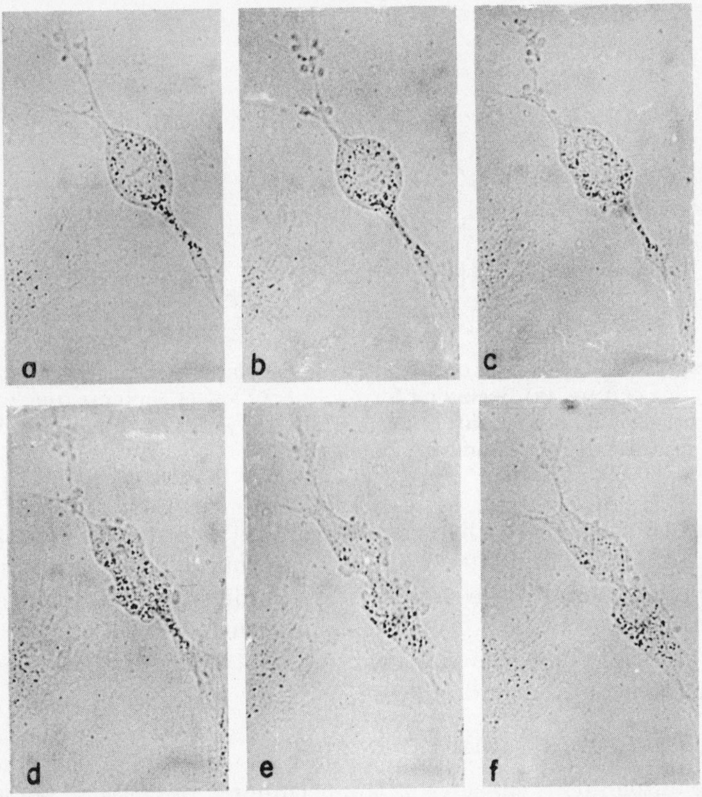

FIGURE 20.1. Chick fibroblast undergoing division in tissue culture. (a–c) Blebs rise and fall on retracting strands in upper part of figure. (d–f) Main body of cell with blebs starting on polar surfaces and spreading over cell except in floor of furrow. (From Chambers, 1938b.)

flattened out and became indistinguishable from their neighbors in the epithelial sheet. Burrows (1927) described a similar phenomenon in a tissue culture cell which was about to divide into two unequal parts. He noted a flow of protoplasm across the equator from the larger to the smaller cell until the two became equal in size. Fission was then completed. Furthermore, it has been shown that in the dividing ameba the fluid endoplasm courses alternately toward one or the other of the daughter cells through the narrowing waist between them.[1]

That the actual cutting through of the furrow is a process largely independent of the spindle and chromosomes, as in the

[1] Dawson, Kessler, and Silberstein (1935); Chalkley (1935, 1951).

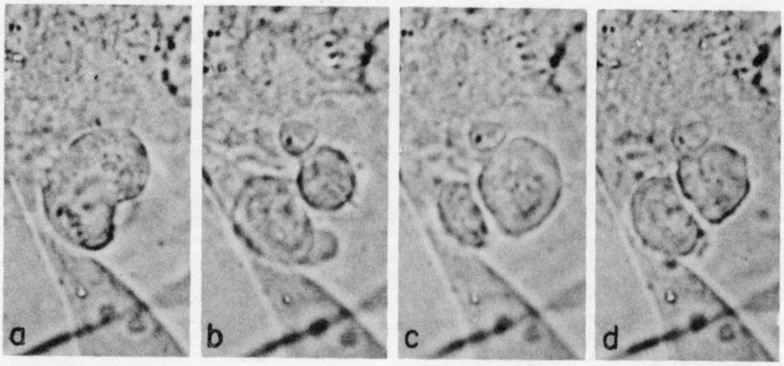

FIGURE 20.2. Four stages of a dividing human epithelial cell (adenocarcinoma) in tissue culture. (a) Beginning furrow. (b, c) Flow of contents from one into the other daughter cell. (d) Completed fission, with two cells approximately equal in size. (From Chambers, 1938b.)

echinoderm egg, has been shown for the neuroblast of the grasshopper by Carlson (1952). Using a microneedle he pushed all the chromosomes and the spindle of the middle anaphase figure over to one pole of the cell. Nonetheless, the furrow formed in the usual position, dividing the cell into one part containing all, and the other none, of the chromosomes.

CELL DIVISION IN YOLK-LADEN EGGS

The autonomous capacity of the peripheral, or cortical, portion of the equatorial region of the dividing cell to form a furrow, clearly shown for several different types of cells (Chapter 19), has also been demonstrated in yolk-laden eggs. A striking series of cutting experiments was performed by Yatsu (1910) on the *Ctenophore* (*Beroë*) egg. Cleavage in this egg is unequal, the furrow sinking in from the animal pole between the two eccentrically placed telophase nuclei. After the furrow had progressed about one third the way through, he separated the nucleated portion from the non-nucleated by a cut at right angles to the cleavage plane, leaving the tip of the advancing furrow intact in the non-nucleated fragment. The furrow continued its advance unimpaired (Figure 20.3). Similar results were obtained by Waddington (1952) for the newt egg. This investigator isolated the furrow region from the interior of the egg and from other cortical areas by inserting cellophane strips between cortex and endoplasm or by cutting into the egg. Despite this, the furrow continued its advance.

The most thorough studies on cell division in yolk-laden eggs are those of Selman and Waddington (1955) carried out on the

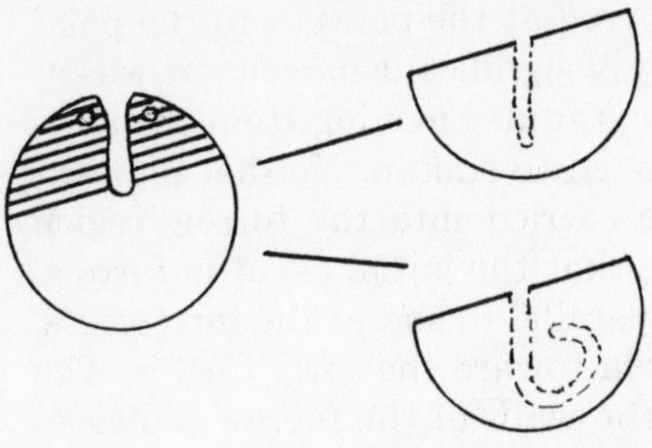

FIGURE 20.3. (Left) cleaving ctenophore egg. Nuclei in upper part. Shaded area shows region transected. (Right, upper) after transection, furrow continues through to eventual completion in non-nucleated fragment. (Right, lower) occasionally after transection furrow spirals back upon itself in non-nucleated fragment. (From Yatsu, 1910.)

newt egg. Like the echinoderm egg, the amphibian egg possesses a definitive gelated cortex containing imbedded pigment granules. The considerable rigidity of this gel layer can readily be demonstrated by attempting to displace the granules with a microneedle or by centrifugal force. The cortex possesses plastic and contractile properties (Holtfreter, 1943). Changes in "rigidity" of the surface layers have been described, based on the amount of sucking force needed to deform the surface of the egg (Selman and Waddington, 1955). An increase in "rigidity" is observed prior to cleavage, attains a maximum as furrowing begins, and thereafter falls off. These results resemble those obtained on the sea urchin egg using a similar method (page 236). Precisely what mechanical property is being measured by this method is not clear. An increase in thickness of the cortex as cleavage approaches might explain the results, but no measurements of width of the peripheral gel layer are available for the various stages of the cycle or for different regions of the egg. The "rigidity" of the surface layers was found to be the same at polar and equatorial regions, as well as at the newly formed furrow surfaces.

As has been generally observed in cells about to divide, the newt egg tends to round up as cleavage approaches. This is particularly conspicious in the denuded egg, which, lying on the bottom of a dish, has a flattened, oval shape. As cleavage approaches the width diminishes and the height increases, reaching a maximum at the moment the furrow begins to sink in. Thereafter, the egg flattens again. The rounding-up process represents a lifting up of the egg contents against gravity and a decrease in surface area. The most probable explanation for this is an increased tension exerted by the cortex on the fluid egg interior.

Movements of the cortex were examined (Selman and Waddington, 1955) by plotting the successive positions of cortical pigment granules or vitally stained areas of the cortex. No significant movements or areal changes occurred at the poles or to the polar sides of the furrow region. The only significant movement was at the site of the future furrow, the granules moving from the sides toward the early furrow as it started to sink in. Neither pigment granules nor stained cortex were carried into the furrow region itself. These observations indicate that the initial event in furrowing is a contractile process acting parallel to and at the furrow site, which pulls the pigment granules in toward the early furrow. The complete absence of pigment in the walls of the furrow indicates that the cortex in this region must have arisen de novo.

Sections through the equatorial region of the newt egg, taken shortly after the dipping in of the furrow, reveal a sheet of yolk-free cytoplasm extending for some distance down from the tip of the furrow, in the cleavage plane. The gel nature of this sheet is inferred from the observation that prior centrifugation caused it to bend.

The extension of the gelated sheet in the presumptive furrow region must occur by a sol–gel transformation at its advancing margin. This implies the presence of subcortical cytoplasmic currents which transfer solated cortical material deep into the equatorial region. This is the very mechanism of furrow formation which Chambers (1937) proposed for the echinoderm egg: "Advance of the wall of the furrow must be considered as a growth phenomenon—material being added progressively to the gelated cortex of the furrow is analogous to the apposition of material along the plane of division of a plant cell." Owing to the opacity of the amphibian egg, the actual presence of currents of flow cannot be checked by visual observation. A gel–sol transformation at the cortex of the egg is inferred (Selman and Waddington, 1955) from the progressive decrease in rigidity of the surface layers (suggesting a decrease in thickness of the cortex) from the time the cleavage furrow first appeared.

Evidently cell division in the newt egg, as indicated also for the echinoderm egg, involves two processes localized at the equatorial region: contraction of the cortex at the equator and inward extension of gel at the furrow tip.

VIII

Micromanipulation

INTRODUCTION TO PART VIII. Protoplasm has been a somewhat hypothetical concept: the physical basis of life, the mysterious complex which constitutes living cells. To ascertain its intimate structure much has been done by analytical methods which involve, at best, destroying protoplasm and examining its disintegrated and isolated remains. A method which retains protoplasm as a living unit is that of the micrurgist. To him protoplasm is a real thing. He can operate on it, dissect it, inject it with substances, and maintain it in a healthy, living condition during his operations. Truly remarkable is the readiness with which micro-operations can be performed on, and solutions injected into, living cells without causing visible indications of injury.

21

The Micromanipulative Method

Micromanipulative studies on the physical nature of protoplasm involve operations performed in the field of the highest available magnifications of the compound microscope. The precision of performance essential to micro-operations on living cells requires that all movements of the microtips, even the finest appreciable to the eye under the highest magnifications, be completely under the control of the operator. The operations are performed in an hanging drop suspended from a coverglass which roofs a moist chamber (Figure 21.1) of adequate height to permit the insertion of the horizontal shafts of the micro-operating instruments. The ends of the needles or pipettes are upturned so that their tips project into the hanging drop. In this way the operations are all performed from below. There being no obstacle above the coverglass, high magnifications can be used.

THE INSTRUMENTS OF MICRURGY

Micrurgy was rendered possible by the adoption of a technique used by M. A. Barber (1904, 1911), bacteriologist at the University of Kansas and later at the Rockefeller Institute in New York. Barber, intent on devising a method for isolating individual micro-organisms, used on the stage of the microscope a moist chamber high enough to permit the insertion into the chamber of the horizontal shafts of micropipettes. By their manipulation, operations could be performed on cells in hanging drops suspended from a coverglass roofing the chamber.

Barber's instrument, called a pipette holder, was based on the principle of a metal block pushed along a groove by a screw. By having a series of three blocks built on one another and each traveling at right angles to each other, movements in any of three directions could be imparted to a needle or pipette clamped to the top block. Barber used micropipettes which depended on capillarity for picking up, and also for depositing, a bacterium. He utilized the fact that the insertion of the pipette tip into a deep drop of fluid caused an inflow, while the insertion of the same tip into a shallower drop caused an outflow.

Later, Barber (1911) developed a method for microinjecting known amounts of fluid into living cells by filling pipettes with a column of mercury. By cooling or heating the mercury, fluid could be sucked into, or ejected from, the micropipette.

The micropipettes Barber used were capillary glass tubing, drawn out in a microflame to rapidly tapering points of extreme fineness. The shaftlets were bent in the flame so that the tips projected upward into the hanging drop of the medium in which the cells to be operated upon were contained.

In 1912 George Lester Kite, a medical student at the University of Chicago, went to Kansas at the instigation of A. P. Mathews, then also of Chicago, to learn the technique developed by Barber. Mathews recognized the possibilities offered by applying Barber's method to the microdissection of living cells. Kite later went to Woods Hole, Massachusetts, the mecca of summer investigators on the Atlantic coast. There he published several preliminary articles[1] which initiated the field of micrurgy.

The term micrurgy (micros = small; ergon = work, Gr.) was introduced by Tibor Péterfi, formerly of Budapest. Péterfi (1923) developed the technique and his instrument, the Péterfi-Zeiss micromanipulator, while he was at the Kaiser Wilhelm Institute, Berlin.

Barber's original purpose was the isolation of individual bacteria. He needed visibility clear enough merely to identify by their outlines the organisms suspended in hanging drops. Hence he was content with the standard substage condenser, which has a working distance equivalent to the thickness of a glass slide. The micrurgist, interested in the fine structure and intimate reactions of living protoplasm, requires the best possible illumination and accuracy of resolution for his micromanipulations. These can be approached by using a substage condenser with a long focal distance, which throws a cone of light high enough so that its apex reaches the undersurface of the coverslip roofing the moist chamber (Figure 21.1).

The Barber pipette holder had a number of deficiencies which introduced much free play. In spite of careful machining, its large

[1] Kite and Chambers (1912); Kite (1913).

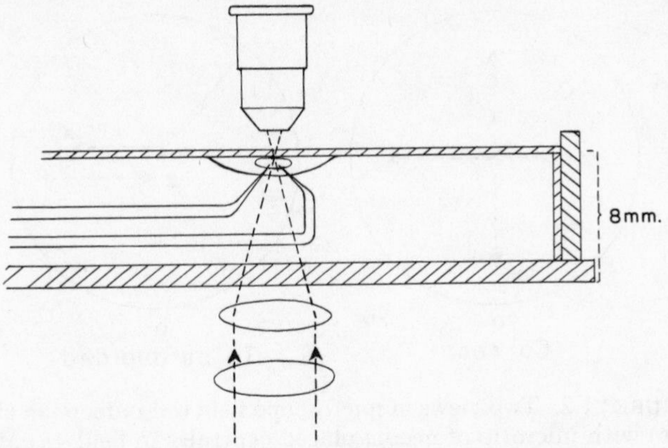

FIGURE 21.1. Side view of moist chamber with objective above and substage condenser having 8 mm. focal length below. Light rays from illuminating source (interrupted lines) are focussed on cells in hanging drop. Two microneedles project into open end of chamber, with their tips in hanging drop suspended from coverslip which roofs chamber.

frictional surfaces (as is true for all machines of this kind) made it impossible to achieve the needed accuracy of movements. Kite's recognition of the possibilities of micro-operations on protoplasm resulted in the development of several types of instruments with greatly improved devices for micromanipulation.[2]

In evaluating the efficiency of micromanipulation it has been generally considered important that the movements be in straight lines at right angles to one another. However, the range and right-angled motions of the tips of the microneedles and pipettes are actually of less consequence than the accuracy of performance of the microtips over short distances. This latter should be the aim of the operator.

Accurate operation under high magnification and within the restricted area of a cell of microscopic dimensions requires that the movements imparted to the microneedles conform to the following conditions. The needles must be mounted so as to be easily maneuverable and perfectly steady. The tips of the microneedles or pipettes must travel true. It must be possible to start and stop them at will with no lag and, when the movement is reversed, to have the tips return along the exact paths they took during their

[2] Chambers (1929a); Chambers and Kopac (1950); de Fonbrune (1949); Kopac (1955, 1959).

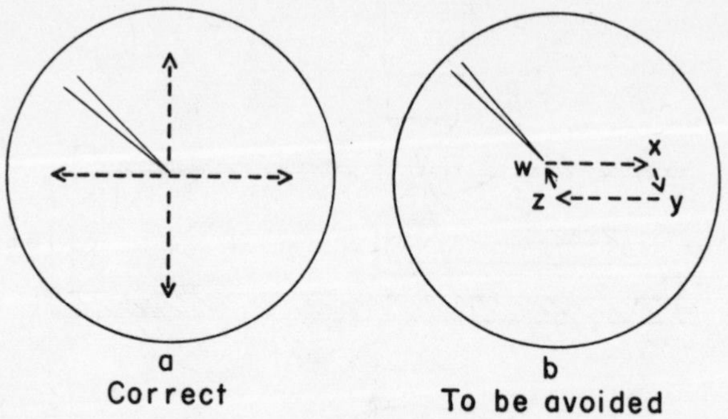

FIGURE 21.2. Two views of microscope field (oil immersion objective) with microtip of needle placed centrally in field. (a) Movements observed with properly constructed micromanipulator. Microtip travels, in one plane, back and forth and side to side with no deviation when direction is changed. (b) Movements of microtip in improperly constructed manipulator. Deviation, x to y and z to w, occurs when direction of movement is changed.

forward motion (Figure 21.2). In horizontal movements the needles must always remain in the same plane, even in the field of an oil immersion objective. In vertical movements there should be no lateral deviation. Every movement, even the slightest which is appreciable to the eye under the compound microscope, must be completely under the control of the operator.

An instrument designed to fulfill these conditions is the Chambers micromanipulator (Figures 21.3, 21.4). The feature of this machine is that friction is minimized because the fine movements depend upon the spreading apart, by micrometer screws, of bars of rigid metal fastened at their ends by resilient metal spring hinges. When the screws are reversed, the bars return to their original positions owing to the spring action at the ends of the bars. Vertical movement is produced by the spread of two bars mounted vertically (Figure 21.4b).

One of the vertical bars is an immovable pillar upon which the other parts are suspended (Figure 21.4b). By this means the microneedles are moved in arcs of circles having radii of 3 to 4 inches. Since the useful range of the movements under the microscope is less than 0.5 mm., the curvature of the arcs is not appreciable when the needles are properly positioned. The movements obtained by this machine are even and smooth, since the metal springs at the ends of the bars are always exerting a positive tension against the machined tips of the micrometer screws. The two horizontal movements at right angles to each other are produced by connecting three bars at their ends with springs to form a Z-like figure (Figure 21.4a).

The microinjection apparatus is used with a micromanipulator to inject aqueous or nonaqueous fluids and suspensions into the cytoplasm, vacuoles, or nuclei of living cells. Chambers' microinjec-

FIGURE 21.3. Chambers micromanipulator and microinjection setup, Gamma Instrument Company, New York.

(a–d) Microinjection setup: (a) syringe; (b) clamp to hold flexible metal tubing (c); (d) micropipette holder connected by adapter to metal tubing (c).

(e–h) Coarse adjustments: (e) screw to fasten holder; (f) controls side-to-side movements; (g) controls height; (h) controls tilt.

(i–k) Fine adjustments: (i) controls vertical movements; (j) controls back-and-forth movements; (k) controls side-to-side movements.

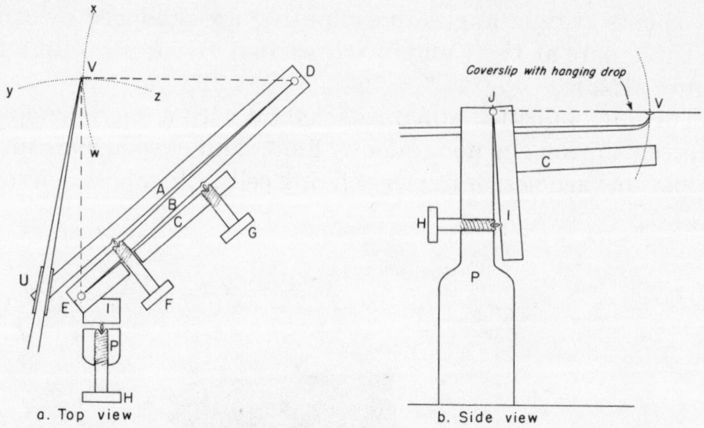

FIGURE 21.4. Diagram showing operative procedure of Chambers micromanipulator. Top and side views.

(a) Top view. Microneedle fastened to the free end of horizontal bar A at U. Needle holder is so adjusted that microtip V is at apex of right-angled triangle (interrupted lines). Base of this triangle is straight line joining springs D and E, which fasten together bars A and B, B and C, respectively. Turning screw F separates bars A and B at D and imparts a back-and-forth arc movement to needle tip V along dotted line W to X. Turning screw G separates bars B and C at E and imparts side-to-side arc movement to microtip V along dotted line Y to Z. Note: horizontal arc movements at V are at right angles to each other. Micromanipulator is set up so that microtip V is at center of the microscope field and at apex of right-angled triangle VED.

(b) Side view. Movement in vertical plane is produced by screw H which is threaded in pillar P. Tip of screw H abuts against vertical bar I, which is connected by spring J to pillar. Horizontal bar C is fastened to vertical bar I. Turning screw H lifts whole combination (bars A, B, C, and I) and imparts an arc movement in vertical plane to tip of needle at V. To procure a movement in vertical part of the arc (dotted line), tip of needle V must be in same horizontal plane with spring J. In practice, level of spring J should be at level of coverslip which roofs moist chamber.

Note: screws F, G, H correspond to screws j, k, i, respectively, of Figure 21.3.

tion apparatus (Figures 21.3, 21.5) consists of a 2 ml. Luer-type syringe connected by means of a metal adapter to a long, flexible metal or plastic capillary tube. This connects, by means of a second adapter, with the pipette holder and pipette. The entire system, except for a small amount of air in the pipette, contains water. The extraordinarily delicate control of the volume of fluid deliverable by this means is due to the great resistance to flow of liquid through the narrow aperture of about 0.5 micron at the microtip.

In addition to the Chambers micromanipulator, with which

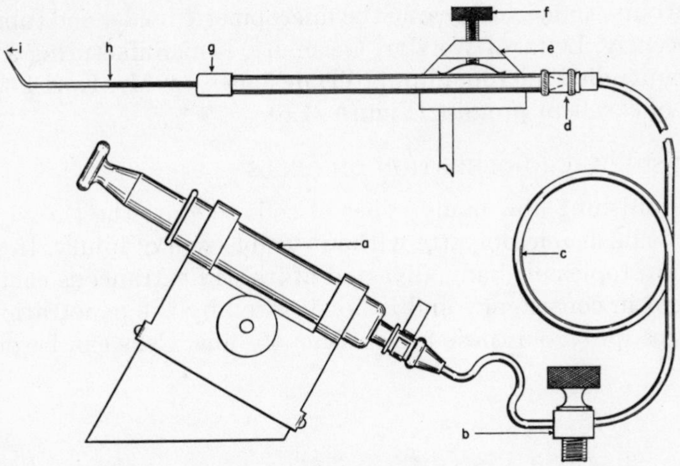

FIGURE 21.5. Microinjection apparatus. (a) Syringe (2 or 2½ ml. Luer) in clamp which is fastened to base of micromanipulator. (b) Clamping screw for fastening tubing (c) to base of micromanipulator. (c) Flexible brass tubing long enough (3 feet) to ensure complete flexibility and nontransmission of vibrations of micropipette. (d) Adapter connecting flexible tubing to pipette holder. (e) Bracket, set into micromanipulator, for pipette holder. (f) Screw for fastening pipette holder, with spring clamp. (g) Screw cap of pipette holder. (h) Pipette shaft filled to (h) with water. (i) Microtip.

almost all of the micrurgical work described in this book was done, there are available several other useful types. In de Fonbrune's (1949) pneumatic micromanipulator,[3] operation of a single control lever pneumatically transmits motion in all planes to the microneedle in the field of the microscope. The specific purpose de Fonbrune had in designing his machine was to get away from the necessity of decomposing three dimensional movements into the rotating of three different screws, one for each plane. The transfer of nuclei from one ameba to another was first successfully accomplished by Commandon and de Fonbrune (1939). It is possible that one reason for their success is the high maneuverability of the de Fonbrune machine in the performance of rapid two- and three-dimensional movements. Investigators who have subsequently carried out nuclear transfer studies in the ameba have used the de Fonbrune machine (this chapter, footnote 6). Microinjections can also be done using either Chambers' microinjection apparatus or de Fonbrune's modification. (The loading capacity of the instru-

[3] Manufactured by Baudouin, 1 et 3 rue Rataud, Paris 5e; also A. S. Aloe Company, St. Louis 12, Missouri.

ment is quite sufficient to carry the micropipette holder and tubing.)

Recently, Leitz, of Wetzlar, Germany, is manufacturing a new lever-controlled micromanipulator, designed by Mr. Karl Frischmann, of excellent promise (Figure 21.6).

EFFECTS OF MICRO-OPERATION ON CELLS

It is astonishing how many types of cells tolerate the thrust of a microneedle or micropipette without visible sign of injury. In general, a protoplasmic body, divested of its stiff extraneous coats, is semifluid in consistency and is not injured by the penetration of microtips, provided these are sufficiently fine. Cells can be punc-

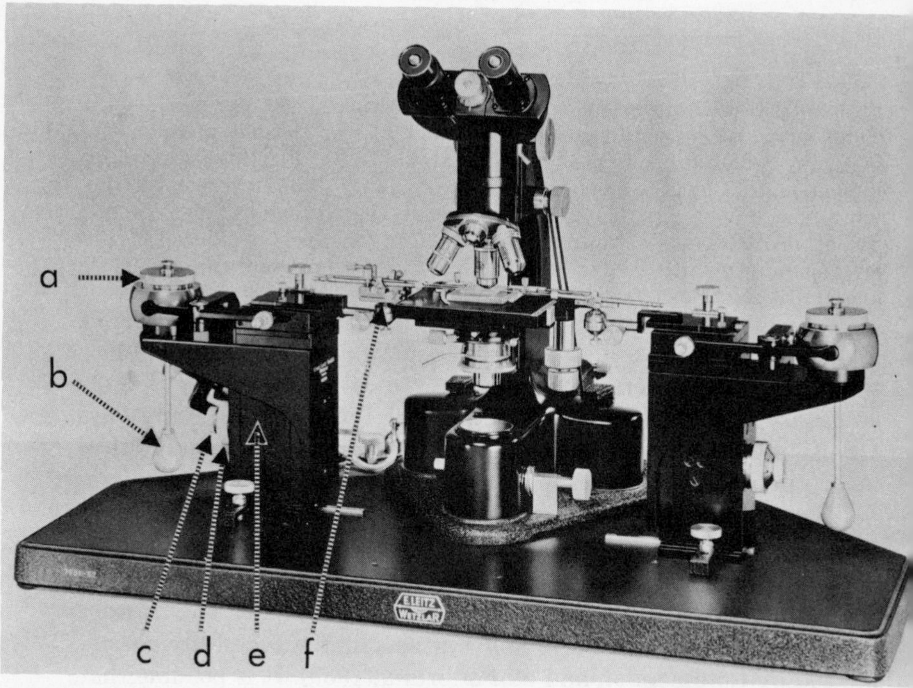

FIGURE 21.6. The new Leitz micromanipulator, E. Leitz, Inc. Two micromanipulat units are mounted on either side of microscope, and two microinstruments are show entering lateral openings of moist chamber placed on microscope stage. Major comp nents are listed below for left-handed unit of micromanipulator: (a) control for adjusti transmission ratio of lever actuated horizontal movements; (b) lever for actuating fi movements in all horizontal coordinates; (c) drive for fine vertical motion; (d) drive f coarse vertical motion; (e) pillar of micromanipulator; (f) ball and socket joint for coar adjustments in all coordinates (four screws controlling coarse horizontal movements i various different directions are not labelled).

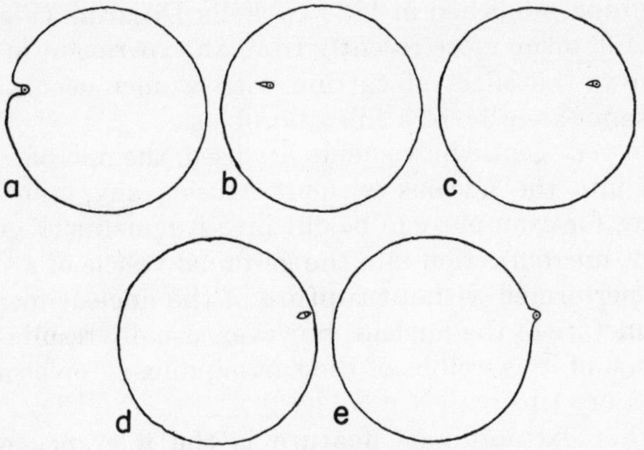

FIGURE 21.7. Successive stages in passage of vertically placed microneedle through cytoplasm of living mature unfertilized echinoderm egg from one side (a) to other (e), with no observable effect other than closure of puncture.

tured and torn repeatedly as long as the successive movements of the needle are performed slowly. Sudden movements may cause disruption of the surface film surrounding the protoplasm, initiating disintegration. This may be localized or general. It is the immediate repair of the surface film upon which the life of any protoplasmic unit absolutely depends. Slowness of the micro-operative movements permits the repair of any break or rupture of the film. For example, the vertically adjusted microneedle can be inserted into an echinoderm egg, and, provided the movement is smooth, the needle can be drawn through the interior of the egg from one side to the other with no observable effect other than a closing of the punctured hole as the needle is moved on (Figure 21.7).

The microtip can be moved from place to place within the cell without altering the normal streaming movements or upsetting the normal sol–gel relations. For example, the microneedle can be inserted into the cortical regions of a cell or into the astral rays of the amphiaster of an echinoderm egg without causing solation of these gelled structures. Solation occurs only when the tip of the needle is moved about rapidly.

The plasticity of protoplasm, in general, is very great. Cells can be readily cut into viable fragments, which tend to round up. This is shown for a protozoan cell, a ciliate, in an old figure taken from a monograph published in 1835 by Félix Dujardin (Figure 21.8).[4] Figure 21.9, taken more recently from an experiment in tissue culture, shows the effect of cutting with a microneedle through a cytoplasmic extension of a chick fibroblast.

When very gentle movements are used, the microneedle can be inserted into the nucleus without causing any injury, and the nucleolus, for example, can be cut into fragments (Figure 21.10). Similarly, microinjection into the germinal vesicle of a starfish egg may be performed without rupture of the nuclear membrane. A rapid puncture of the nucleus, however, usually results in the disintegration of it as well as of the surrounding cytoplasmic constituents (Figure 1.15).

Another extraordinary feature is the tolerance many cells exhibit to the microinjection of a number of different aqueous solutions which do not coagulate proteins. These solutions readily

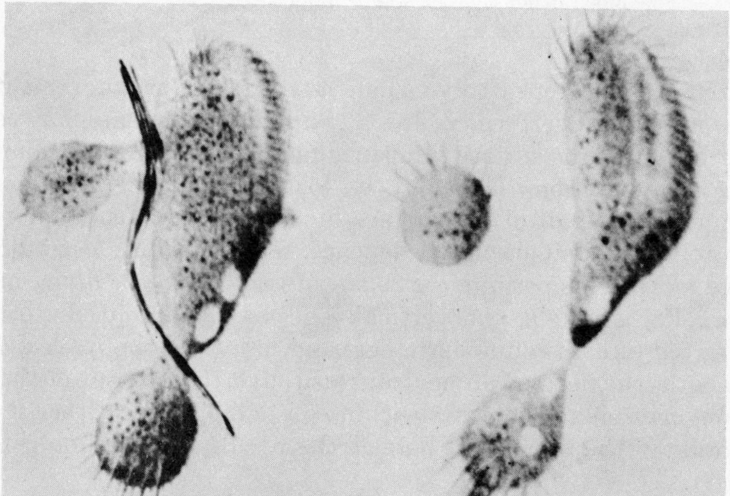

FIGURE 21.8. Micro-operation performed in 1835 by Dujardin, probably among the first to transect a living protozoan. Cotton fibers were placed in drop of water containing several organisms, and drop was covered with coverglass. When one of swimming ciliates moved under a fiber, pressure was exerted on coverslip with the result that the fiber, lying over organism as shown in (a), was forced down sufficiently to pinch off fragment from organism (b).

[4] From observations of the kind depicted here, Dujardin (1835) developed his famous description of sarcode, the semifluid material composing the bodies of protozoa, as a "substance glutinous, perfectly homogenous, elastic, contractile, diaphanous . . . ."

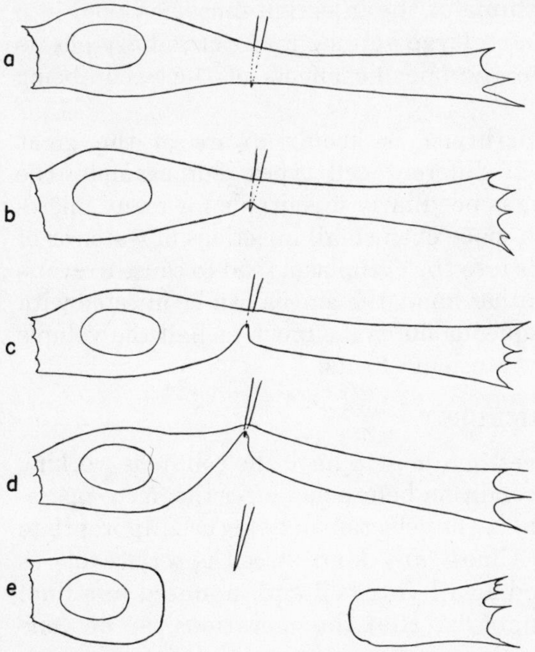

FIGURE 21.9. (a–e) Cutting off extended cytoplasmic process of epithelial cell in tissue culture. Note elasticity of extended cytoplasmic process as indicated by retraction of cut ends in (e). (From Chambers and Fell, 1931.)

spread through the protoplasmic interior. For successful microinjection experiments, the operator must be able to control the delivery of amounts of solution which are minute with reference to the volume of the cell being injected. This can be readily accomplished by using micropipettes with a sufficiently fine bore (bore at tip 0.5 to 0.3 micron in diameter). Controlling the rate of delivery is important because the reaction of the protoplasm to variations

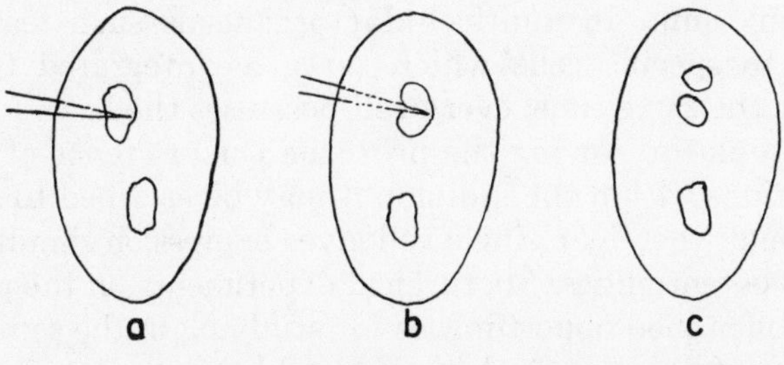

FIGURE 21.10. Manipulation of one of nucleoli of fibroblast nucleus in tissue culture. Only nucleus shown. Needle inserted gradually so that no injury occurs. One of two nucleoli transected. (From Chambers and Fell, 1931.)

in the mere rate and volume of the injection must be taken into account. A rapid rate and a large volume may per se have effects out of all proportion to the specific effects of the agent being injected.

The experienced micrurgist is keenly aware of the great variability in response of different cell types. For example, the *Arbacia* (sea urchin) egg is peculiarly unsuitable for many micromanipulative procedures, since even small injections of water or of monovalent salt solutions into the cytoplasm tend to cause irreversible breakdown. On the other hand, the ameba can be injected with large amounts of such aqueous fluids (as much as half the volume of the ameba) with only transient effects.

CELLS AMENABLE TO MICRURGY

The chief aim of the investigator is to have the cell he is working with living and in good condition before and after the micro-operation. Size is not a problem in the selection of living cells appropriate for micromanipulation. Almost any kind of cell is serviceable as long as it can be isolated as a living cell and mounted in a fluid medium in a moist chamber so that the operations can be done under the high magnifications of the microscope. It is important, however, for the micrurgist to realize that different cell types, or the same cell type in different species, vary greatly in their susceptibility to micro-operative procedures. Micrurgical studies on protoplasm have been made on cells which fall within three main categories: protozoa, plant cells, and cells of the metazoa.

The protozoa

Some of the protozoa are highly differentiated, and many contain specialized organellae. These serve as skeletal, contractile, assimilatory, or nerve-like structures, all within the body of the same protoplasmic unit. In multicellular organisms such features are allocated to specific cells which serve as integrated functional groups. At the same time, every cell possesses the basic features of living protoplasm. Among the protozoa the existence of localized specializations within the individual may be ascribed to an evolutionary trend whereby a single cell gives expression simultaneously to various potentialities. Micrurgical experiments on the protozoan organism offer good opportunities for studying in the same cell the many functions that protoplasm is capable of.

A frequent feature in protozoa is the presence within the protoplasm of temporary vacuoles possessing various metabolic functions. Some serve as gastric, others as excretory, vacuoles. In some protozoa the cytoplasm is so highly vacuolated that when micro-injections are made the microtip of the pipette always enters a vacuole. An example is *Actinosphaerium* (Figure 21.11), which in reality is a colonial aggregate containing many nuclei distributed throughout a frothy cytoplasm. In one group of protozoa, the *Suctoria,* there are intraprotoplasmic vacuoles which serve as "brood pouches." During the reproductive period of this organism, when nuclear division is to take place, the "brood pouch" takes form as a cytoplasmic vacuole adjacent to the cell nucleus. Into this vacuole an offshoot of the nucleus protrudes, surrounded by a small amount of cytoplasm, which is pinched off. The nucleated offshoot within the vacuole develops into a ciliated, single-cell off-spring which is eventually liberated when the "brood pouch" opens out to the exterior of the parent organism.

The structural complexities met with among the protozoa give indications of the multiplicity of reactions which can occur within a single body of protoplasm without the need of cellular partitions. For those concerned with the phenomena of gelation and solation in relation to ameboid movement, good examples for study are to be found among the *Rhizopoda.* In these, of which the fresh water ameba is a familiar form, morphological differentiation is at a

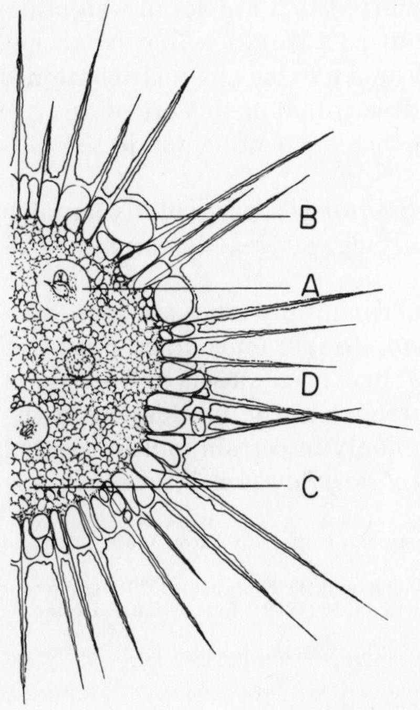

FIGURE 21.11. *Actinosphaerium eichhorni.* This is a multinucleated, highly vacuolated unicellular organism. Its surface is adhesive, and individuals readily fuse on contact. Gastric vacuole (A); contractile vacuole (B); cytoplasmic vacuoles (C); intervacuolar cytoplasm (D). (After Howland, 1930.)

minimum.[5] In some cases, as in *Amoeba dubia,* even the enveloping pellicle is so tenuous as to be imperceptible. The weakness of the external coat makes *A. dubia* ideal for microinjection studies, since the micropipette readily penetrates the protoplasmic interior. This species should prove suitable for nuclear transfer studies,[6] since the tenuous pellicle offers negligible resistance to the nucleus' being pushed across membranes (and coats) of adjacent amebae. Other types of amebae, like *Amoeba proteus,* have relatively firm pellicles, and these individuals are more difficult to microinject. The pellicle of *Amoeba verrucosa* is very tough (Figure 2.2), and several rhizopods are enclosed in coats stiffened with silicious or calcareous material. This shell possesses one or several apertures through which ameboid processes of the protoplasmic body can be extruded, as in *Difflugia.* The latter organism can form filament-like, gelled pseudopodia which, when stimulated with the microneedle, exhibit prominent contractile properties.

Excellent material for microinjection studies, especially with relation to sol–gel transformations, is the streaming, multinucleated plasmodium of the slime mold *Physarum.* The plasmodium does not possess a pellicle, but is surrounded by a jelly coat, which the micropipette readily penetrates. Small pieces can be cut away from a large plasmodial mass, and the surface film immediately regenerates at the cut edge. Transferred to a droplet of water on a coverslip, the fragment spreads out as a thin, fan-like expansion. The protoplasm streams in numerous narrow channels contained within gelated borders.[7] Further description of this organism with reference to its use in the study of the effects of microinjected salts is to be found on page 137.

Among the ciliated protozoa, *Stentor* is a particularly suitable organism, since the repair of the surface very readily occurs following cutting or tearing operations.

Interesting possibilities for experimental studies are those protozoan organisms which can be made to fuse on contact. Fusion of contiguous protoplasmic surfaces has been observed mostly in cases when this is a natural occurrence in the life history of the organism. Fusion occurs very commonly in certain multinucleated organisms such as the heliozoan, *Actinosphaerium.* Whenever these

[5] An excellent mass culture method for *A. proteus* is that of Prescott (1956); it is also applicable to the culture of *A. dubia.*

[6] The nuclear transfer studies of Commandon and de Fonbrune (1939), Lorch and Danielli (1950), Danielli et al. (1955), Danielli (1958, 1959), were performed using *A. sphaeronucleus, A. proteus,* and *A. discoides.*

[7] The reader is specially referred to the articles by A. R. Moore (1935, 1945), also Winer and Moore (1941).

organisms touch, either of their own accord or by being pushed together with needles, they immediately fuse at their surfaces of contact. The same phenomenon occurs with the plasmodia of the myxomycetes in their ameboid stage. In this regard it is to be specially noted that pseudopodia of two different species of plasmodia have never been seen to coalesce. In *Pelomyxa,* the giant multinucleated ameba, a micro-operative procedure is required to achieve fusion of two organisms or of previously cut apart portions of an individual. The fusion is accomplished by approximating the organisms and tearing across the intervening membranes with microneedles.[8]

Plant cells

Most plant cells are characterized by being enclosed in a more or less rigid cellulose wall and by possessing a large vacuole located centrally within the body of protoplasm. The protoplasmic body, the protoplast, is morphologically well defined and consists of a fine, granular, actively streaming, fluid body containing the cell nucleus. As the cell grows in size, the enlarging vacuole expands the protoplast until it is converted into a thin layer lying against the inner surface of the wall. The cellulose wall is a secretion product of the underlying protoplast and is so intimately associated with it that whenever the protoplast becomes separated from the cellulose, the wall deteriorates and becomes brittle. This occurs when plant tissues are exposed to freezing and to agents which induce plasmolysis, or shrinkage, of the protoplast away from its cellulose wall. The relatively large size of many plant cells is due mainly to the dimensions of the vacuole of sap.

The cellulose wall is an obstacle to micromanipulative studies. However, the walls of young unicellular root hairs or the hairs of the unopened buds of many young flowering water plants are thin and can be readily punctured (Figure 13.4). In the case of cells with rigid walls, one end can be cut away and the microneedles inserted through the open end (Figure 21.12).

> Naked, living protoplasts can be obtained from plant cells by the following method. Thin slices of the epidermis of an onion bulb scale are immersed in a plasmolyzing solution of 1.0 M sucrose until the protoplasts have shrunk well away from their enclosing walls of cellulose. The sheets of the epidermis are then cut with a sharp knife at right angles

[8] Okada (1930). Daniels (1952, 1958) in interesting studies has combined this procedure with differential centrifugation of the intraprotoplasmic components of irradiated and nonirradiated organisms.

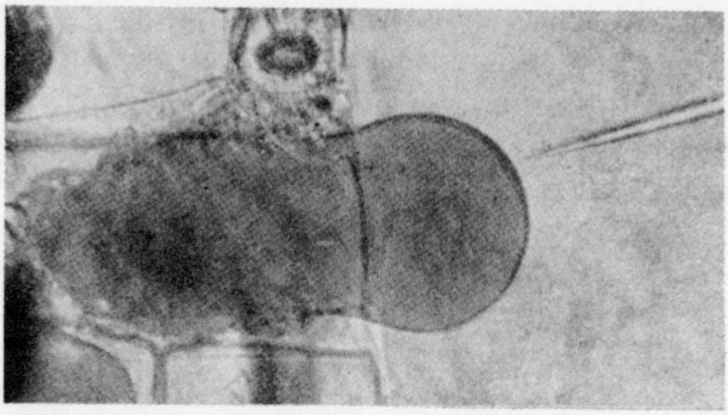

FIGURE 21.12. Bulb scale of *Allium cepa* immersed in solution containing 1.2 M sucrose and cut transversely with razor. Plasmolyzed, healthy protoplast partly protrudes through cut-open end of epidermal cell. (From Chambers and Höfler, 1931.)

to the long axis of the oval plasmolyzed cells in the sheet. By this means many cells along the edge have their end walls cut away, leaving their plasmolyzed protoplasts exposed and still intact. The strips of the epidermis are then mounted in a shallow hanging drop of the sucrose solution in the micrurgical moist chamber. Micro-operations can be done on the protoplast, the cytoplasm still streaming, by inserting the tips of microneedles and pipettes through the cut-open end of the cell wall (Figure 21.12). Moreover, tissues prepared in this way give access, in the neighborhood of the cut, to unopened cells, the noncuticularized walls of which can be readily punctured. Frequently, isolated protoplasts are found floating in the medium. These have "fallen out" through the cut-open ends. An interesting feature of these preparations is that the protoplasm of the protoplast can be dispersed, while the integrity of the central vacuole and its bounding membrane, the tonoplast, is still retained (Chambers and Höfler, 1931).

Cells of the metazoa

Within the body of the multicellular animal cells exist either isolated (like the various types of blood cells and the cells of the connective tissue) or grouped in closely adherent columns (like glands) or sheets (like epithelial or endothelial cells). In general, the protoplasmic portion of these animal cells, except for dispersed vacuoles and minute granules, occupies almost the entire volume of the cell. Actually, the animal cell includes one or more extraneous coats which are closely adherent to the surface of the protoplasm. These coats have been a source of confusion in attempts at determining the true surface of protoplasm. It is the protoplasmic surface film upon which the physiologic integrity of the protoplasm depends.

Objects favorable for experimental study are cells which can be obtained and collected in quantity as isolated, living cells, such as the ova of echinoderms and of other marine invertebrates. These are exceptionally valuable for a variety of micrurgical studies on the physical nature and other properties of protoplasm. In the full-grown state echinoderm eggs are spherical and average, in different species, from 75 to 150 micra in diameter. When obtained from a single ripe female, they are uniform in size. On being inseminated, they undergo cleavage and further development. They contain a relatively small amount of dispersed yolk. The specific gravity of the yolk bodies in these eggs is only slightly different from that of the cytoplasmic matrix. This and the continual streaming movements inherent in living protoplasm keep the yolk bodies evenly distributed. The inclusions can be conveniently segregated within the egg interior by centrifugation. When the cell is ready for cleavage, the nucleus occupies a relatively central position, and the early cleavages are symmetrical. The ova of the sand dollar and of the sea urchin *Lytechinus* are exceptionally favorable for micrurgical study because of their high translucency.

Micro-operative procedures frequently require the prior removal of the extraneous coats which envelop the eggs, using the methods described in Chapter 2. This is to ensure the penetration of the microinstruments into the cell interior. Attempts, for example, to insert a micropipette into a cell whose vitelline or fertilization membrane is intact frequently accomplishes nothing more than an inpocketing of this extraneous coat, pushed ahead of the microtip as it is advanced.

The eggs of different species of invertebrates vary greatly in their susceptibility to micro-operative procedures. For example, the cytoplasm of the mature *Arbacia* egg is peculiarly intolerant to the microinjection of aqueous solutions and quite susceptible to puncture by the microneedle. On the other hand, the cytoplasm of the immature ovum of the starfish *Asterias* is extraordinarily resistant to these same micro-operative procedures. The developing ovum of the worm *Cerebratulus* has proved especially valuable in micro-operative studies on cell division, since this egg is peculiarly resistant to solation of gelled structures in response to mechanical trauma, a minimum amount of which is necessarily involved in even the most delicately performed micro-operations.

Giant nerve fibers dissected out of crustacean or cephalopod tissues, and striated muscle fibers of the frog or other species, are, of course, excellent material. A good feature of the frog muscle fiber is that if in the course of the micro-operations injury occurs, this becomes immediately evident by the appearance of longitudinal striations in the interior.

Other highly desirable materials for micrurgy are the outgrowing cellular sheets in tissue cultures of explanted fragments of tissues from both plants and animals. The cultures may be of any fragment of tissue, aseptically excised from the body of an animal, preferably of young animals or their embryonic forms. The culture medium of animal cells usually consists of a sterile clot of blood plasma. Outgrowths of the cells spread through the medium from the edges of fragments of endothelial or epithelial tissues as sheets, one cell thick. An example of this in the case of the epithelial cells of the proximal tubules of the kidney is described on page 192. Plant cells which grow out in sheets may be obtained from excised growing tips of plant roots, cultured in fluid suitable for the growth of plant tissues.

Appendices

A

The Technique of Micromanipulation

The following description of the micromanipulative technique is added to provide helpful hints to the potential micrurgist.

CHOICE OF MICROSCOPE AND ACCESSORIES

The binocular, mono-objective microscope with quadruple nosepiece is preferable. The tube must be focusable. It is advantageous to have, in addition, a stage focusable by fine adjustment. Centered and parfocal microscope objectives greatly facilitate the micromanipulative work, which involves constantly adjusting the needle tips under different magnifications.

A great deal of the micro-operative work is done by means of the mechanical stage, the movements of which must be true, steady, and fully controllable even when the work is done under high magnifications, including the oil immersion objectives. In practice, movements of the mechanical stage are often used in preference to those controlling the horizontally moving microneedles. This keeps the microtips in the central zone of the field of vision of the microscope. Micro-operations can be carried out mostly by using the horizontal movements of the mechanical stage and the vertical movements of the micromanipulator.

The stage of the microscope should be deep and the mechanical stage of a sufficiently long range (50 × 75 mm.) to allow the front edge of the moist chamber to be moved back beyond the center of the field of the microscope. This enables one manually to adjust the position of freshly mounted needles and pipettes without danger of striking the needle tips against the operating chamber. With a moist, or an oil, chamber open at one or both ends, the shafts of the needles and pipettes are operated with their microtips inserted in drops of aqueous media suspended from a coverslip which roofs the chamber.

A condenser with a working distance equal at least to the height of the operating chamber is an essential component of the micromanipulative equipment. A moist chamber 8 to 10 mm. in height is ample for operations on cells in hanging drops. Long-working-distance condensers with this focal length for both bright field and phase contrast are commercially available.[1] The numerical aperture of these condensers is about 0.7, which is entirely adequate for objectives of at least the same (and less) aperture. A moist or oil chamber only 4 or 5 mm. in height can be used if the microinstruments are made by means of a microforge, such as that of de Fonbrune. Condensers of approximately this working distance are also obtainable, and even for dark field work.

An inverted microscope (Figure A.1) with focusable tube can be used with the micromanipulator. The objective occupies a position

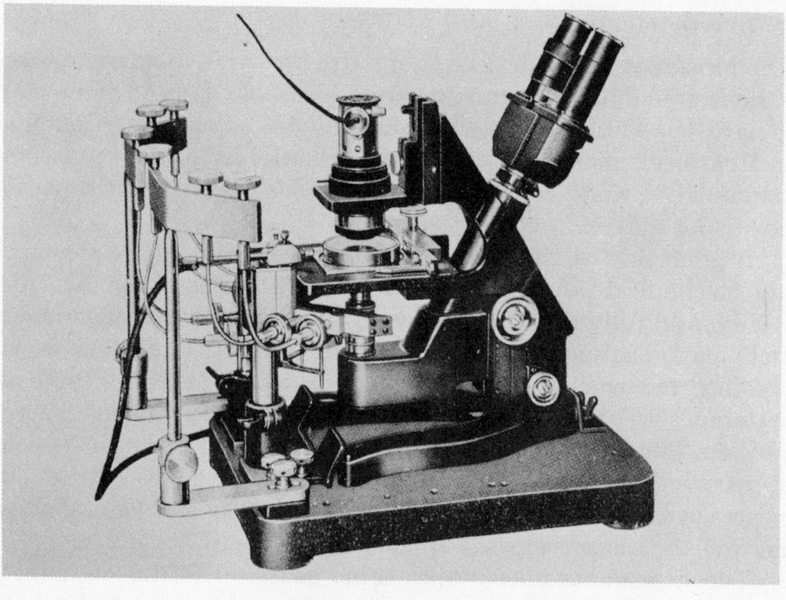

FIGURE A.1. Inverted microscope and Chambers micromanipulator. Specially constructed microscope with objective *below,* and condenser with source of illumination *above,* microscope stage. Operating microneedles and pipettes are passed from front of stage and operate on floor of moist chamber. In order to focus on the object the microscope tube is moved; this feature is essential in micromanipulative work. (From Chambers and Kopac, 1950.)

[1] For example, Bausch and Lomb Optical Co., Rochester, New York, or Reichert, Vienna, Austria. A triple-lens, achromatic condenser with numerical aperture of about 1.2 can frequently be converted to a condenser with focal length of 8 to 10 mm. (aperture of 0.6 or 0.7) by removing the top lens.

below the stage of the microscope. The moist chamber is also inverted, the floor of the chamber being the coverglass on top of which the organisms to be operated upon are placed. A large amount of fluid may then be used, thereby offsetting the tendency to rapid evaporation that occurs in the usual type of moist chamber. Another advantage of this arrangement is the fact that the organisms, mounted in a drop of water, fall through the drop to rest on the glass floor of the chamber. Micro-operations can, therefore, be performed without necessarily having to resort to various measures to hold a cell in place. With the inverted microscope, the microneedles and micropipettes are manipulated above the coverslip and object. Above these is the superstage consisting of a long-working-distance condenser.

In spite of the advantages of the inverted microscope, we do not know of any machines presently available commercially which are satisfactory for the work. Most inverted microscopes lack a focusable tube. The supporting pillars of the micromanipulator units would then have to be adapted to fasten directly to the focusable stage. This is feasible only with some types of micromanipulators. Furthermore, the inverted instruments lack a rotating nosepiece for changing objectives. The inconvenience is compounded by the importance of having a set of objectives all centered to coincident optical axes.

The lower power objectives should be avoided for micromanipulative work, since their greater depth of focus renders it impossible to determine accurately the vertical level of the tip of a microneedle with respect to that part of an object which the needle tip is to touch. The shallow depth of focus of high power objectives enables one to place the microtip into a cell with far greater precision than is possible with lower power objectives.

A serviceable lighting system is one in which the source is a broad, flat, structureless, incandescent surface. This can be had by means of the recently developed ribbon filament electric light bulbs.

Poor phase contrast optics frequently result when one attempts to observe cells in hanging drops; this is due to the curvature of the air–water interface. The difficulty can be diminished by using flat, shallow droplets; also by using a chamber of low height and filling the air-space between roofing coverslip and floor with oil (de Fonbrune's oil chamber).

Emphasis must be laid upon the fact that study of living cellular phenomena, especially in micromanipulative work in which the elaborate methods of fixing, staining, and clearing tissues have to be discarded, requires an acute appreciation of the limits and possibilities of critical microscopic vision.

OPERATING CHAMBER AND COVERSLIPS

A moist chamber is commonly used with Chambers' instruments. It is open at one end only (Figure A.2). With this chamber, the two micromanipulator units are placed so that both of the microneedles or pipettes pass into the chamber from the one end. The coverglass which roofs the chamber is held in place with vaseline.

A convenient height for micro-operative purposes is 8 to 10 mm., and the substage condenser must have approximately the same focal length. A greater height than this is to be avoided, since resolution is impaired by the necessity of using a condenser with inadequate numerical aperture.

The height of the space between roofing coverslip and floor of the moist chamber is determined by the vertical distance which must be allowed for the insertion of the microinstruments. The investigator who bends his needles by hand in a microflame finds that a height of less than 8 mm. is inconvenient. A minimum workable height is 6 to 7 mm. (The total height of the chamber can be minimized by using a coverslip for the floor of the chamber, rather than a glass slide.) When, however, needles are fashioned by means

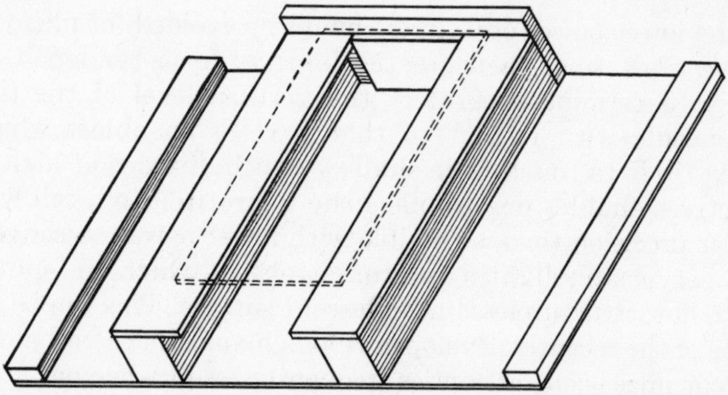

FIGURE A.2. Moist chamber for micromanipulation, open at front to allow insertion of shafts of microneedle and micropipette from double micromanipulator placed in front of microscope. Coverglass (interrupted line) roofs chamber and is held in place with vaseline. Moist chamber is 2 cm. deep and has walls of glass or bakelite, 8 to 10 mm. high. Inner end of chamber is closed by strip of glass or bakelite of same height as chamber and backed by another strip, 1 to 2 mm. higher. Along inner walls of chamber wet strips of filter paper are placed. Floor may be flooded with water by applying vaseline across floor at open end. (From Chambers, 1929a.)

of an incandescent filament in a microscopic field (de Fonbrune's microforge) needles can be made for convenient insertion into a space 3 mm. or less in height. This permits using a condenser of short working distance and large aperture.

A moist chamber is purposely made rather long because of the evaporation which occurs at the open end. At a certain distance from the open end is a zone where the water condensation on the coverglass is at an optimum for maintaining the hanging drop which contains the cells or tissues. Evaporation can be minimized by closing the open end off with soft vaseline and flooding the floor of the chamber with water.

Using a hanging drop of inert paraffin oil to cover the drop of aqueous solution containing the material for micro-operations is an efficient method for obviating evaporation. This method was used by Chambers and Hale (1932) in their studies on the effects of sub-cooling amebae and tissue cells. The hanging drop of paraffin oil can also be used for microchemical experiments as a "micro-test-tube" into which two or more measured spherical drops of aqueous solutions are introduced with micropipettes and then caused to mix by approaching the drops till they fuse.

An operating chamber, open at both ends, can be made into an oil chamber (de Fonbrune, 1949) by filling the space (2.5 to 3 mm. vertical distance) between roofing coverslip and floor with heavy paraffin oil (viscosity of 335 to 350); capillarity holds the oil in place. Cells to be operated upon are introduced by inserting the tip of a mouth pipette through the oil and applying a droplet of fluid to the undersurface of the coverslip. If the coverslip is scrupulously clean, the water droplet adheres to the glass. By withdrawing excess water, the drop can be made extremely shallow, and cells as small as nucleated red cells can be held in place and even flattened against the coverslip by the oil–water interface. In the oil chamber evaporation of water is virtually eliminated. Cells may continue to live in a tiny droplet for several days, since the oil is entirely innocuous and freely permeable to carbon dioxide and oxygen. According to de Fonbrune, the optical properties of the oil filled chamber are superior to those of the moist chamber due to the minimizing of the diffraction which occurs at air–water and air–glass interfaces.

A special type of hermetic chamber has been developed for micro-operations in atmospheres of various gases (Cohen, Chambers, and Reznikoff, 1928). The open end of a moist chamber is closed by a trough of mercury through which hollow metal tubes with U-bends pass into the hermetic chamber (Figure A.3). Short micro-

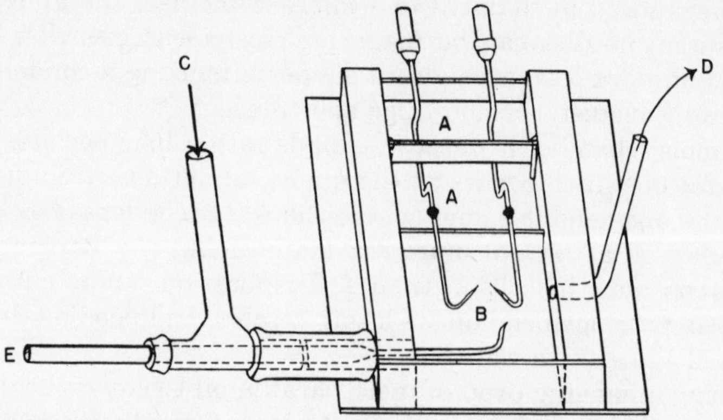

FIGURE A.3. Hermetic chamber. Moist chamber, roofed by large coverglass (30 × 40 mm.) and sealed with vaseline, is divided into two troughs, A and B. Trough A, filled with mercury, is traversed by crossbar which, along its upper surface, is flush with undersurface of roofing coverglass. Crossbar extends part way down in trough A to allow micropipettes, because of U-bend of copper parts of their shafts, to pass through mercury seal under the crossbar and traverse trough A so as to extend into trough B. Trough B and adjacent half of trough A are sealed off (with vaseline) portions of chamber and are filled, through inlet C, with a specific gas. Pipette E is for delivering solution to be microinjected. Outlet is at D. Glass portions of micropipettes are connected to U-shaped copper tubes by seal of cement, to be replaced as required. (From Cohen, Chambers, and Reznikoff, 1928.)

needles or pipettes are cemented into the metal tubes by means of a quick-hardening cement. The chamber is supplied with two inlets for operative purposes: one to flood the chamber with the gas, and the other for a delivery pipette to deposit a hanging drop of a desired solution after the chamber has been filled with the gas.

The cleaning of glass is a most important feature, since the required small, flat, shallow droplets can be suspended from a coverslip only if it is scrupulously clean. (For further hints, see Chambers and Kopac, 1950; de Fonbrune, 1949.)

MAKING MICRONEEDLES AND MICROPIPETTES

The needles and pipettes are made from either soft or hard glass rods or tubing, 0.8 to 1.0 mm. in diameter.[2] The microtips can be readily made by free hand using a gas microburner made out of a fine hypodermic needle with the beveled end ground off.[3]

[2] The glass stock should be on hand in lengths of several feet and of uniform diameter or bore. This is available from James D. Graham (glassblower to the Marine Biological Laboratory, Woods Hole, Mass.), 11 Mountwell Ave., Haddonfield, N. J.

[3] For further description of the technique, see Chambers (1929a); Chambers and Kopac (1950).

Needles and pipettes can also be made by means of specially constructed mechanical apparatus, the heating elements of which are incandescent platinum loops. The apparatus depends upon the lifting of the shaft of the capillary away from the heat source just as softening of the glass occurs. An excellent apparatus is Livingston's needle puller, shown in Figure A.4. In another instrument of this type the heat-softened glass tubing is pulled out electromagnetically.[4] The great advantage of these automatic instruments is that a microtip of any desired type can be made and faithfully duplicated. When doing microinjection work many micropipettes are used in an afternoon, and generally only one particular kind of microtip is optimal for the specific job at hand.

FIGURE A.4. Livingston's needle puller. A glass rod or tubing, 0.8 to 1.0 mm. in diameter, for making microneedles or pipettes, is held horizontally, under tension, between incandescent platinum loops (within a slot) on vertically adjustable pillar. When glass rod or tubing is heat softened, tension of stretched spring holding wheels is released. This release causes rod or tubing to be drawn apart, whereupon microtips are produced at ends. (Made by Otto Hebel, Swarthmore College, Swarthmore, Pennsylvania.)

[4] Alexander and Nastuk (1953), available from Industrial Science, 63-15 Forest Avenue, Ridgewood 27, N. Y.

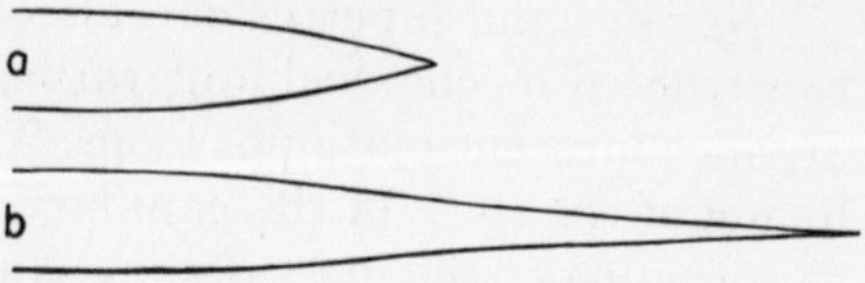

FIGURE A.5. Views of serviceable microneedles and pipettes under high power objective. (a) Quick taper. (b) Gradual taper.

Serviceable microneedles and pipettes are classified into two general types: the quick taper (Figure A.5a), in which the diameter decreases rapidly to a microtip, and the gradual taper (Figure A.5b).[5] Too slow a taper results in a lack of rigidity and the rapid clogging of a pipette. If the taper is too rapid, the microtip is apt to be too blunt or the orifice of the pipette too wide when opened.

After a suitable microtip is made, the shaft 2 to 4 mm. from the microtip is bent at an angle over the microflame. The height of the portion of the needle beyond the bend should not be over 3 or 4 mm., to fit into the moist chamber. In Figure A.6 several different types of bends are shown. A rigid needle results if the bend is on the shaft just on the wider end of the slender extension (Figure A.6a). This type of needle, especially if the bend approaches a right angle, is easily broken when the tip meets the coverslip. A more resilient and yielding type results when the bend is made on the slender extension (Figure A.6b), especially when the angle of the bend is obtuse. A type of needle in which the shaftlet is bent upon itself (Figure A.6c) is well adapted for moving a cell about or for cutting cells in two (Figure A.7).

P. de Fonbrune (1949) has described thoroughly methods for making glass microinstruments by means of his microforge.[6] This

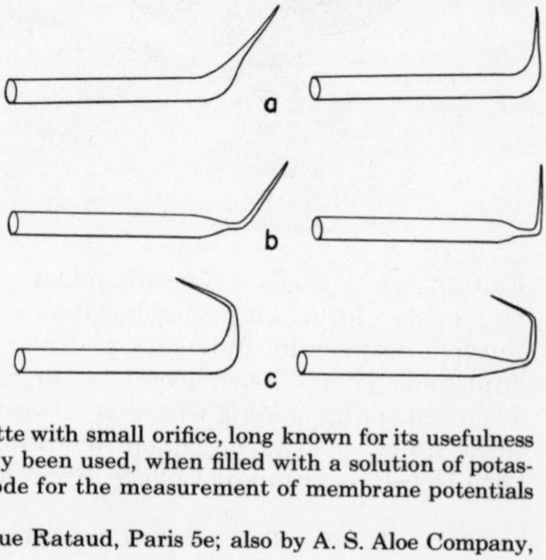

FIGURE A.6. Shafts of microneedles or pipettes showing different types of bends of microtips. (a) Bend on heavy part of shaft. (b) Bend on slender part of shaft. (c) Double bend.

[5] The gradual taper type of micropipette with small orifice, long known for its usefulness in microinjection work, has more recently been used, when filled with a solution of potassium chloride, as an intracellular electrode for the measurement of membrane potentials (Ling and Gerard, 1949).

[6] Manufactured by Baudouin, 1 et 3 rue Rataud, Paris 5e; also by A. S. Aloe Company, St. Louis 12, Missouri.

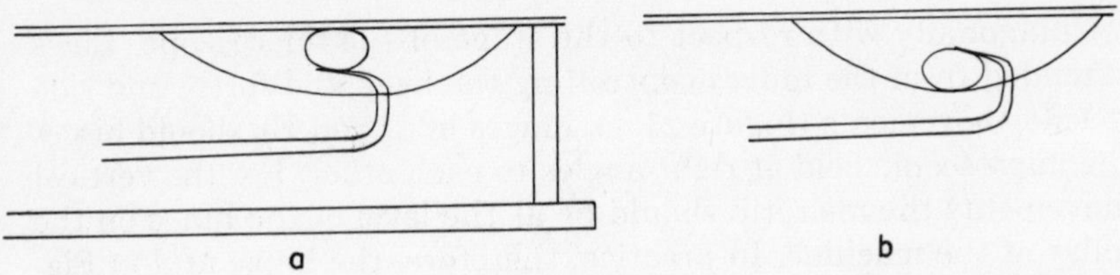

FIGURE A.7. Methods for cutting echinoderm egg in two. (a) Side view of moist chamber showing needle in position with its microtip placed to compress egg between it and coverslip. (b) Cutting egg by bringing end limb of double bend of microtip down on egg so as to press egg against lower surface of hanging drop.

consists of two mechanically movable parts, a platinum filament and a holder for glass rods or tubes. The two parts are mounted in the field of a microscope. The microforge possesses the important advantage that microneedles and micropipettes can be appropriately bent very short distances from their tips. This permits the use of operating chambers of minimal working height. By making a double bend, a short horizontal shaftlet extending to the microtip can be constructed for microdissection or microinjection of cells from the side. Micropipettes horizontal a short distance back from the microtip are useful for quantitative microinjections, since the distance from the microtip to the meniscus of the fluid being injected can be easily measured. Furthermore, microhooks can be made, which are useful to hold cells in place during micro-operative procedures.

POSITIONING THE CHAMBERS MICROMANIPULATOR

Chambers' micromanipulator depends for its movements on metal bars which are so located that extensions from them move along arcs of large circles.

The advantage of the large arc principle is that the fine movements are imparted by the spreading apart of the bars against a spring hinge. The frictional surfaces are at a minimum, since they are the two tips of small spindle-shaped pins, lying in pointed depressions, at the end of each turning screw. The frictional points should be adequately greased.

Proper positioning of the micromanipulator with reference to the position of the microscope is essential. For horizontal movements the microtips should operate along arcs at the apex of an imaginary right-angled triangle, as shown in Figure 21.4a. This is achieved by having the micromanipulator with its horizontal bars set diagonally with respect to the stage of the microscope. Lines extended from the hinges controlling the back and forth, and side to side, movements (Figure 21.4a, hinges at D and E), should bisect the microscopic field at right angles to each other. For the vertical movements the microtip should be at the level of the hinge on the pillar of the machine. In practice, therefore, the hinge at J in Figure 21.4b is set at the level of the coverglass which roofs the moist chamber. This position ensures that the movements of the microtip will take place along the most vertical part of the arc within the field of microscopic vision, thereby minimizing lateral deviation.

MOUNTING THE MICRONEEDLES

A microneedle is mounted by inserting its shaft into a specially designed metal holder (Figure A.8). This arrangement, with minute rubber washer, firmly clamps the glass shaft of the needle or pipette in place and is also leakproof to the water used in the microinjection system. With the needle mounted in the holder, the holder is then clamped to the carrier of the micromanipulator. The microtip is then centered within the field of the microscope using the coarse adjustments of the micromanipulator (Figure 21.3).

HOLDING CELLS IN PLACE

A feature of importance is to have adequate means to hold in place the cell to be microdissected or injected. If amebae are being used, a good method is to cover the floor of a petri dish with moist filter paper, and place in the dish a small horizontal rack. A grease-free coverslip is set on the rack, a series of small drops containing amebae is deposited by mouth pipette, and the petri dish is covered. After half an hour the coverslip is removed and inverted over the micro-operating chamber. The amebae are now found attached to

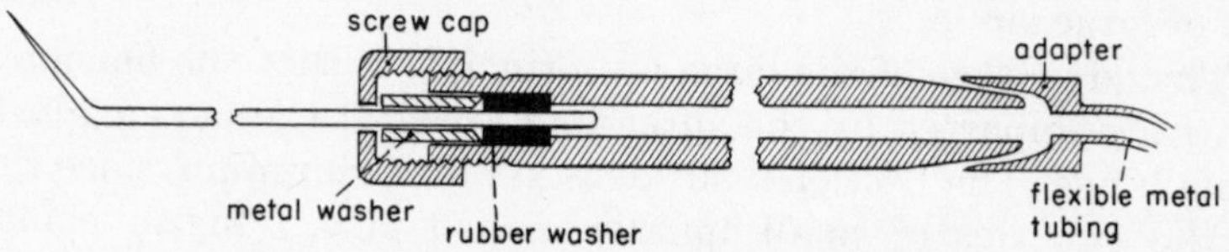

FIGURE A.8. Micropipette holder, enlarged, with parts untightened to show construction. At right of figure is end of flexible metal tubing connected to holder by water-tight adapter. At left of figure, glass shaft of micropipette is held in place by screw cap and metal and rubber washers. Rubber washer fits against shelf in holder. Tightening of screw cap causes concave end of metal washer to press against rubber, sealing glass shaft. Flexible tubing and holder must be filled with air-free water.

and spread out on the undersurface of the coverslip. In this condition they can easily be operated upon or microinjected. The amebae release their attachment only if they have been deleteriously affected by a micromanipulative procedure.

Frequently, a cell can be immobilized by dragging it with a microneedle to the edge of a shallow drop where the air–water interface tends to compress the cell against the undersurface of the coverslip. A cell can be held in the crook of a microhook, or by being sucked into the mouth of a pipette with a flattened orifice (a "sucker"); both of these are glass instruments which can be made with de Fonbrune's (1949) microforge.

Another method is to use a drop sufficiently shallow so that the contained cells are held against the undersurface of the coverslip by the air–water interface. When a moist chamber is being used, difficulties with evaporation make this method applicable only to large cells such as amebae, echinoderm eggs, and striated muscle fibers of the frog. If small cells are to be held stationary by this means, the amount of water in the drop must be greatly reduced. To minimize evaporation, the required extremely shallow drops are best deposited under oil or against a coverslip roofing an oil-filled chamber. If, in addition to making very shallow drops, the diameter of the drop is diminished to constrain the cell from the sides also, even better immobilization is achieved. In his nuclear transfer experiments de Fonbrune used an oil chamber and tiny water droplets hardly larger than the two contained amebae (one of the amebae was held with a microhook during the actual transfer).

MICROINJECTION

The ability to microinject depends not only on the fineness of the micropipettes but also on the accurate control of pressure and the surface tension conditions of the fluids to be injected. The diameter of the bore of the micropipette may be anywhere from less than half a micron to more, according to the type of microinjection required.

The syringe and lower end of the flexible tube are fastened firmly at the base of the microscope (Figures 21.3, 21.5). This is to prevent vibrations caused by touching the syringe from being transmitted to the micropipette. The length of the metal tube, curved into a large loop, is to prevent constraint on the movements of the manipulator. The entire system, from the syringe through the metal tubing and to the pipette holder, is filled with air-free water. The plunger should be pushed at least half way into the syringe. To mount the micropipette, its shaft is inserted into the water-filled pipette holder, which is then fastened to the micromanipulator. The vertical positioning of the microtip of the pipette should be as shown in Figure 21.4b.

It should be noted that the only air in the injection system is in the glass shaft of the micropipette. This small amount of air seems to be necessary; otherwise the system would be too rigid. An amount of air in excess of this, however, makes control of the microinjections impossible. The injection system must be water-tight when tested by exerting pressure on the plunger of the syringe.

The tip of a properly made micropipette is always closed; it is converted into an open pipette only just before actual use. This is done by raising the tip into the hanging drop of the solution to be injected and barely touching the tip against the under surface of the coverslip. This procedure should be done under the same magnification as that used for injecting.

The filling of the pipette is done by capillarity aided by pulling on the plunger of the syringe. When the filling is completed, the pipette is lowered out of the drop and is ready for an injection. For expelling fluid from the pipette the plunger of the syringe should be gently pushed with a rotating action.

For microinjection it is preferable to use a double micromanipulator, one holding a microneedle and the other the micropipette. The microneedle is for holding the cells to be microinjected. Both needle and pipette should be relatively stiff and rapidly tapering, with the shafts bent so that the tips are directed upward toward the coverslip. The tip of the micropipette, after being opened, should average about 0.2 to 0.5 micron in diameter. The following represents a typical injection experiment on an echinoderm egg.

> A shallow hanging drop of sea water containing the cells to be injected is mounted on the under surface of a coverslip alongside an adjacent drop containing the fluid to be injected (for example, a pH indicator solution). The tip of the still closed micropipette is brought into focus, well below the level of the coverslip; the drop of the solution to be microinjected is brought into position; and the micropipette is raised into the hanging drop. The tip of the pipette is broken open by gently touching the microtip to the surface of the coverslip, and the pipette fills by capillarity with aid from the syringe. The solution-filled micropipette is lowered out of and well below the drop. The moist chamber is then moved to bring into position the cell to be

injected. A microneedle may be used to hold the cell to be operated upon. The filled micropipette is now brought into position with its microtip just below the surface of the shallow hanging drop containing the cell to be microinjected. The positioning of the tip of the micropipette is aided by having a fine pointer (for example, an eyelash) fixed in an ocular. By moving the mechanical stage, the operator first places the tip of the pointer over a given region of the cell to be microinjected. Then the operator lowers the microscope tube until the focus is located just below the surface of the hanging drop. The microtip is moved until its position coincides with the tip of the pointer. The operator refocuses on the cell and the micropipette, now in position, is raised steadily and carefully until its open tip penetrates the cell being injected in the region originally selected. If a displacement of the microtip with reference to the position of the pointer has occurred during the raising, this is a result of the vertical arc movement. The occurrence of a displacement indicates that the spring controlling the vertical movement of the micromanipulator may not have been at its proper level (Figure 21.4b); the hanging drop may have been too deep; or the pipette may have been too far below the hanging drop when moved to a position which coincided with the tip of the pointer in the ocular.

When the micropipette is inserted into the cell the solution to be injected flows out either spontaneously or when pressure is exerted on the plunger of the syringe. (An excessive, uncontrollable outflow of fluid indicates that the microtip is too large). The amount of the injection is controlled either by varying the pressure or by suddenly dropping the pipette out of the cell. Care must be taken, when raising the micropipette into the cell, to avoid the entry of extraneous fluid from the hanging drop; otherwise one may unknowingly inject only hanging drop fluid. This pitfall is best avoided by always adequately filling the pipette with the fluid to be injected, that is, to the bend in the shaft.

With use, micropipettes rapidly become clogged because of adherence of cytoplasmic material at their tips; this difficulty can be minimized by siliconizing their inner and outer surfaces (Kopac, 1953).

The operator must be sure that the micropipette actually enters the protoplasmic interior. When a pipette is pushed into a cell, especially a cell surrounded by an elastic extraneous coat, or if the tip of the pipette is more than a micron in diameter, the surface of the cell frequently invaginates, forming a deep pocket

(Chambers, 1922c). Even if the cell surface is penetrated, the tip is apt to be separated from the interior by the formation of a new surface film. This film may or may not be continuous with the original surface of the cell. Attention has again been drawn to this feature with reference to the measurement of membrane potentials in the starfish egg by insertion of a micropipette (microelectrode) filled with potassium chloride solution (Tyler et al., 1956).

B

Technique of Determining Intracellular pH by the Microinjection Method

In using the microinjection technique it is essential that the protoplasm be maintained in a normal state and under normal conditions throughout the procedure. For this reason the investigator must have available many individual cells of the same type. This permits microinjecting one cell after another with a series of overlapping indicators.

The indicators used are those of the Clark and Lubs series, as the sodium, or preferably the potassium, salts, covering the pH range from 4.4 to 8.8 (Table 13.1). Their properties are discussed in Chapters 13 and 15. The only basic dye indicator used is neutral red, as the chloride salt.

> The pH indicators of the Clark and Lubs series include chlor phenol red, brom cresol purple, brom thymol blue, phenol red, and cresol red, prepared according to Clark (1928) in 0.4 per cent aqueous solution with a molecular equivalent of sodium hydroxide or potassium hydroxide. In our experience the potassium salts of the indicators are not as toxic as the sodium salts when microinjected. Of them, phenol red is the least toxic and brom thymol blue the most toxic. All the other indicators are relatively nontoxic except brom cresol purple, which gives evidence of toxicity only when injected into the cell nucleus. Variations in toxicity of the indicators may occur in preparations obtained from different manufacturers.
>
> Brom cresol purple, in its alkaline range, has the disadvantage of exhibiting dichromatism. Accordingly, chlor phenol red, having nearly the same transformation range, is to be preferred.
>
> In order to rule out the possibility that the introduction

of the indicator, in itself, might impose a pH change, the solutions microinjected should not only be adjusted to neutral pH but should also be acidified or alkalinized to the extent just necessary to secure the acid or alkaline colors.

The important feature of the sulfonphthalein indicators (as sodium or potassium salts) is that when microinjected they promote the fluid state of the cell interior and diffuse evenly throughout the protoplasmic matrix, as previously discussed. Neutral red is a valuable indicator, but its use requires special precautions. Only when microinjected in low concentrations does this dye spread evenly and diffusely through the cytoplasmic matrix and nuclear sap. If introduced in higher concentrations, coagulation occurs due to the formation of insoluble neutral red proteinates. The investigator must also remember that this basic dye, after being microinjected into the cytoplasm, rapidly passes from the matrix to segregate in acidic vacuoles, and if lipid deposits are present, these will assume the yellow color of the free base. The importance of using the basic dye is that, if protein or salt errors were serious, they would cause deviations in the colorimetric estimations, opposite in direction for the basic as compared with the acidic indicators.

Under no circumstances should the quantity of the microinjected solution cause any appreciable increase in volume of the cell. In general, 0.2 to 0.4 per cent aqueous solutions of the Clark and Lubs indicators have been found to be sufficiently concentrated so that an injected minimal quantity gives an appreciable color to the cell. If the indicator solution is too dilute, a larger amount of fluid must be injected. This, however, is to be avoided, since many cells tolerate only a limited volume of injected solution. During and after the injection, signs should be looked for which might indicate injury. For echinoderm eggs the most frequently observed sign of injury is the development of an acid of injury of pH 5.2 to 5.4, generally localized and, if not too extensive, reversible. Generally speaking, cells in a normal state are characterized by the glairy, hyaline appearance of their cytoplasm. A frequent sign of irreversible injury is the appearance of reticulated structure in the nucleus, which under normal circumstances is usually optically homogeneous.

A good way of evaluating the color tint of a microinjected indicator in terms of the color tint of the indicator in one or another of a series of buffered solutions was proposed by Pantin (1923a, b). A series of test tubes containing the given indicator in standard Clark and Lubs buffer solutions of varying pH values is projected one tube after the other into the microscopic field in which lies the microinjected cell. The intensity of color projected in the microscopic field may be approximated to that within the cell by using buffer solutions containing different concentrations of the indicator. If the color in the microinjected cell is the same as that of the background, and if the illumination is sufficiently intense to minimize structural details, the cell, being of the same color, will almost disappear.

Bibliography

Bibliography

BIBLIOGRAPHY OF THE WORKS OF ROBERT CHAMBERS

BARON, H., AND CHAMBERS, R. (1936). A micromanipulative study on the migration of blood cells in frog capillaries. Am. J. Physiol. 114:700–708.

BECK, L. V., AND CHAMBERS, R. (1935). Secretion in tissue culture; II—Effect of Na iodoacetate on the chick kidney. J. Cell. & Comp. Physiol. 6:441–455.

BECK, L. V., AND CHAMBERS, R. (1937). Permeability of sand dollar ova and of *Modiolus* and *Mytilus* gills to various acids and their salts. J. Exper. Biol. 14:278–289.

BOUCEK, R. J., CAMERON, GLADYS, SAURINO, V. R., AND CHAMBERS, R. (1954). A tissue culture study of the chick fibroblast in relation to streptococcal filtrates and rheumatic heart disease. Circulation Research 2:84–89.

BUNO, W., AND CHAMBERS, R. (1947). Direct determination of internal hydrostatic pressure of *Fundulus* eggs. Biol. Bull. 93:190.

CAMERON, GLADYS, AND CHAMBERS, R. (1937). Neoplasm studies; III—Organization of cells of human tumors in tissue culture. Am. J. Cancer 30:115–129.

CAMERON, GLADYS, AND CHAMBERS, R. (1938). Direct evidence of function in kidney of an early human fetus. Am. J. Physiol. 123:482–485.

CAMERON, GLADYS, KENSLER, C. J., AND CHAMBERS, R. (1944). The action of heptanal sodium bisulfite methylsalicylate and of 2, 4, 6-trimethylpyridine on tissue cultures of human and mouse carcinoma and rat lymphosarcoma. Cancer Research 4:495–501.

CAMERON, GLADYS, KOPAC, M. J., AND CHAMBERS, R. (1943). Neoplasm studies; IX—The effects in tissue culture of N, N-dimethyl-p-phenylenediamine on rat liver tumors induced by p-dimethylaminoazobenzene. Cancer Research 3:281–289.

CHAMBERS, E. L., AND CHAMBERS, R. (1938). The resistance of fertilized *Arbacia* eggs to immersion in KCL and NaCl solutions. Biol. Bull. 75:356.

CHAMBERS, E. L., AND CHAMBERS, R. (1949). Ion exchanges and fertilization in echinoderm eggs. Am. Naturalist 83:269–284.

CHAMBERS, E. L., WHITELEY, A., CHAMBERS, R., AND BROOKS, S. C. (1948). Distribution of radioactive phosphate in the eggs of the sea urchin, *Lytechinus pictus*. Biol. Bull. 95:263.

CHAMBERS, R. (1908). Einfluss der Eigrösse und der Temperatur auf das Wachstum und die Grösse des Frosches und dessen Zellen. Arch. f. mikr. Anat. u. Entwcklungsmechn. 72:607–661.

Chambers, R. (1912a). Egg maturation, chromosomes, and spermatogenesis in *Cyclops*. University of Toronto Studies, Biological Series, No. 14, pp. 5–37.

Chambers, R. (1912b). A discussion of *Cyclops viridis* Jurine. Biol. Bull. 22:291–296.

Chambers, R. (1913). The spermatogenesis of a daphnid, *Simocephalus vetulus*. Biol. Bull. 25:134–140.

Chambers, R. (1914a). Linkage of the factor of bifid wing; the bifid wing and other sex-linked factors in *Drosophila*. Biol. Bull. 27:151–164.

Chambers, R. (1914b). Some physical properties of the cell nucleus. Science 40:824–827.

Chambers, R. (1915a). Microdissection studies on the germ cell. Science 41:290–293.

Chambers, R. (1915b). Microdissection studies on the physical properties of protoplasm. Lancet-Clinic 113:363–369.

Chambers, R. (1917a). Microdissection studies; i—The visible structure of cell protoplasm and death changes. Am. J. Physiol. 43:1–12.

Chambers, R. (1917b). Microdissection studies; ii—The cell aster; a reversible gelation phenomenon. J. Exper. Zool. 23:483–504.

Chambers, R. (1918). The microvivisection method. Biol. Bull. 34:121–136.

Chambers, R. (1919a). Changes in protoplasmic consistency and their relation to cell division. J. Gen. Physiol. 2:49–68.

Chambers, R. (1919b). Some studies on the surface layer in the living egg cell. Proc. Soc. Exper. Biol. & Med. 17:41–43.

Chambers, R. (1920a). Dissection and injection studies on the amoeba. Proc. Soc. Exper. Biol. & Med. 18:66–68.

Chambers, R. (1920b). A note on the structure of the contractile vacuole in *Amoeba proteus*. Anat. Rec. 18:225.

Chambers, R. (1920c). The ineffectiveness of churning the contents of an egg on its subsequent development. Anat. Rec. 18:226.

Chambers, R. (1921a). The formation of the aster in artificial parthenogenesis. J. Gen. Physiol. 4:33–39.

Chambers, R. (1921b). Studies on the organization of the starfish egg. J. Gen. Physiol. 4:41–44.

Chambers, R. (1921c). A simple apparatus for micromanipulation under the highest magnifications of the microscope. Science 54:411–413.

Chambers, R. (1921d). The effect of experimentally induced changes in consistency on protoplasmic movements. Proc. Soc. Exper. Biol. & Med. 19:87–88.

Chambers, R. (1921e). Microdissection studies; iii—Some problems in the maturation and fertilization of the echinoderm egg. Biol. Bull. 41:318–350.

Chambers, R. (1922a). New apparatus and methods for the dissection and injection of living cells. Anat. Rec. 24:1–19.

Chambers, R. (1922b). New apparatus and methods for the dissection and injection of living cells. J. Roy. Micr. Soc., pp. 373–388.

Chambers, R. (1922c). A microinjection study on the permeability of the starfish egg. J. Gen. Physiol. 5:189–193.

Chambers, R. (1922d). Permeability of the cell; the surface as contrasted with the interior. Proc. Soc. Exper. Biol. & Med. 20:72–74.

Chambers, R. (1923a). Some changes in the dying cell. Proc. Soc. Exper. Biol. & Med. 20:367–368.

CHAMBERS, R. (1923b). The mechanism of the entrance of sperm into the starfish egg. J. Gen. Physiol. 5:821–829.

CHAMBERS, R. (1924a). The physical structure of protoplasm as determined by microdissection and injection. *In* Cowdry, E. V. (ed.). General Cytology, pp. 237–309. Univ. Chicago Press, Chicago.

CHAMBERS, R. (1924b). Etudes de microdissection; IV—Les structures mitochondriales et nucléaires dans les cellules germinales mâles chez la Sauterelle. Cellule 35:107–124.

CHAMBERS, R. (1926). The nature of the living cell as revealed by microdissection. Harvey Lectures, Series 22, pp. 41–58. Williams and Wilkins, Baltimore.

CHAMBERS, R. (1927). Physical properties of protoplasm. Lectures on the Biologic Aspects of Colloid and Physiologic Chemistry (Mayo Foundation Lectures, 1925–1926), pp. 113–132. Saunders, Philadelphia.

CHAMBERS, R. (1928a). The nature of the living cell as revealed by micromanipulation. *In* Alexander, J. (ed.). Colloid Chemistry: Theoretical and Applied, Vol. II, pp. 467–486. Chemical Catalog Co., New York.

CHAMBERS, R. (1928b). Intracellular hydrion concentration studies; I—The relation of the environment to the pH of protoplasm and of its inclusion bodies. Biol. Bull. 55:369–376.

CHAMBERS, R. (1928c). Micromanipulative technique. *In* Gatenby, J. B., and Cowdry, E. V. (eds.). Bolles Lee's Microtomist's Vade Mecum (Ed. 9), pp. 390–407. Churchill, London.

CHAMBERS, R. (1929a). Methods for the study of fresh material. Physical agents: microdissection, micro-injection. *In* McClung, C. E. (ed.). Handbook of Microscopical Technique (Ed. 1), pp. 39–73. Hoeber, New York.

CHAMBERS, R. (1929b). Hydrogen ion concentration studies of protoplasm. National Research Council, Bulletin No. 69, pp. 37–47.

CHAMBERS, R. (1929c). The oxidation–reduction potential of protoplasm. National Research Council, Bulletin No. 69, pp. 48–50.

CHAMBERS, R. (1930a). Vital staining with methyl red. Proc. Soc. Exper. Biol. & Med. 27:809–811.

CHAMBERS, R. (1930b). La membrane semi-perméable de la cellule. Ann. de physiol. 6:233–239.

CHAMBERS, R. (1930c). The manner of sperm entry in the starfish egg. Biol. Bull. 58:344–369.

CHAMBERS, R. (1932). Intracellular hydrion concentration studies; V—The pH of the protoplasm of the *Fundulus* egg. J. Cell. & Comp. Physiol. 1:65–70.

CHAMBERS, R. (1933a). The manner of sperm entry in various marine ova. J. Exper. Biol. 10:130–141.

CHAMBERS, R. (1933b). An analysis of oxidation and reduction of indicators in the living cell. Cold Spring Harbor Symposia on Quantitative Biology, Vol. 1, pp. 205–213.

CHAMBERS, R. (1933c). Intracellular hydrion concentration studies; IV—Gastric epithelium of the frog and nerve cells of *Lophius*. Ztschr. f. wissensch. Mikr. 50:239.

CHAMBERS, R. (1933d). The relation of the nucleus to ameboid movement in *Amoeba dubia*. Anat. Rec. 57 (Suppl.) :93.

CHAMBERS, R. (1935a). The living cell. *In* Harrow, B., and Sherwin, C. P. Text Book of Biochemistry, pp. 17–39. Saunders, Philadelphia.

CHAMBERS, R. (1935b). Disposal of dyes by proximal tubule cells of chick mesonephros in tissue culture. Proc. Soc. Exper. Biol. & Med. 32:1199–1200.

CHAMBERS, R. (1937). The physical state of the wall of the furrow in a dividing cell. Biol. Bull. 73:367–368.

CHAMBERS, R. (1938a). The physical state of protoplasm with special reference to its surface. Am. Naturalist 72:141–159.

CHAMBERS, R. (1938b). Structural and kinetic aspects of cell division. J. Cell. & Comp. Physiol. 12:149–165.

CHAMBERS, R. (1938c). The sperm aster and protoplasmic movement. Anat. Rec. 72 (Suppl.):63.

CHAMBERS, R. (1938d). Cytoplasmic inclusions and matrix of the *Arbacia* egg. Biol. Bull. 75:350–351.

CHAMBERS, R. (1940a). The micromanipulation of living cells. The Cell and Protoplasm. American Association for the Advancement of Science, Publication No. 14, pp. 20–30.

CHAMBERS, R. (1940b). Recent developments of the micromanipulative technique and its application. J. Roy. Micr. Soc. 60:113–127.

CHAMBERS, R. (1940c). The relation of extraneous coats to the organization and permeability of cellular membranes. Cold Spring Harbor Symposia on Quantitative Biology, Vol. 8, pp. 144–153.

CHAMBERS, R. (1941). Blue nuclei in tulip petals. Am. J. Botany 28:445–446.

CHAMBERS, R. (1942a). The intrinsic expansibility of the fertilization membrane of echinoderm ova. J. Cell. & Comp. Physiol. 19:145–150.

CHAMBERS, R. (1942b). Experimental studies on protoplasm. Proceedings of the VIII American Scientific Congress, Section 11, Biological Science, Vol. 3, pp. 25–34.

CHAMBERS, R. (1943a). Electrolytic solutions compatible with the maintenance of protoplasmic structures. Biological Symposia 10:91–109.

CHAMBERS, R. (1943b). Tissue culture method in recent investigations. Tr. New York Acad. Sc., Series II, 5:92–96.

CHAMBERS, R. (1944). Some physical properties of protoplasm. *In* Alexander, J. (ed.). Colloid Chemistry: Theoretical and Applied, Vol, V, pp. 864–875. Reinhold, New York.

CHAMBERS, R. (1946). Karyokinetic lengthening and cleavage. Anat. Rec. 94:373.

CHAMBERS, R. (1947). The shape of oil drops injected into the axoplasm of the giant nerve of the squid. Biol. Bull. 93:191.

CHAMBERS, R. (1948a). Vasomotion in the hemodynamics of the blood capillary circulation. Ann. New York Acad. Sc. 49:549–552.

CHAMBERS, R. (1948b). Blood capillary circulation under normal conditions and in traumatic shock. Nature 162:835.

CHAMBERS, R. (1949a). Micrurgical studies on protoplasm. Biol. Rev. 24:246–265.

CHAMBERS, R. (1949b). Sir D'Arcy Wentworth Thompson, C. B., F. R. S., 1860–1948. Science 109:138–139.

CHAMBERS, R. (1950a). The cell as an integrated functional body. Ann. New York Acad. Sc. 50:817–823.

CHAMBERS, R. (1950b). Micromanipulation of protoplasm. *In* Glasser, O. (ed.). Medical Physics, Vol. II, pp. 503–511. Year Book Publishers, Chicago.

CHAMBERS, R. (1951). Micrurgical studies on the kinetic aspects of cell division. Ann. New York Acad. Sc. 51:1311–1326.

CHAMBERS, R., AND BAEZ, S. (1947). A method for subcooling cells under microscopic observation at room temperature. Biol. Bull. 93:226.

CHAMBERS, R., BECK, L. V., AND BELKIN, M. (1935). Secretion in tissue cultures; I—Inhibition of phenol red accumulation in the chick kidney. J. Cell. & Comp. Physiol. 6:425–439.

CHAMBERS, R., BECK, L. V., AND GREEN, D. E. (1933). Intracellular oxidation-reduction studies; V—A comparison of intact and cytolyzed starfish eggs by the immersion method. J. Exper. Biol. 10:142–152.

CHAMBERS, R., AND BLACK, M. (1941). Electrolytes and nuclear structure of the cells of the onion bulb epidermis. Am. J. Botany 28:364–371.

CHAMBERS, R., AND BORQUIST, MAY (1928). In vitro observations of living monocytes and clasmatocytes with reference to phagocytosis and physical properties. Nat. Tuberc. A. Tr. 24:259–263.

CHAMBERS, R., AND CAMERON, GLADYS (1932). Intracellular hydrion concentration studies; VII–The secreting cells of the mesonephros in the chick. J. Cell. & Comp. Physiol. 2:99–103.

CHAMBERS, R., AND CAMERON, GLADYS (1941). The reaction of kidney tubules in tissue culture to roentgen rays. Radiology 37:186–193.

CHAMBERS, R., AND CAMERON, GLADYS (1943). The effect of l-ascorbic acid on epithelial sheets in tissue culture. Am. J. Physiol. 139:21–25.

CHAMBERS, R., AND CAMERON, GLADYS (1944). Adrenal cortical compounds and l-ascorbic acid on secreting kidney tubules in tissue culture. Am. J. Physiol. 141:138–142.

CHAMBERS, R., CAMERON, GLADYS, AND GRAND, C. G. (1949). Calcium and intercellular cement of normal and cancerous epithelium in tissue culture. Acta Unio Internat. contra Cancrum 6:696–699.

CHAMBERS, R., CAMERON, GLADYS, AND KOPAC, M. J. (1943). Neoplasm studies; XI—The effects in tissue culture of N, N, N′, N′-tetramethyl-o-phenylenediamine and other compounds on malignant lymph nodes. Cancer Research 3:293–295.

CHAMBERS, R., AND CHAMBERS, E. L. (1949). Nuclear and cytoplasmic interrelations in the fertilization of the *Asterias* egg. Biol. Bull. 96:270–282.

CHAMBERS, R., AND CHAMBERS, E. L. (1953). Die Wirkung von Na-, K- und Ca-Salzen auf den physikalischen Zustand des Protoplasmas. Arzneimittel-Forsch. 3:322–325.

CHAMBERS, R., CHAMBERS, E. L., AND KAO, C. Y. (1951). The internal hydrostatic pressure of the unfertilized activated *Fundulus* egg exposed to various experimental conditions. Biol. Bull. 101:206.

CHAMBERS, R., CHAMBERS, E. L., AND LEONARD, L. M. (1949). Rhythmic alterations in certain properties of the fertilized *Arbacia* egg. Biol. Bull. 97:233.

CHAMBERS, R., COHEN, B., AND POLLACK, H. (1931). Intracellular oxidation-reduction studies; III—Permeability of echinoderm ova to indicators. J. Exper. Biol. 8:1–8.

CHAMBERS, R., COHEN, B., AND POLLACK, H. (1932). Intracellular oxidation-reduction studies; IV—Reduction potentials of European marine ova and *Amoeba proteus* as shown by indicators. Protoplasm 17:376–387.

CHAMBERS, R., AND DAWSON, J. A. (1925). The structure of the undulating membrane in the ciliate *Blepharisma*. Biol. Bull. 48:240–242.

Chambers, R., and Fell, H. B. (1931). Micro-operations on cells in tissue cultures. Proc. Roy. Soc., London, s. B. 109:380–403.

Chambers, R., and Grand, C. G. (1933). The cortex of the *Arbacia* egg during segmentation. Anat. Rec. 54 (Suppl.):75.

Chambers, R., and Grand, C. G. (1936). The chemotactic reaction of leucocytes to foreign substances in tissue culture. J. Cell. & Comp. Physiol. 8:1–17.

Chambers, R., and Grand, C. G. (1937a). Neoplasm studies; ii—The effect of injecting starch grains into transplanted tumors. Am. J. Cancer 29:111–115.

Chambers, R., and Grand, C. G. (1937b). Leucocytic infiltration of irradiated mouse Sarcoma 180. Proc. Soc. Exper. Biol. & Med. 36:673–675.

Chambers, R., and Grand, C. G. (1939). Neoplasm studies; v—The effect of carbohydrate on the accumulation of granulocytes in various mouse tumors. Am. J. Cancer 36:369–382.

Chambers, R., and Grand, C. G. (1948). Micromanipulation. Encyclopaedia Britannica, Vol. 15, pp. 428–430.

Chambers, R., and Hale, H. P. (1932). The formation of ice in protoplasm. Proc. Roy. Soc., London, s. B. 110:336–352.

Chambers, R., and Höfler, K. (1931). Micrurgical studies on the tonoplast of *Allium cepa.* Protoplasma 12:338–355.

Chambers, R., and Howland, Ruth B. (1930). Micrurgical studies in cell physiology; vii—The action of the chlorides of Na, K, Ca, and Mg on vacuolated protoplasm. Protoplasma 11:1–18.

Chambers, R., and Kao, C. Y. (1952). The effect of electrolytes on the physical state of the nerve axon of the squid and of *Stentor,* a protozoon. Exper. Cell Research 3:564–573.

Chambers, R., and Kempton, R. T. (1933). Indications of function of the chick mesonephros in tissue culture with phenol red. J. Cell. & Comp. Physiol. 3:131–167.

Chambers, R., and Kempton, R. T. (1937). The elimination of neutral red by the frog's kidney. J. Cell. & Comp. Physiol. 10:199–221.

Chambers, R., and Kerr, T. (1932). Intracellular hydrion concentration studies; viii—Cytoplasm and vacuole of *Limnobium* root-hair cells. J. Cell. & Comp. Physiol. 2:105–119.

Chambers, R., and Kopac, M. J. (1937a). The coalescence of living cells with oil drops; i—*Arbacia* eggs immersed in sea water. J. Cell. & Comp. Physiol. 9:331–343.

Chambers, R., and Kopac, M. J. (1937b). The coalescence of sea urchin eggs with oil drops. Annual Report of the Tortugas Laboratory, Carnegie Institution of Washington, Yearbook No. 36, pp. 88–90.

Chambers, R., and Kopac, M. J. (1950). Micrurgical technique for the study of cellular phenomena. *In* Jones, Ruth McClung (ed.). McClung's Handbook of Microscopical Technique (Ed. 3), pp. 492–543. Hoeber, New York.

Chambers, R., and Ludford, R. J. (1932a). Microdissection studies on malignant and non-malignant tissue cells. Arch. f. exper. Zellforsch. 12:555–569.

Chambers, R., and Ludford, R. J. (1932b). Intracellular hydrion concentration studies; vi—Colorimetric pH of malignant cells in tissue culture. Proc. Roy. Soc., London, s. B. 110:120–124.

Chambers, R., and Pollack, H. (1926). The hydrogen ion concentration of the nucleus and cytoplasm of the egg cell. Proc. Soc. Exper. Biol. & Med. 24:42–43.

Chambers, R., and Pollack, H. (1927a). Micrurgical studies in cell physiology; iv—Colorimetric determinations of the nuclear and cytoplasmic pH in the starfish egg. J. Gen Physiol. 10:739–755.

Chambers, R., and Pollack, H. (1927b). The pH of the blastocoele of echinoderm embryos. Biol. Bull. 53:233–238.

Chambers, R., Pollack, H., and Cohen, B. (1929). Intracellular oxidation-reduction studies; ii—Reduction potentials of marine ova as shown by indicators. Brit. J. Exper. Biol. 6:229–247.

Chambers, R., Pollack, H., and Hiller, S. (1927). The protoplasmic pH of living cells. Proc. Soc. Exper. Biol. & Med. 24:760–761.

Chambers, R., and Renyi, G. S. (1925). The structure of the cells in tissues as revealed by microdissection; i—The physical relationships of the cells in epithelia. Am. J. Anat. 35:385–402.

Chambers, R., and Reznikoff, P. (1926). Micrurgical studies in cell physiology; i—The action of the chlorides of Na, K, Ca, and Mg on the protoplasm of *Amoeba proteus.* J. Gen. Physiol. 8:369–401.

Chambers, R., and Sands, H. C. (1923). A dissection of the chromosomes in the pollen mother cells of *Tradescantia virginica* L. J. Gen. Physiol. 5:815–819.

Chambers, R., and Stern, K. G. (1948). Protoplasm. Encyclopaedia Britannica, Vol. 18, pp. 617–619.

Chambers, R., and Zweifach, B. W. (1940). Capillary endothelial cement in relation to permeability. J. Cell. & Comp. Physiol. 15:255–272.

Chambers, R., and Zweifach, B. W. (1944). Topography and function of the mesenteric capillary circulation. Am. J. Anat. 75:173–205.

Chambers, R., and Zweifach, B. W. (1946). Functional activity of the blood capillary bed, with special reference to visceral tissue. Ann. New York Acad. Sc. 46:683–695.

Chambers, R., and Zweifach, B. W. (1947a). Intercellular cement and capillary permeability. Physiol. Rev. 27:436–463.

Chambers, R., and Zweifach, B. W. (1947b). Blood-borne vasotropic substances in experimental shock. Am. J. Physiol. 150:239–252.

Chambers, R., Zweifach, B. W., and Lowenstein, B. E. (1943). Circulatory reactions of rats traumatized in the Noble-Collip drum. Am. J. Physiol. 139:123–128.

Cohen, B., Chambers, R., and Reznikoff, P. (1928). Intracellular oxidation-reduction studies; i—Reduction potentials of *Amoeba dubia* by micro-injection of indicators. J. Gen. Physiol. 11:585–612.

Gordon, H. K., and Chambers, R. (1941). The particle size of acid dyes and their diffusibility into living cells. J. Cell. & Comp. Physiol. 17:97–108.

Grand, C. G., and Chambers, R. (1936). The chemotactic reaction of leucocytes to irritated tissues. J. Cell. & Comp. Physiol. 9:165–175.

Grand, C. G., and Chambers, R. (1940). Neoplasm studies; vii—Granulocytes in regressing transplants of spontaneous mouse tumors. Am. J. Cancer 39:211–219.

Grand, C. G., Chambers, R., and Cameron, Gladys (1935). Neoplasm studies; i—Cells of melanoma in tissue culture. Am. J. Cancer 24:36–50.

Grundfest, H., Nachmansohn, D., Kao, C. Y., and Chambers, R. (1952). Mode of blocking of axonal activity by curare and inhibitors of anticholinesterases. Nature 169:190–191.

HERSHEY, S. G., ZWEIFACH, B. W., CHAMBERS, R., AND ROVENSTINE, E. A. (1945). Peripheral circulatory reactions as a basis for evaluating anesthetic agents. Anesthesiology 6:362–375.

HIBBARD, H., AND CHAMBERS, R. (1935). Micromanipulation of egg and nurse cells in *Bombyx mori.* Biol. Bull. 69:332.

HYMAN, C., AND CHAMBERS, R. (1943). Effect of adrenal cortical compounds on edema formation of frog's hind limbs. Endocrinology 32:310–318.

KAO, C. Y., AND CHAMBERS, R. (1954). Internal hydrostatic pressure of the *Fundulus* egg; I—The activated egg. J. Exper. Biol. 31:139–149.

KITE, G. L., AND CHAMBERS, R. (1912). Vital staining of chromosomes and the function and structure of the nucleus. Science 36:639–641.

KOPAC, M. J., CAMERON, GLADYS, AND CHAMBERS, R. (1943). Neoplasm studies; X—The effects in tissue culture of some split products of p-dimethylaminoazobenzene on rat liver tumors. Cancer Research 3:290–292.

KOPAC, M. J., AND CHAMBERS, R. (1937). The coalescence of living cells with oil drops; II—*Arbacia* eggs immersed in acid or alkaline calcium solutions. J. Cell. & Comp. Physiol. 9:345–361.

MARTENS, P., AND CHAMBERS, R. (1932). Etudes de microdissection; V—Les poils staminaux de *Tradescantia.* Cellule 41:131–143.

MORITA, Y., AND CHAMBERS, R. (1929). Permeability differences between nuclear and cytoplasmic surfaces in *Amoeba dubia.* Biol. Bull. 56:64–67.

PANDIT, C. G., AND CHAMBERS, R. (1932). Intracellular hydrion concentration studies; IX—The pH of the egg of the sea urchin, *Arbacia punctulata.* J. Cell. & Comp. Physiol. 2:243–249.

REZNIKOFF, P., AND CHAMBERS, R. (1927). Micrurgical studies in cell physiology; III—The action of CO_2 and some salts of Na, Ca, and K on the protoplasm of *Amoeba dubia.* J. Gen. Physiol. 10:731–738.

RUDZINSKA, MARIA A., AND CHAMBERS, R. (1951a). The activity of the contractile vacuole in a suctorian (*Tokophrya infusionum*). Biol. Bull. 100:49–58.

RUDZINSKA, MARIA A., AND CHAMBERS, R. (1951b). An abnormal type of development in the life cycle of a suctorian, *Tokophrya infusionum.* Tr. Am. Micr. Soc. 70:168–172.

RUDZINSKA, MARIA A., AND CHAMBERS, R. (1952). The contractile vacuole in enucleated *Amoeba dubia.* Proc. Soc. Protozoologists 3:9.

SPEK, J., AND CHAMBERS, R. (1933). Neue experimentelle Studien über das Problem der Reaktion des Protoplasmas. Protoplasma 20:376–406.

WILBUR, K. M., AND CHAMBERS, R. (1942). Cell movements in the healing of micro-wounds in vitro. J. Exper. Zool. 91:287–302.

ZWEIFACH, B. W., ABELL, R. G., CHAMBERS, R., AND CLOWES, G. H. A. (1945). Role of the decompensatory reactions of peripheral blood vessels in tourniquet shock. Surg. Gynec. & Obst. 80:593–608.

ZWEIFACH, B. W., AND CHAMBERS, R. (1950). The action of hyaluronidase extracts on the capillary wall. Ann. New York Acad. Sc. 52:1047–1051.

ZWEIFACH, B. W., CHAMBERS, R., LEE, R. E., AND HYMAN, C. (1948). Reactions of peripheral blood vessels in experimental hemorrhage. Ann. New York Acad. Sc. 49:553–570.

ZWEIFACH, B. W., HERSHEY, S. G., ROVENSTINE, E. A., AND CHAMBERS, R. (1944). Influence of anesthesia on circulatory changes in dogs subjected to graded hemorrhage. Proc. Soc. Exper. Biol. & Med. 56:73–77.

ZWEIFACH, B. W., HERSHEY, S. G., ROVENSTINE, E. A., LEE, R. E., AND

CHAMBERS, R. (1945). Anesthetic agents as factors in circulatory reactions induced by hemorrhage. Surgery 18:48–65.
ZWEIFACH, B. W., LEE, R. E., HYMAN, C., AND CHAMBERS, R. (1944). Omental circulation in morphinized dogs subjected to graded hemorrhage. Ann. Surg. 120:232–250.
ZWEIFACH, B. W., LOWENSTEIN, B. E., AND CHAMBERS, R. (1944). Responses of blood capillaries to acute hemorrhage in the rat. Am. J. Physiol. 142:80–93.

GENERAL BIBLIOGRAPHY

ABELSON, P. H., AND DURYEE, W. R. (1949). Radioactive sodium permeability and exchange in frog eggs. Biol. Bull. 96:205–217.
AFZELIUS, B. A. (1956). The ultrastructure of the cortical granules and their products in the sea urchin egg as studied with the electron microscope. Exper. Cell Research 10:257–285.
ALEXANDER, J. T., AND NASTUK, W. L. (1953). An instrument for the production of microelectrodes used in electrophysiological studies. Review of Scientific Instruments 24:528–531.
ALLEN, R. D. (1954). Fertilization and activation of sea urchin eggs in glass capillaries. Exper. Cell Research 6:403–424.
ALLEN, R. D., AND ROWE, E. C. (1958). The dependence of pigment granule migration on the cortical reaction in the eggs of *Arbacia punctulata*. Biol. Bull. 114:113–117.
ATKINS, W. R. G. (1922a). Dibromthymolsulphonephthalein as a reagent for determining the hydrogen ion concentration of living cells. J. Marine Biol. Assoc., United Kingdom 12:781–784.
ATKINS, W. R. G. (1922b). The hydrogen ion concentration of the cells of some marine Algae. J. Marine Biol. Assoc., United Kingdom 12:785–788.
BAIRATI, A., AND LEHMANN, F. E. (1953). Structural and chemical properties of the plasmalemma of *Amoeba proteus*. Exper. Cell Research 5:220–233.
BAJER, A. (1957). Ciné-micrographic studies on mitosis in endosperm; III—The origin of the mitotic spindle. Exper. Cell Research 13:493–502.
BAJER, A. (1958). Ciné-micrographic studies on mitosis in endosperm; IV—The mitotic contraction stage. Exper. Cell Research 14:245–256.
BALDWIN, E. (1937). An Introduction to Comparative Biochemistry. Cambridge Univ. Press, London.
BALINSKY, B. I. (1959). An electron microscopic investigation of the mechanisms of adhesion of the cells in a sea urchin blastula and gastrula. Exper. Cell Research 16:429–433.
BARBER, M. A. (1904). A new method for isolating micro-organisms. J. Kansas M. Soc. 4:487.
BARBER, M. A. (1911). A technic for the inoculation of bacteria and other substances into living cells. J. Infect. Dis. 8:348–360.
BARTLEY, W., AND DAVIES, R. E. (1954). Active transport of ions by subcellular particles. Biochem. J. 57:37–49.
BARY, H. A. DE (1859). Myxomyceten. Ztschr. f. wissensch. Zool.
BEAMS, H. W., AND EVANS, T. C. (1940). Some effects of colchicine upon the first cleavage in *Arbacia punctulata*. Biol. Bull. 79:188–198.
BECK, L. V., AND CHAMBERS, R. (1935). Secretion in tissue culture; II—Effect

of Na iodoacetate on the chick kidney. J. Cell. & Comp. Physiol. 6:441–455.

Beck, L. V., and Chambers, R. (1937). Permeability of sand dollar ova and of *Modiolus* and *Mytilus* gills to various acids and their salts. J. Exper. Biol. 14:278–289.

Bělăr, K. (1929). Beiträge zur Kausalanalyse der Mitose; Untersuchungen an den Spermatocyten von *Chorthippus* (*Stenobothrus*) *lineatus* Panz. Arch. f. Entwcklungsmechn. d. Organ. 118:359–484.

Bělăr, K., and Huth, W. (1933). Zur Teilungsautonomie der Chromosomen. Ztschr. f. Zellforsch. u. mikr. Anat. 17:51–66.

Bennett, H. S., Luft, J. H., and Hampton, J. C. (1959). Morphological classification of vertebrate blood capillaries. Am. J. Physiol. 196:381–390.

Bennett, M. C., and Rideal, E. A. (1954). Membrane behaviour in *Nitella*. Proc. Roy. Soc., London, s. B. 142:483–496.

Berkeley, E. (1948). Spindle development and behaviour in the giant amoeba. Biol. Bull. 94:169–175.

Bichat, F. M. X. (1802). Traité des membranes (Ed. 2). Paris.

Blinks, L. R., and Nielsen, J. P. (1940). The cell sap of *Hydrodictyon*. J. Gen. Physiol. 23:551–559.

Bodine, J. H. (1926). Quantitative potentiometric measurements on intracellular pH values of single *Fundulus* egg cells. Science 64:532.

Boné, G. J. (1944). Le rapport sodium/potassium dans la liquide coelomique des insects. Ann. soc. roy. zool. Belg. 75:123–132.

Bonner, J. (1935). Pectins. Chemisch weekblad 32:118–120.

Boveri, T. (1918). Zwei Fehlerquellen bei Merogonieversuchen und die Entwicklungsfähigkeit merogonischer und partiellmerogonischer Seeigelbastarde. Arch. f. Entwcklungsmechn. d. Organ. 44:417–471.

Brachet, A. (1910). La polyspermie expérimentale comme moyen d'analyse de la fécondation. Arch. f. Entwcklungsmechn. d. Organ. 30:261–303.

Brachet, J. (1955). Recherches sur les interactions biochimiques entre le noyau et le cytoplasme chez les organismes unicellulaires; i—*Amoeba proteus*. Biochim. et Biophys. Acta 18:247–268.

Brachet, J. (1959). Cytoplasmic dependence in Amoebae. Ann. New York Acad. Sc. 78:688–695.

Brachet, J., Chantrenne, H., and Vanderhaeghe, F. (1955). Recherches sur les interactions biochimiques entre le noyau et le cytoplasme chez les organismes; ii—*Acetabularia mediterranea*. Biochim. et Biophys. Acta 18:544–563.

Bradford, N. M., and Davies, R. E. (1950). The site of hydrochloric acid production in the stomach as determined by indicators. Biochem. J. 46:414–420.

Brandt, P. W. (1958). A study of the mechanism of pinocytosis. Exper. Cell Research 15:300–313.

Brooks, S. C. (1937). Selective accumulation with reference to ion exchange by the protoplasm. Tr. Faraday Soc. 33:1002–1006.

Brooks, S. C. (1940). The intake of radioactive isotopes by living cells. Cold Spring Harbor Symposia on Quantitative Biology, Vol. 8, pp. 171–180.

Brooks, S. C., and Chambers, E. L. (1954). The penetration of radioactive phosphate into marine eggs. Biol. Bull. 106:279–296.

Brown, D. E. S. (1934). The pressure coefficient of viscosity in the eggs of *Arbacia punctulata*. J. Cell. & Comp. Physiol. 5:335–346.

BROWN, D. E. S., AND MARSLAND, D. A. (1936). The viscosity of *Amoeba* at high hydrostatic pressure. J. Cell. & Comp. Physiol. 8:159–165.

BROWN, R. (1831). On the organs and mode of fecundation in *Orchidaceae* and *Asclepiadeae*. Tr. Linnean Soc., London 16:685–745 (published in 1835).

BUCK, R. C. (1958). The fine structure of endothelium of large arteries. J. Biophys. & Biochem. Cytol. 4:187–190.

BÜTSCHLI, O. (1894). Investigations on Microscopic Foams and on Protoplasm (E. A. Minchin, trans.). Black, London.

BURROWS, T. M. (1927). The mechanism of cell division. Am. J. Anat. 39:83–134.

BUYTENDYK, F. J. J., AND WOERDEMAN, M. W. (1927). Die physico-chemischen Erscheinungen während der Eientwicklung; I—Die Messung der Wasserstoffionenkonzentration. Arch. f. Entwcklungsmechn. d. Organ. 112:387–410.

CALDWELL, P. C. (1954). An investigation of the intracellular pH of crab muscle fibres by means of micro-glass and micro-tungsten electrodes. J. Physiol. 126:169–180.

CALDWELL, P. C. (1955). Studies of ionic mobilities in the giant axon of the squid by means of an intracellular electrode system. J. Physiol. 129:16 P.

CALDWELL, P. C. (1956). Intracellular pH. Internat. Rev. Cytology 5:229–277.

CAMERON, GLADYS (1953). Secretory activity of the chorioid plexus in tissue culture. Anat. Rec. 117:115–126.

CAMERON, GLADYS, AND CHAMBERS, R. (1938). Direct evidence of function in kidney of an early human fetus. Am. J. Physiol. 123:482–485.

CARLSON, J. G. (1952). Microdissection studies of the dividing neuroblast of the grasshopper *Chortophaga viridifasciata* (De Geer). Chromosoma 5:199–220.

CARNOT, P., GLÉNARD, R., AND MME. GRUZEWSKA (1925). Les colorations vitales au rouge neutre comme indices de la concentration ionique des organes vivants. Compt. rend. Soc. de biol. 92:865–868.

CARRÉ, M. H. (1922). An investigation of the changes which occur in the pectic constituents of stored fruit. Biochem. J. 16:704–712.

CASTILLO, J. DEL, AND KATZ, B. (1956). Biophysical aspects of neuro-muscular transmission. Progress in Biophysics and Biophysical Chemistry 6:122–170.

CHALKLEY, H. W. (1935). The mechanism of cytoplasmic fission in *Amoeba proteus*. Protoplasma 24:607–621.

CHALKLEY, H. W. (1951). Control of fission in *Amoeba proteus* as related to the mechanism of cell division. Ann. New York Acad. Sc. 51:1303–1310.

CHAMBERS, E. L. (1939). The movement of the egg nucleus in relation to the sperm aster in the echinoderm egg. J. Exper. Biol. 16:409–424.

CHAMBERS, E. L. (1949). The uptake and loss of K in the unfertilized and fertilized eggs of *S. purpuratus* and *A. punctulata*. Biol. Bull. 97:251–252.

CHAMBERS, E. L., AND CHAMBERS, R. (1938). The resistance of fertilized *Arbacia* eggs to immersion in KCl and NaCl solutions. Biol. Bull. 75:356.

CHAMBERS, E. L., AND CHAMBERS, R. (1949). Ion exchanges and fertilization in echinoderm eggs. Am. Naturalist 83:269–284.

CHAPMAN-ANDRESEN, C., AND HOLTER, H. (1955). Studies on the ingestion of C^{14} glucose by pinocytosis in the amoeba *Chaos chaos*. Exper. Cell Research 3 (Suppl.):52–63.

CHASE, H. Y. (1935). The origin and nature of the fertilization membrane in various marine ova. Biol. Bull. 69:159–184.

CHOLODNY, N. (1923). Zur Frage über die Beeinflussung des Protoplasmas durch mono- und bivalente Metallionen. Beihefte zum botan. Centralbl., Abt. 1, 39:231–238.

CHOLODNY, N. (1924). Über Protoplasmaveränderung bei Plasmolyse. Biochem. Ztschr. 147:22–29.

CLARK, W. M. (1928). The Determination of Hydrogen Ions (Ed. 3). Williams & Wilkins, Baltimore.

CLELAND, K. W., AND SLATER, E. C. (1953). The sarcosomes of heart muscle; their isolation, structure, and behaviour under various conditions. Quart. J. Micr. Sc. 94:329–346.

COHEN, B., CHAMBERS, R., AND REZNIKOFF, P. (1928). Intracellular oxidation-reduction studies; I—Reduction potentials of *Amoeba dubia* by micro-injection of indicators. J. Gen. Physiol. 11:585–612.

COLE, K. S., AND HODGKIN, A. L. (1939). Membrane and protoplasm resistance in the squid giant axon. J. Gen. Physiol. 22:671–687.

COMMANDON, J., AND DE FONBRUNE, P. (1939). Greffe nucléaire totale, simple ou multiple, chez une amibe. Compt. rend. Soc. de biol. 130:744–748.

CONKLIN, E. G. (1902). Karyokinesis and cytokinesis in the maturation, fertilization, and cleavage of *Crepidula* and other *Gasteropoda*. J. Acad. Nat. Sc. 12:1–121.

CONKLIN, E. G. (1908). The habits and early development of *Linerges mercurius*. Carnegie Institution of Washington, Publication No. 103, pp. 155–170.

CONKLIN, E. G. (1917). Effects of centrifugal force on the structure and development of the eggs of *Crepidula*. J. Exper. Zool. 22:311–419.

CONWAY, E. J., AND DOWNEY, M. (1950). pH values of the yeast cell. Biochem. J. 47:355–360.

CONWAY, E. J., AND FEARON, P. J. (1944). The acid-labile CO_2 in mammalian muscle and the pH of the muscle fibre. J. Physiol. 103:274–289.

COWAN, S. L. (1933). The carbon dioxide dissociation curves and the buffering of crab's muscle and nerve preparations. J. Exper. Biol. 10:401–411.

CURTIS, H. J., AND COLE, K. S. (1938). Transverse electric impedance of the squid giant axon. J. Gen. Physiol. 21:757–765.

DAN, K. (1943). Behaviour of the cell surface during cleavage; VI—On the mechanism of cell division. Journal of the Faculty of Science, Imperial University of Tokyo, Section IV, 6:323–368.

DAN, K. (1952). Cyto-embryological studies in sea urchins; II—Blastula stage. Biol. Bull. 102:74–89.

DAN, K. (1954a). Further study on the formation of the "new membrane" in the eggs of the sea urchin, *Hemicentrotus* (*Strongylocentrotus*) *pulcherrimus*. Embryologia 2:99–114.

DAN, K. (1954b). The cortical movement in *Arbacia punctulata* eggs through cleavage cycles. Embryologia 2:115–122.

DAN, K. (1956). SH-protein and the cell aster. Cytologia Suppl. Proc. International Genetics Symposia, 1956, pp. 216–218.

DAN, K., AND DAN, J. C. (1940). Behaviour of the cell surface during cleavage; III—On the formation of new surface in the eggs of *Strongylocentrotus pulcherrimus*. Biol. Bull. 78:486–501.

DAN, K., AND NAKAJIMA, T. (1956). On the morphology of the mitotic apparatus isolated from echinoderm eggs. Embryologia 3:187–200.

DAN, K., AND ONO, T. (1952). Cyto-embryological studies of sea urchins; I— The means of fixation of the mutual positions among the blastomeres of sea urchin larvae. Biol. Bull. 102:58–73.

DAN, K., AND ONO, T. (1954). A method of computation of the surface area of the cell. Embryologia 2:87–98.

DAN, K., YANAGITA, T., AND SUGIYAMA, M. (1937). Behaviour of the cell surface during cleavage. I. Protoplasma 28:66–81.

D'ANGELO, E. G. (1946). Micrurgical studies on *Chironomus* salivary gland chromosomes. Biol. Bull. 90:71–87.

D'ANGELO, E. G. (1950). Salivary gland chromosomes. Ann. New York Acad. Sc. 50:815–1012.

DANIELLI, J. F. (1937). The relations between surface pH, ion concentrations and interfacial tension. Proc. Roy. Soc., London, s. B. 122:155–174.

DANIELLI, J. F. (1958). Studies of inheritance in amoebae by the technique of nuclear transfer. Proc. Roy. Soc., London, s. B. 148:321–331.

DANIELLI, J. F. (1959). The cell-to-cell transfer of nuclei in amoebae and a comprehensive cell theory. Ann. New York Acad. Sc. 78:675–687.

DANIELLI, J. F., LORCH, I. J., ORD, M. J., AND WILSON, E. G. (1955). Nucleus and cytoplasm in cellular inheritance. Nature 176:1114–1115.

DANIELS, E. W. (1952). Cell division in the giant amoeba, *Pelomyxa carolinensis,* following x-irradiation. J. Exper. Zool. 120:525–545.

DANIELS, E. W. (1959). Micrurgical studies on irradiated *Pelomyxa.* Ann. New York Acad. Sc. 78:662–674.

DAWSON, J. A., KESSLER, W. R., AND SILBERSTEIN, J. K. (1935). Mitosis in *Amoeba dubia.* Biol. Bull. 69:447–461.

DELAGE, Y. (1901). Etudes expérimentales sur la maturation cytoplasmique et sur la parthénogénèse artificielle chez les échinodermes. Arch. de zool. expér. et gén., 3e série, 9:285–326.

DELAGE, Y. (1907). Les vrais facteurs de la parthénogénèse expérimentale. Elevage de larves parthénogénétiques jusqu'à la forme parfaite. Arch. de zool. expér. et gén., 4e série, 7:445–506.

DEMPSEY, E. W., AND WISLOCKI, G. B. (1955). An electron microscopic study of the blood–brain barrier in the rat, employing silver nitrate as a vital stain. J. Biophys. & Biochem. Cytol. 1:245–256.

DEUEL, H., AND SOLMS, J. (1954). Observations on pectic substances. Natural Plant Hydrocolloids. American Chemical Society, Advances in Chemistry Series, No. 11, pp. 62–67.

DICK, D. A. T. (1959). Osmotic properties of living cells. Internat. Rev. Cytology 8:387–448.

DOBELL, C. (1932). Antony van Leeuwenhoek and His "Little Animals." Harcourt, New York.

DODSON, E. O. (1948). A morphological and biochemical study of the lampbrush chromosomes of vertebrates. University of California Publications in Zoology, Vol. 53, No. 8, pp. 281–314.

DOLE, M. (1941). The Glass Electrode. Wiley, New York.

DORFMAN, W. A. (1936). A simple type of micro-electrode for the determination of pH and eH. Protoplasma 25:465–468.

DORFMAN, W. A., AND GRODSENSKY, D. E. (1937). The oxidation–reduction and pH gradient of the unfertilized amphibian egg and the technic of determination of these values in situ. Bull. biol. et méd. expér. U.R.S.S. 4:265–268.

DUBUISSON, M. (1950). Some chemical and physical aspects of muscle contraction and relaxation. Proc. Roy. Soc., London, s. B. 137:63–71.

DUJARDIN, F. (1835). Recherches sur les organismes inférieurs. Ann. sc. nat., 2e série, Zool. 4:343–377.

DURYEE, W. R. (1941). The chromosomes of the amphibian nucleus. Cytology, Genetics and Evolution, pp. 129–141. Univ. of Pennsylvania Press, Philadelphia.

DURYEE, W. R. (1950). Chromosomal physiology in relation to nuclear structure. Ann. New York Acad. Sc. 50:920–953.

DURYEE, W. R. (1954). Microdissection studies on human ovarian eggs. Tr. New York Acad. Sc., Series II, 17:103–108.

DURYEE, W. R., AND DOHERTY, J. K. (1954). Nuclear and cytoplasmic organoids in the living cell. Ann. New York Acad. Sc. 58:1210–1230.

DUTROCHET, R. J. H. (1824). Recherches anatomiques et physiologiques sur la structure intime des animaux et des végétaux, et sur leur motilité. Paris.

EDWARDS, J. G., AND MARSHALL, E. K. (1924). Microscopic observations of the living kidney after the injection of phenolsulphonephthalein. Am. J. Physiol. 70:489–495.

EKHOLM, R., AND SJÖSTRAND, F. S. (1957). The ultrastructural organization of the mouse thyroid gland. J. Ultrastructure Research 1:178–199.

ENDO, Y. (1952). The role of the cortical granules in the formation of the fertilization membrane in eggs from Japanese sea urchins. Exper. Cell Research 3:406–418.

FAURÉ-FREMIET, E. (1923). Variation de l'alcalinité de l'oeuf de *Sabellaria* pendant la maturation. Compt. rend. Soc. de biol. 88:863–866.

FAURÉ-FREMIET, E. (1924). L'oeuf de *Sabellaria alveoleta* L. Arch. d'anat. micr. 20:211–342.

FAURÉ-FREMIET, E. (1925). La cinétique du développement. Presses Universitaires de France, Paris.

FAWCETT, D. W., AND ITO, S. (1958). Observations on the cytoplasmic membranes of testicular cells, examined by phase contrast and electron microscopy. J. Biophys. & Biochem. Cytol. 4:135–142.

FENN, W. O., AND COBB, D. M. (1934). The potassium equilibrium in muscle. J. Gen. Physiol. 17:629–656.

FENN, W. O., COBB, D. M., MANERY, J. F., AND BLOOR, W. R. (1938). Electrolytic changes in cat muscle during stimulation. Am. J. Physiol. 121:595–608.

FENN, W. O., AND MAURER, F. W. (1935). The pH of muscle. Protoplasma 24:337–345.

FISCHER, M. H., AND OSTWALD, W. (1905). Zur physikalisch-chemischen Theorie der Befruchtung. Arch. f. d. ges. Physiol. 106:229–66.

FOL, H. (1873). Die erste Entwickelung des *Geryonideneies*. Jenaische Zeitschrift für Medizin und Naturwissenschaft 7:471–492.

FOL, H. (1879). Recherches sur la fécondation et le commencement de l'hénogénie chez divers animaux. Mémoires de la Societé de physique et d'histoire naturelle de Genève 26:89–397.

FONBRUNE, P. DE (1949). Technique de Micromanipulation. Masson, Paris.

FREY-WYSSLING, A. (1952). Growth of plant cell walls. Symposia of the Society for Experimental Biology, No. 6, pp. 320–328.

FREY-WYSSLING, A., AND MÜLLER, H. R. (1957). Submicroscopic differentiation of plasmodesmata and sieve plates in *Cucurbita*. J. Ultrastructure Research 1:38–48.

Gordon, H. K., and Chambers, R. (1941). The particle size of acid dyes and their diffusibility into living cells. J. Cell. & Comp. Physiol. 17:97–108.

Grand, C. G. (1938). Intracellular pH studies on the ova of *Mactra solidissima*. Biol. Bull. 75:369.

Gray, J. (1926). The properties of an intercellular matrix and its relation to electrolytes. J. Exper. Biol. 3:167–187.

Gray, J. (1927). The mechanism of cell division; iv—The effect of gravity on the eggs of *Echinus*. J. Exper. Biol. 5:102–111.

Gray, J. (1931). A Text-Book of Experimental Cytology. Cambridge Univ. Press, London.

Green, P. B., and Chapman, G. B. (1955). On the development and structure of the cell wall in *Nitella*. Am. J. Botany 42:685–693.

Grossfeld, H., Meyer, K., Godman, G., and Linker, A. (1957). Mucopolysaccharides produced in tissue culture. J. Biophys. & Biochem. Cytol. 3:391–396.

Gruber, A. (1885). Über künstliche Teilung bei Infusorien. Biol. Centralbl. 4:717–722; 5:137–141.

Guyénot, E., and Danon, M. (1953). Chromosomes et ovocytes de batraciens. Rev. suisse Zool. 60:1–129.

Hämmerling, J. (1934). Über formbildende Substanzen bei *Acetabularia mediterranea,* ihre räumliche und zeitliche Verteilung und ihre Herkunft. Arch. f. Entwcklungsmechn. d. Organ. 131:1–81.

Hämmerling, J. (1953). Nucleo-cytoplasmic relationships in the development of *Acetabularia*. Internat. Rev. Cytology 2:475–497.

Hämmerling, J., Clauss, H., Keck, H., Richter, G., and Werz, G. (1959). Growth and protein synthesis in nucleated and enucleated cells. Exper. Cell Research 6 (Suppl.):210–226.

Hansteen, B. (1910). Über das Verhalten der Kulturplanzen zu den Bodensalzen. Jahrb. wiss. Botan. 47:289–376.

Harris, D. L. (1943). The osmotic properties of cytoplasmic granules of the sea urchin egg. Biol. Bull. 85:179–192.

Harris, E. J. (1957). Permeation and diffusion of K ions in frog muscle. J. Gen. Physiol. 41:169–195.

Harvey, E. B. (1935). The mitotic figure and cleavage plane in the egg of *Parechinus microtuberculatus,* as influenced by centrifugal force. Biol. Bull. 69:287–297.

Harvey, E. N. (1911). Studies on the permeability of cells. J. Exper. Zool. 10:507–556.

Harvey, E. N. (1954). Tension at the cell surface. Protoplasmatologia, Handbuch der Protoplasmaforschung, Band 2, E. 5, pp. 1–30. Springer, Vienna.

Harvey, E. N., and Marsland, D. A. (1932). The tension at the surface of *Amoeba dubia* with direct observations on the movement of cytoplasmic particles at high centrifugal speeds. J. Cell. & Comp. Physiol. 2:75–97.

Harwood, F. C. (1923). The colloidal electrolyte extracted from Carrageen (*Chondrus crispus*). J. Chem. Soc. 123 (Part 2):2254–2258.

Hassid, W. Z. (1936). Carbohydrates in *Irideae laminarioides*. Plant Physiol. 11:461–463.

Heilbrunn, L. V. (1923). The colloid chemistry of protoplasm: i—General considerations; II—The electrical charges of protoplasm. Am. J. Physiol. 64:481–498.

HEILBRUNN, L. V. (1926a). The centrifuge method of determining protoplasmic viscosity. J. Exper. Zool. 43:313–320.

HEILBRUNN, L. V. (1926b). The physical structure of the protoplasm of sea-urchin eggs. Am. Naturalist 60:143–156.

HEILBRUNN, L. V. (1930). The action of various salts on the first stage of the surface precipitation reaction in *Arbacia* egg protoplasm. Protoplasma 11:558–573.

HEILBRUNN, L. V. (1952). An Outline of General Physiology (Ed. 3). Saunders, Philadelphia.

HEILBRUNN, L. V. (1956). The Dynamics of Living Protoplasm. Academic Press, New York.

HEILBRUNN, L. V., AND DAUGHERTY, K. (1931). The action of the chlorides of sodium, potassium, calcium, and magnesium on the protoplasm of *Amoeba dubia.* Physiol. Zool. 4:635–651.

HEILBRUNN, L. V., AND DAUGHERTY, K. (1932). The action of sodium, potassium, calcium, and magnesium ions on the plasmagel of *Amoeba proteus.* Physiol. Zool. 5:254–274.

HEILBRUNN, L. V., AND DAUGHERTY, K. (1933). The action of ultraviolet rays on ameba protoplasm. Protoplasma 18:596–619.

HENDEE, E. C. (1931). Formed components and fertilization in the egg of the sea urchin, *Lytechinus variegatus.* Carnegie Institution of Washington, Publication No. 413, pp. 99–105.

HERBST, C. (1895). Experimentelle Untersuchungen über den Einfluss der veränderten chemischen Zusammensetzung des umgebenden Mediums auf die Entwicklung der Thiere. Mitt. Zool. Station Neapel 11:136–220.

HERBST, C. (1900). Über das Auseinandergehen von Furchungs- und Gewebezellen in kalkfreien Medium. Arch. f. Entwcklungsmechn. d. Organ. 9:424–463.

HERTWIG, R. (1903). Ueber Korrelation von Zell- und Kerngrösse und ihre Bedeutung für die geschlechtliche Differenzierung und die Teilung der Zelle. Biol. Centralbl. 23:49–62, 108–119.

HIBBARD, H., AND CHAMBERS, R. (1935). Micromanipulation of egg and nurse cells in *Bombyx mori.* Biol. Bull. 69:332.

HILL, A. V. (1955). The influence of the external medium on the internal pH of muscle. Proc. Roy. Soc., London, s. B. 144:1–22.

HILL, A. V., AND KUPALOV, P. S. (1930). The vapour pressure of muscle. Proc. Roy. Soc., London, s. B. 106:445–477.

HILLIER, J., AND HOFFMAN, J. F. (1953). On the ultrastructure of the plasma membrane as determined by the electron microscope. J. Cell. & Comp. Physiol. 42:203–248.

HIRAMOTO, Y. (1956). Cell division without mitotic apparatus in sea urchin eggs. Exper. Cell Research 11:630–636.

HIRAMOTO, Y. (1957). The thickness of the cortex and the refractive index of the protoplasm in sea urchin eggs. Embryologia 3:361–374.

HIRSHFIELD, H. I. (1959). Nuclear control of cytoplasmic activities. Ann. New York Acad. Sc. 78:647–654.

HOAGLAND, D. R., AND DAVIS, A. R. (1923). The composition of the cell sap of the plant in relation to the absorption of ions. J. Gen. Physiol. 5:629–646.

HOBSON, A. D. (1932). On the vitelline membrane of the egg of *Psammechinus miliaris* and of *Teredo norvegica*. J. Exper. Biol. 9:93–106.

HODGKIN, A. L., AND KATZ, B. (1949). The effect of calcium on the axoplasm of giant nerve fibres. J. Exper. Biol. 26:292–294.

HODGKIN, A. L., AND KEYNES, R. D. (1953). The mobility and diffusion coefficient of potassium in giant axons from *Sepia*. J. Physiol. 119:513–528.

HODGKIN, A. L., AND KEYNES, R. D. (1956). Experiments on the injection of substances into squid giant axons by means of a microsyringe. J. Physiol. 131:592–616.

HODGKIN, A. L., AND KEYNES, R. D. (1957). Movements of labelled calcium in squid giant axons. J. Physiol. 138:253–281.

HÖBER, R. (1898). Ueber Resorption im Dünndarm. Arch. f. d. ges. Physiol. 70:624–642.

HÖBER, R. (1899). Ueber Resorption im Dünndarm. Arch. f. d. ges. Physiol. 74:246–271.

HÖBER, R. (1901). Ueber Resorption im Darm. Arch. f. d. ges. Physiol. 86:199–214.

HÖBER, R., AND HÖBER, J. (1937). Experiments on the absorption of organic solutes in the small intestine of rats. J. Cell. & Comp. Physiol. 10:401–422.

HOFER, B. (1890). Experimentelle Untersuchungen über den Einfluss des Kerns auf des Protoplasma. Jenaische Zeitschrift für Medizin und Naturwissenschaft 24:105–176.

HOLTFRETER, J. (1943a). Properties and functions of the surface coat in amphibian embryos. J. Exper. Zool. 93:251–319.

HOLTFRETER, J. (1943b). A study of the mechanics of gastrulation. J. Exper. Zool. 94:261–318.

HOWLAND, RUTH B. (1924). Dissection of the pellicle of *Amoeba verrucosa*. J. Exper. Zool. 40:263–270.

HOWLAND, RUTH B. (1928). The pH of gastric vacuoles. Protoplasma 5:127–134.

HOWLAND, RUTH B. (1930). Micrurgical studies on the contractile vacuole; III—The pH of the vacuolar fluid in *Actinosphaerium eichhorni*. J. Exper. Zool. 55:53–62.

HUGHES, A. F., AND SWANN, M. M. (1948). Anaphase movements in the living cell; a study with phase contrast and polarized light on chick tissue cultures. J. Exper. Biol. 25:45–70.

HUXLEY, T. H. (1868). On the physical basis of life. (A lay sermon delivered in Edinburgh in 1868.) *In* Huxley, T. H. Lay Sermons, Addresses and Reviews, Ch. 7. Appleton, New York, 1876.

INOUÉ, S. (1953). Polarization optical studies of the mitotic spindle; I—The demonstration of spindle fibres in living cells. Chromosoma 5:487–500.

INOUÉ, S., AND DAN, K. (1952). Birefringence of the dividing cell. J. Morphol. 89:423–451.

JUST, E. E. (1922). Studies on cell division; I—The effect of dilute sea-water on the fertilized egg of *Echinarachnius parma* during the cleavage cycle. Am. J. Physiol. 61:505–515.

KARRER, H. E. (1956). An electronmicroscopic study of the fine structure of pulmonary capillaries and alveoli of the mouse. Bull. Johns Hopkins Hosp. 98:65–82.

KEOSIAN, J. (1938). Secretion in tissue cultures; III—Tonicity of fluid in chick mesonephric cysts. J. Cell. & Comp. Physiol. 12:23–37.

KERR, T. (1933). The injection of certain salts into the protoplasm and vacuoles of the root hairs of *Limnobium spongia*. Protoplasma 18:420–440.

KERR, T., AND BAILEY, I. W. (1934). The cambium and its derivative tissues; x—Structure, optical properties and chemical composition of the so-called middle lamella. J. Arnold Arboretum 15:327–349.

KEYNES, R. D. (1951). The ionic movements during nervous activity. J. Physiol. 114:119–150.

KEYNES, R. D., AND LEWIS, P. R. (1956). The intracellular calcium contents of some invertebrate nerves. J. Physiol. 134:399–407.

KITE, G. L. (1913). Studies on the physical properties of protoplasm; I—The physical properties of the protoplasm of certain animal and plant cells. Am. J. Physiol. 32:146–164.

KITE, G. L., AND CHAMBERS, R. (1912). Vital staining of chromosomes and the function and structure of the nucleus. Science 36:639–641.

KOPAC, M. J. (1935). Intracellular pH determinations on marine ova. Carnegie Institution of Washington, Year Book No. 34, pp. 85–86.

KOPAC, M. J. (1937). The coalescence of a plant cell with oil drops. Biol. Bull. 73:363.

KOPAC, M. J. (1938a). The Devaux effect at oil-protoplasm interfaces. Biol. Bull. 75:351.

KOPAC, M. J. (1938b). Micro-estimation of protein adsorption at oil-protoplasm interfaces. Biol. Bull. 75:372.

KOPAC, M. J. (1940). The physical properties of the extraneous coats of living cells. Cold Spring Harbor Symposia on Quantitative Biology, Vol. 8, pp. 154–170.

KOPAC, M. J. (1950). The surface chemical properties of cytoplasmic proteins. Ann. New York Acad. Sc. 50:870–909.

KOPAC, M. J. (1953). Submicro methods in enzymatic cytochemistry. Tr. New York Acad. Sc., Series II, 15:290–297.

KOPAC, M. J. (1955). Cytochemical micrurgy. Internat. Rev. Cytology 4:1–29.

KOPAC, M. J. (1959). Micrurgical studies on living cells. *In* Brachet, J., and Mirsky, A. E. (eds.). The Cell, Vol. 1, pp. 161–191. Academic Press, New York.

KOPAC, M. J., AND CHAMBERS, R. (1937). The coalescence of living cells with oil drops; II—*Arbacia* eggs immersed in acid or alkaline calcium solutions. J. Cell. & Comp. Physiol. 9:345–361.

KRISZAT, G. (1950). Die Wirkung von Adenosintriphosphat und Calcium auf Amöben (*Chaos chaos*). Arkiv för Zoologi, S.2, 1:477–485.

KUHL, W., AND KUHL, G. (1949). Neue Ergebnisse zur Cytodynamik der Befruchtung und Furchung des Eies von *Psammechinus miliaris*. Zoologische Jahrbücher Abteilung für Anatomie und Otogenie der Tiere 70:1–59.

KUNO, M. (1954). On the nature of the egg surface during cleavage of the sea urchin egg. Embryologia 2:33–41.

KUNO-KOJIMA, M. (1957). On the regional difference in the nature of the cortex of the sea urchin egg during cleavage. Embryologia 3:279–293.

LANDIS, E. M. (1934). Capillary pressure and capillary permeability. Physiol. Rev. 14:404–481.

LANDIS, E. M. (1946). Capillary permeability and the factors affecting the composition of capillary filtrate. Ann. New York Acad. Sc. 46:713–727.

LEITCH, J. L. (1934). The water exchanges of living cells; II—Non-solvent volume determinations from swelling and analytical data. J. Cell. & Comp. Physiol. 4:457–473.

LENHER, S., AND SMITH, J. E. (1936). A diffusion study of dyes. J. Phys. Chem. 40:1005–1020.

LEPESCHKIN, W. W. (1930). My opinion about protoplasm. Protoplasma 9:269–297.

LEWIS, W. H. (1931). Pinocytosis. Bull. Johns Hopkins Hosp. 49:17–27.

LILLIE, R. S. (1906). The relation of ions to contractile processes; I—The action of salt solutions on the ciliated epithelium of *Mytilus edulis*. Am. J. Physiol. 17:89–141.

LILLIE, R. S. (1921). A simple case of salt antagonism in starfish eggs. J. Gen. Physiol. 3:783–794.

LING, G., AND GERARD, R. W. (1949). The normal membrane potential of frog sartorius fibers. J. Cell. & Comp. Physiol. 34:383–396.

LOCONTI, J. D., AND KERTESZ, Z. I. (1941). Identification of calcium pectate as the tissue firming compound formed by the treatment of tomatoes with calcium chloride. Food Research 6:499–508.

LOEB, J. (1906). The Dynamics of Living Matter. Columbia Univ. Press, New York.

LOEB, J. (1908). Über die osmotischen Eigenschaften und die Entstehung der Befruchtungs-membran beim Seeigelei. Arch. f. Entwcklungsmechn. d. Organ. 26:82–88.

LOEB, J. (1914). Cluster formation of spermatozoa caused by specific substances from eggs. J. Exper. Zool. 17:123–140.

LORCH, I. J., AND DANIELLI, J. F. (1950). Transplantation of nuclei from cell to cell. Nature 166:329–330.

MACALLUM, A. B. (1926). The palaeochemistry of the body fluids and tissues. Physiol. Rev. 6:316–357.

MACFARLANE, M. G., AND SPENCER, A. G. (1953). Changes in the water, sodium, and potassium content of rat-liver mitochondria during metabolism. Biochem. J. 54:569–575.

MANGIN, L. (1891, 1892). Etude historique et critique sur la présence des composés pectiques dans les tissus des végétaux. J. de botanique 5:400–413, 440–448; 6:12–19.

MANGIN, L. (1892, 1893). Propriétés et réactions des composés pectiques. J. de botanique 6:206–212, 235–244, 363–368; 7:37–47, 121–131, 325–343.

MANGIN, L. (1893). Sur l'emploi du rouge de ruthénium en anatomie végétale. Compt. rend. Acad. d. sc. 116:653–656.

MARSLAND, D. A. (1939). The mechanism of cell division; hydrostatic pressure effects upon dividing egg cells. J. Cell. & Comp. Physiol. 13:15–22.

MARSLAND, D. (1956). Protoplasmic contractility in relation to gel structure; temperature–pressure experiments on cytokinesis and amoeboid movement. Internat. Rev. Cytology 5:199–227.

MARSLAND, D. (1957). Temperature–pressure studies on the role of sol–gel reactions in cell division. *In* Johnson, F. H. (ed.). Influence of Temperature on Biological Systems (incorporating papers presented at a symposium held at the University of Connecticut, August 27–28, 1956). Americal Physiological Society, Washington.

MARTENS, P., AND CHAMBERS, R. (1932). Etudes de microdissection; v—Les poils staminaux de *Tradescantia.* Cellule 41:131–143.

MAST, S. O. (1926). Structure, movement, locomotion and stimulation in *Amoeba.* J. Morphol. & Physiol. 41:347–425.

MAST, S. O. (1931). Effect of salts, hydrogen-ion concentration, and pure water on length of life in *Amoeba proteus.* Physiol. Zool. 4:58–71.

MATHEWS, A. P. (1907). A contribution to the chemistry of cell division, maturation and fertilization. Am. J. Physiol. 18:89–111.

MAZIA, D. (1955). The organization of the mitotic apparatus. Symposia of the Society for Experimental Biology, No. IX, pp. 335–357.

MAZIA, D. (1958). The production of twin embryos in *Dendraster* by means of mercaptoethanol (monoethylene glycol). Biol. Bull. 114:247–254.

MAZIA, D., AND DAN, K. (1952). The isolation and biochemical characterization of the mitotic apparatus of dividing cells. Proc. Nat. Acad. Sc. 38:826–838.

MAZIA, D., AND PRESCOTT, D. M. (1955). The role of the nucleus in protein synthesis in *Amoeba.* Biochim. et Biophys. Acta 17:23–33.

MAZIA, D., AND ZIMMERMAN, A. M. (1958). II—The effect of mercaptoethanol on the structure of the mitotic apparatus in sea urchin eggs. Exper. Cell Research 15:138–153.

MEYEN, J. (1839). Beiträge zur Bildungsgeschichte verschiedener Pflanzentheile. Arch. Anat. Physiol. wissenschaftliche Med., pp. 255–279.

MICHAELIS, L., AND KRAMSZTYK, A. (1914). Die Wasserstoffionenkonzentration der Gewebassäfte. Biochem. Ztschr. 62:180–185.

MIRBEL, BRISSEAU DE (1809). Exposition de la théorie de l'organisation végétale (Ed. 2). Paris.

MITCHISON, J. M. (1953). Microdissection experiments on sea urchin eggs at cleavage. J. Exper. Biol. 30:515–524.

MITCHISON, J. M. (1956). The thickness of the cortex of the sea urchin egg and the problem of the vitelline membrane. Quart. J. Micr. Sc. 97:109–121.

MITCHISON, J. M., AND SWANN, M. M. (1955). The mechanical properties of the cell surface; III—The sea urchin egg from fertilization to cleavage. J. Exper. Biol. 32:734–750.

MOHL, H. VON (1835). Ueber die Vermehrung der Pflanzenzellen durch Theilung. Dissert. Tübingen. Flora 20:1–31 (published 1837).

MOHL, H. VON (1846). Ueber die Saftbewegung im inneren der Zellen. Botan. Zeitung 4:89–94.

MOLISCH, H. (1913). Mikrochemie der Pflanze. Fischer, Jena.

MONGAR, I. L., AND WASSERMANN, A. (1952). Absorption of electrolyte by alginate gels without and with cation exchange. J. Chem. Soc., pp. 492–497.

MONNÉ, L. (1935). Permeability of the nuclear membrane to vital stains. Proc. Soc. Exper. Biol. & Med. 32:1197–1199.

MONNÉ, L., AND HÅRDE, S. (1951). On the formation of the blastocoele and similar embryonic cavities. Arkiv för Zoologi, S.2, 1:463–469.

MONNÉ, L., AND SLAUTTERBACK, D. B. (1950). Differential staining of various polysaccharides in sea urchin eggs. Exper. Cell Research 1:477–491.

MONNÉ, L., AND SLAUTTERBACK, D. B. (1952). On the staining of the cytoplasm with the Schiff reagent during the development of the eggs of *Paracentrotus lividus.* Arkiv för Zoologi, S.2, 3:349–356.

MOORE, A. R. (1912). On the nature of the cortical layer in sea urchin eggs. University of California, Publications in Physiology, No. 4, pp. 89–90.

MOORE, A. R. (1928). On the hyaline membrane and hyaline droplets of the fertilized egg of the sea urchin, *Strongylocentrotus purpuratus*. Protoplasma 3:524–530.

MOORE, A. R. (1930a). Fertilization and development without membrane formation in the egg of the sea urchin, *Strongylocentrotus purpuratus*. Protoplasma 9:9–17.

MOORE, A. R. (1930b). Fertilization and development without the fertilization membrane in the egg of *Dendraster eccentricus*. Protoplasma 9:18–24.

MOORE, A. R. (1932). The dependence of cytoplasmic structures in the egg of the sea urchin on the ionic balance of the environment. J. Cell. & Comp. Physiol. 2:41–51.

MOORE, A. R. (1935). On the significance of cytoplasmic structure in plasmodium. J. Cell. & Comp. Physiol. 7:113–129.

MOORE, A. R. (1937). On the centering of the nuclei in centrifuged eggs as a result of fertilization and artificial parthenogenesis. Protoplasma 27:544–551.

MOORE, A. R. (1940). Osmotic and structural properties of the blastular wall in *Dendraster excentricus*. J. Exper. Zool. 84:73–83.

MOORE, A. R. (1945). The Individual in Simpler Forms. University of Oregon Monographs, Studies in Psychology, No. 2. University of Oregon Press, Eugene.

MOORE, A. R. (1949). The relation of ions to the appearance and persistence of the fertilization and hyaline membranes in the eggs of the sea urchin. Am. Naturalist 83:233–247.

MOORE, A. R. (1952). The process of gastrulation in trypsin embryos of *Dendraster excentricus*. J. Exper. Zool. 119:37–46.

MOORE, A. R., AND BURT, A. S. (1939). On the locus and nature of the forces causing gastrulation in the embryos of *Dendraster excentricus*. J. Exper. Zool. 82:159–171.

MOORE, A. R., AND MOORE, M. M. (1931). Fertilization and development without membrane formation in the egg of the sea urchin (*Paracentrotus lividus*). Arch. de. biol. 42:375–388.

MOORE, D. H., AND RUSKA, H. (1957). The fine structure of capillaries and small arteries. J. Biophys. & Biochem. Cytol. 3:457–462.

MORGAN, T. H. (1900). Further studies on the action of salt solutions and other agents on the eggs of *Arbacia*. Arch. f. Entwcklungsmechn. d. Organ. 10:489–524.

MORGAN, T. H. (1901). Regeneration of proportionate structures in *Stentor*. Biol. Bull. 2:311–328.

MORGAN, T. H. (1910). Cytological studies of centrifuged eggs. J. Exper. Zool. 9:593–655.

MORITA, Y., AND CHAMBERS, R. (1929). Permeability differences between nuclear and cytoplasmic surfaces in *Amoeba dubia*. Biol. Bull. 56:64–67.

MOSCONA, A. (1952). Cell suspensions from organ rudiments of chick embryos. Exper. Cell Research 3:535–539.

MOSER, F. (1939). Studies on a cortical layer response to stimulating agents in the *Arbacia* egg. J. Exper. Zool. 80:423–445.

MOSER, F. (1940). Studies on a cortical layer response to stimulating agents in the *Arbacia* egg; III—Response to non-electrolytes. Biol. Bull. 78:68–79.

MOTOMURA, I. (1934). On the mechanism of fertilization and development without membrane formation in the sea urchin egg, with notes on a new method of artificial parthenogenesis. Science Report. Tôhoku Imperial University, Series 4, 9:33–45.

MOTOMURA, I. (1935). Determination of the embryonic axis in the eggs of amphibia and echinoderms. Science Report, Tôhoku Imperial University, Series 4, 10:211–245.

MOTOMURA, I. (1941). Materials of the fertilization membrane in the eggs of echinoderms. Science Report, Tôhoku Imperial University, Series 4, 16:345–363.

MOTOMURA, I. (1957). On the nature and localization of the third factor for toughening of the fertilization membrane of the sea urchin egg. Science Report, Tôhoku University, Series 4, 23:167–181.

MÜHLETHALER, K. (1950). Electronenmikroskopische Untersuchungen über den Feinbau und das Wachstum der Zellmembranen in Mais- und Hafer-koleoptilen. Ber. schweiz. botan. Ges. 60:614–628.

NAKANO, E. (1956). Physiological studies on refertilization of the sea urchin egg. Embryologia 3:139–165.

NEEDHAM, J., AND NEEDHAM, D. M. (1925). The hydrogen-ion concentration and oxidation–reduction potential of the cell-interior: a micro-injection study. Proc. Roy. Soc., London, s. B. 98:259–286.

NEEDHAM, J., AND NEEDHAM, D. M. (1926a). The hydrogen-ion concentration and oxidation–reduction potential of the cell-interior before and after fertilization and cleavage: a microinjection study on marine eggs. Proc. Roy. Soc., London, s. B. 99:173–199.

NEEDHAM, J., AND NEEDHAM, D. M. (1926b). Further microinjection studies on the oxidation–reduction potential of the cell-interior. Proc. Roy. Soc., London, s. B. 99:383–397.

NORTHEN, H. T., AND NORTHEN, R. T. (1939). Effects of cations and anions on protoplasmic elasticity. Plant Physiol. 14:539–547.

ODOR, L. (1956). Uptake and transfer of particulate matter from the peritoneal cavity of the rat. J. Biophys. & Biochem. Cytol. 2 (Suppl.):105–108.

OGAWA, J. (1929). Ueber die Reaktion der Gewebe; IV—Studien über intrazelluläre Wasserstoffionenkonzentration der *Entamoeba histolytica* und *Entamoeba coli.* Zentralbl. f. Bakteriol., Abt. 1, 114:68–81.

OKADA, Y. K. (1930). Über den Bau und die Bewegungsweise von *Pelomyxa.* Arch. f. Protistenk. 70:131–154.

O'KELLEY, J. C., AND CARR, P. H. (1954). An electron micrographic study of the cell walls of elongating cotton fibers, root hairs, and pollen tubes. Am. J. Botany 41:261–264.

OKEN, L. (1810). Lehrbuch der Naturphilosophie (Ed. 1). Jena.

OSTERHOUT, W. J. V. (1936). The absorption of electrolytes in large plant cells. Botanical Rev. 2:283–315.

OVERTON, W. (1904). Beiträge sur allgemeinen Muskel- und Nervenphysiologie. Arch. f. d. ges. Physiol. 105:176–290.

OWENS, H. S., LOTZKAR, H., SCHULTZ, T. H., AND MACLAY, W. D. (1946). Shape and size of pectinic acid molecules deduced from viscosimetric measurements. J. Am. Chem. Soc. 68:1628–1632.

PALADE, G. E. (1953a). An electron microscope study of the mitochondrial structure. J. Histochem. 1:188–211.

Palade, G. E. (1953b). Fine structure of blood capillaries. J. Appl. Physics 24:1424.

Palade, G. E. (1956). The endoplasmic reticulum. J. Biophys. & Biochem. Cytol. 2 (Suppl.):85–97.

Palade, G. E., and Porter, K. R. (1954). Studies on the endoplasmic reticulum; i—Its identification in cells *in situ*. J. Exper. Med. 100:641–656.

Pandit, C. G., and Chambers, R. (1932). Intracellular hydrion concentration studies; ix—The pH of the egg of the sea urchin *Arbacia punctulata*. J. Cell. & Comp. Physiol. 2:243–249.

Pantin, C. F. A. (1923a). The determination of pH of microscopic bodies. Nature 111:81.

Pantin, C. F. A. (1923b). On the physiology of amoeboid movement. J. Marine Biol. Assoc., United Kingdom 13:1–69.

Pappas, G. D. (1954). Structural and cytochemical studies of the cytoplasm in the family *Amoebidae*. Ohio J. Sc. 54:195–222.

Pappas, G. D. (1956). The fine structure of the nuclear envelope of *Amoeba proteus*. J. Biophys. & Biochem. Cytol. 2 (Suppl.):431–434.

Pappenheimer, J. R. (1953). Passage of molecules through capillary walls. Physiol. Rev. 33:387–423.

Pease, D. C. (1946). Hydrostatic pressure effects upon the spindle figure and chromosome movement; ii—Experiments on the meiotic divisions of *Tradescantia* pollen mother cells. Biol. Bull. 91:145–169.

Pease, D. C. (1955). Fine structures of the kidney seen by electron microscopy. J. Histochem. 3:295–308.

Personius, C. J., and Sharp, P. F. (1939). Adhesion of potato tissue cells as influenced by pectic solvents and precipitants. Food Research 4:299–307.

Péterfi, T. (1923). Mikrurgische Methodik. Handbuch der biologischen Arbeitsmethoden (Abderhalden), Abt. 5, Teil 2, 1:479–516.

Pfeffer, W. (1899). The Physiology of Plants (Ed. 3; A. J. Ewart, trans.). Clarendon Press, Oxford.

Pfeiffer, H. H. (1940). Mikrurgische Versuche in polariseiertem Lichte zur Analyse des Feinbaues der Riesenchromosomen von *Chironomus*. Chromosoma 1:26–30.

Pollack, H. (1927). Action of picric acid on living protoplasm. Proc. Soc. Exper. Biol. & Med. 25:145.

Pollack, H. (1928a). Micrurgical studies in cell physiology; vi—Calcium ions in living protoplasm. J. Gen. Physiol. 11:539–545.

Pollack, H. (1928b). Intracellular hydrion concentration studies; iii—The buffer action of the cytoplasm of *Amoeba dubia* and its use in measuring the pH. Biol. Bull. 55:383–385.

Prescott, D. M. (1956). Mass and lone culturing of *Amoeba proteus* and *Chaos chaos*. Comptes rendus des travaux du Laboratoire Carlsberg, Série chimique 30:1–12.

Preston, R. D. (1952). The Molecular Architecture of Plant Cell Walls. Wiley, New York.

Purkinje, J. (1840). Ueber die Analogieen in den Strukturelementen des pflanzlichen und tierischen Organismus. Uebersicht der Arbeiten und Veränderungen der schlesischen Gesellschaft für vaterländische Cultur im Jahre, 1893. Dresden.

Rapkine, L., and Wurmser, R. (1928). On intracellular oxidation-reduction potential. Proc. Roy. Soc., London, s. B. 102:128–137.

Ray, D. L. (1956). Nucleolar activity during encystment in *Hartmanella astronyxis* N. Sp. (Amoeba). Proceedings of the XIV International Congress of Zoology (Copenhagen, 1953), pp. 180–181.

Reed, H. S. (1907). The value of certain nutritive elements to the plant cell. Ann. Botany 21:501–543.

Reiss, P. (1926). Le pH intérieur cellulaire. Presses Universitaires de France, Paris.

Remak, R. (1841). Theilung rother Blutzellen beim Embryo. Med. Vereinszeitung.

Reznikoff, P. (1926). Micrurgical studies in cell physiology; II—The action of the chlorides of lead, mercury, copper, iron, and aluminum on the protoplasm of *Amoeba proteus.* J. Gen. Physiol. 10:9–21.

Reznikoff, P. (1928). Micrurgical studies in cell physiology; V—The antagonism of cations in their actions on the protoplasm of *Amoeba dubia.* J. Gen. Physiol. 11:221–232.

Reznikoff, P., and Chambers, R. (1927). Micrurgical studies in cell physiology; III—The action of CO_2 and some salts of Na, Ca, and K on the protoplasm of *Amoeba dubia.* J. Gen. Physiol. 10:731–738.

Reznikoff, P., and Pollack, H. (1928). Intracellular hydrion concentration studies; II—The effect of injection of acids and salts on the cytoplasmic pH of *Amoeba dubia.* Biol. Bull. 55:377–382.

Rhumbler, L. (1898). Physikalische Analyse von Lebenserscheinungen der Zelle. Arch. f. Entwcklungsmechn. d. Organ. 7:103–350.

Ringer, S. (1890). Concerning experiments to test the influence of lime, sodium and potassium salts on the development of ova and growth of tadpoles. J. Physiol. 11:79–84.

Ris, H. (1949). The anaphase movement of chromosomes in the spermatocytes of the grasshopper. Biol. Bull. 96:90–106.

Roberts, H. S. (1955). The mechanism of cytokinesis in neuroblasts of *Chortophaga viridifasciata* (De Geer). J. Exper. Zool. 130:83–105.

Robertson, J. D. (1957). New observations on the ultrastructure of the membranes of frog peripheral nerve fibres. J. Biophys. & Biochem. Cytol. 3:1043–1047.

Robertson, J. D. (1959). The ultrastructure of cell membranes and their derivatives. Biochemical Society Symposium, No. 16, pp. 1–43. Cambridge Univ. Press, London.

Rohde, K. (1917). Untersuchungen über den Einfluss der freien H-Ionen im innern lebender Zellen auf den Vorgang der vitalen Färbung. Arch. f. d. ges. Physiol. 168:411–433.

Rothstein, A. (1954). The enzymology of the cell surface. Protoplasmatologia, Handbuch der Protoplasmaforschung, Band 2, E.4, pp. 1–86. Springer, Vienna.

Rous, P. (1925a). The relative reaction within living mammalian tissues; I—General features of vital staining with litmus. J. Exper. Med. 41:379–397.

Rous, P. (1925b). The relative reaction within living mammalian tissues; II—Mobilization of acid material within cells and reaction as influenced by cell state. J. Exper. Med. 41:399–411.

Rous, P. (1925c). The relative reaction within living mammalian tissues; III—Indicated differences in reaction of blood and tissues on vital staining with phthaleins. J. Exper. Med. 41:451–470.

Rous, P. (1925d). The relative reaction within living mammalian tissues; IV—Indicated differences in reaction of organs on vital staining with phthaleins. J. Exper. Med. 41:739–759.

Rudzinska, Maria A., and Chambers, R. (1951). The activity of the contractile vacuole in a suctorian (*Tokophrya infusionum*). Biol. Bull. 100:49–58.

Runnström, J. (1935). An analysis of the action of lithium on sea urchin development. Biol. Bull. 68:378–384.

Runnström, J. (1948). On the action of trypsin and chymotrypsin on the unfertilized sea urchin egg; a study concerning the mechanism of formation of the fertilization membrane. Arkiv för Zoologi, Vol. 40A, No. 17, pp. 1–16.

Runnström, J., Hagström, B. E., and Perlman, P. (1959). Fertilization. *In* Brachet, J., and Mirsky, A. E. (eds). The Cell, pp. 327–397. Academic Press, New York.

Runnström, J., and Kriszat, G. (1950). On the effect of adenosine triphosphoric acid and of Ca on the cytoplasm of the egg of the sea urchin, *Psammechinus miliaris*. Exper. Cell Research 1:284–303.

Runnström, J., and Kriszat, G. (1952). The cortical propagation of the activation impulse in the sea urchin egg. Exper. Cell Research 3:419–426.

Runnström, J., and Monné, L. (1945). On some properties of the surface layers of immature and mature sea urchin eggs, especially the changes accompanying nuclear and cytoplasmic maturation. Arkiv för Zoologi, Vol. 36A, No. 18, pp. 1–26.

Runnström, J., Monne, L., and Broman, L. (1944). On some properties of the surface layers in the sea urchin egg and their changes upon activation. Arkiv för Zoologi, Vol. 35A, No. 3, pp. 1–32.

Runnström, J., Monné, L., and Wicklund, E. (1944). Mechanism of formation of the fertilization membrane in the sea urchin egg. Nature 153:313–314.

Sabin, F. R. (1923). Studies of living human blood-cells. Bull. Johns Hopkins Hosp. 34:277–288.

Sabin, F. R. (1937). Chemical agents: supravital stains. *In* McClung, C. E. (ed.). Handbook of Microscopical Technique (Ed. 2), pp. 117–131. Hoeber, New York.

Salm-Horstmar (1856). Versuche und Resultate über die Nahrung der Pflanzen. Braunschweig.

Saric, S. P., and Schofield, R. K. (1946). The dissociation constants of the carboxyl and hydroxyl groups in some insoluble and sol-forming polysaccharides. Proc. Roy. Soc., London, s. A. 185:431–447.

Scarth, G. W. (1924). Colloidal changes associated with protoplasmic contraction. Quart. J. Exper. Physiol. 14:99–113.

Schaede, R. (1924). Über die Reaktion des lebenden Plasmas. Ber. deutsche botan. Ges. 42:219–224.

Schechtman, A. M. (1937). Localized cortical growth as the immediate cause of cell division. Science 85:222–223.

Schleiden, M. J. (1838). Beiträge zur Phytogenesis. Arch. Anat. Physiol. und wiss. Med., pp. 137–176.

Schmidt, W. J. (1939). Doppelbrechung der Kernspindel und Zugfasertheorie der Chromosomenbewegung. Chromosoma 1:253–264.

Schmidtmann, M. (1924). Uber eine Methode zur Bestimmung der Wasserstoffzahl im Gewebe und in einzelnen Zellen. Biochem. Ztschr. 150:253–255.

Schmidtmann, M. (1925). Über die intracelluläre Wasserstoffionenkonzentration unter physiologischen und einigen pathologischen Bedingungen. Ztschr. f. d. ges. exper. Med. 45:714–742.

Schultze, M. (1861). Ueber Muskelkörperchen und das, was man eine Zelle zu nennen habe. Arch. f. Anat. u. Physiol., pp. 1–27.

Schultze, M. (1863). Das protoplasma der Rhizopoden und der Pflanzenzelle. Ein Beitrag zur Theorie der Zelle. Engelmann, Leipzig.

Schumaker, V. N. (1958). Uptake of protein from solution by *Amoeba proteus*. Exper. Cell Research 15:314–331.

Schwann, T. (1839). Mikroskopische Untersuchungen über die Uebereinstimmung in der Struktur und dem Wachstum der Tiere und Pflanzen. Berlin.

Scott, F. M., Hamner, K. C., Baker, E., and Bowler, E. (1956). Electron microscope studies of cell wall growth in the onion root. Am. J. Botany 43:313–324.

Seifriz, W., and Zetzmann, M. (1935). A slime mould pigment as indicator of acidity. Protoplasma 23:175–179.

Selenka, E. (1878). Befruchtung des Eies von *Toxopneustes variegatus*. Engelmann, Leipzig.

Selman, G. G., and Waddington, C. H. (1955). The mechanism of cell division in the cleavage of the newt's egg. J. Exper. Biol. 32:700–733.

Shanes, A. M. (1958). Electrochemical aspects of physiological and pharmacological action in excitable cells; I—The resting cell and its alteration by extrinsic factors. Pharmacological Reviews 10:59–164.

Shapiro, N. N. (1927). The cycle of hydrogen-ion concentration in the food vacuoles of *Paramecium, Vorticella,* and *Stylonychia*. Tr. Am. Micr. Soc. 46:45–50.

Shimamura, T. (1940). Studies on the effect of the centrifugal force upon nuclear division. Cytologia 11:186–216.

Sichel, F. J. M., and Burton, A. C. (1936). A kinetic method of studying surface forces in the egg of *Arbacia*. Biol. Bull. 71:397–398.

Small, J. (1929). Hydrogen-Ion Concentration in Plant Cells and Tissues. Protoplasma-monographien, No. 2. Borntraeger, Berlin.

Small, J. (1955). The pH of plant cells. Protoplasmatologia, Handbuch der Protoplasmaforschung, Band 2, 2, pp. 1–116. Springer, Vienna.

Smith, H. W. (1932). Water regulation and its evolution in the fishes. Quart. Rev. Biol. 7:1–26.

Spector, W. G. (1953). Electrolyte flux in isolated mitochondria. Proc. Roy. Soc., London, s. B. 141:268–279.

Spek, J., and Chambers, R. (1933). Neue experimentelle Studien über das Problem der Reaktion des Protoplasmas. Protoplasma 20:376–406.

Spooner, G. B. (1911). Embryological studies with the centrifuge. J. Exper. Zool. 10:23–49.

Steiner, A. B., and McNeely, H. (1954). Algin in review. Natural Plant Hydrocolloids. American Chemical Society, Advances in Chemistry Series, No. 11, pp. 68–82.

Stella, G. (1929). The combination of carbon dioxide with muscle; its heat of neutralization and its dissociation curve. J. Physiol. 68:49–66.

Steward, F. C., and Mühlethaler, K. (1953). The structure and development of the cell wall in the *Valoniaceae* as revealed by the electron microscope. Ann. Botany, N. S. 17:295–316.

Strangeways, T. S. P. (1924). Tissue Culture in Relation to Growth and Differentiation. Heffer, Cambridge.

Sugiyama, M. (1951). Re-fertilization of the fertilized eggs of the sea urchin. Biol. Bull. 101:335–344.

Sugiyama, M. (1956). Physiological analysis of the cortical response of the sea urchin egg. Exper. Cell Research 10:364–376.

Swann, M. M. (1951a). Protoplasmic structure and mitosis; I—The birefringence of the metaphase spindle and asters of the living sea urchin egg. J. Exper. Biol. 28:417–433.

Swann, M. M. (1951b). Protoplasmic structure and mitosis; II—The nature and cause of birefringence changes in the sea-urchin egg at anaphase. J. Exper. Biol. 28:434–444.

Swann, M. M., and Mitchison, J. M. (1953). Cleavage of sea-urchin eggs in colchicine. J. Exper. Biol. 30:506–514.

Tartar, V. (1956). Pattern and substance in *Stentor*. *In* Rudnick, D. (ed.). Cellular Mechanisms in Differentiation and Growth, pp. 73–100. Princeton Univ. Press, Princeton.

Taylor, C. V. (1925). Microelectrodes and micromagnets. Proc. Soc. Exper. Biol. & Med. 23:147–150.

Taylor, C. V., and Whitaker, D. M. (1927). Potentiometric determinations in the protoplasm and cell-sap of *Nitella*. Protoplasma 3:1–6.

Teichmann, E. (1903). Über die Beziehung zwischen Astrosphären und Furchen. Arch. f. Entwcklungsmechn. d. Organ. 16:243–327.

Tennent, D. H., Taylor, C. V., and Whitaker, D. M. (1929). An investigation on organization in a sea urchin egg. Carnegie Institution of Washington, Publication No. 391. Papers from the Tortugas Laboratory, Vol. 26, pp. 1–104.

Teorell, T. (1952). Permeability properties of erythrocyte ghosts. J. Gen. Physiol. 35:669–701.

Thiele, H., and Andersen, G. (1955). Ionotrope Gele von Polyuronsäuren; III—Alginat und Pektin in nativen organisierten Gelen. Kolloid Ztschr. 143:21–31.

Tobias, J. M. (1948). The high potassium and low sodium in the body fluid and tissues of a phytophagous insect, the silkworm *Bombyx mori,* and the change before pupation. J. Cell. & Comp. Physiol. 31:143–148.

Townsend, C. O. (1897). Der Einfluss des Zellkerns auf die Bildung der Zellhaut. Jahrb. f. wissensch. Botan. 30:484–510.

True, R. H. (1922). The significance of calcium for higher green plants. Science 55:1–6.

Tyler, A. (1941). The role of fertilizin in the fertilization of eggs of the sea-urchin and other animals. Biol. Bull. 81:190–204.

Tyler, A., Monroy, A., Kao, C. Y., and Grundfest, H. (1956). Membrane potential and resistance of the starfish egg before and after fertilization. Biol. Bull. 111:153–177.

Vasseur, E. (1952). Periodate oxidation of the jelly coat substance of *Echinocardium cordatum.* Acta chem. scandinav. 6:376–384.

Vellinger, E. (1926). Recherches potentiométriques sur le pH intérieur de l'oeuf d'Oursin. Compt. rend. Soc. de. biol. 94:1371–1373.

Verworn, M. (1888). Biologische Protisten-Studien. Ztschr. f. wissensch. Zoöl. 46:455–470.

Vlès, F. (1924). Recherches sur le pH intérieur cellulaire. Arch. de physique biol. 4:1–20.

Vlès, F., and Vellinger, E. (1928). Recherches sur le pigment de l'oeuf d' *Arbacia* envisagé comme indicateur de pH intracellulaire. Bull. Institut océanographique 513:1–16.

Waddington, C. H. (1952). Preliminary observations on the mechanism of cleavage in the amphibian egg. J. Exper. Biol. 29:484–489.

Walker, C. E., and Tozer, F. M. (1909). Observations on the history and possible function of the nucleoli in the vegetative cells of various animals and plants. Quart. J. Exper. Physiol. 2:187–200.

Wallace, W. M., and Hastings, A. B. (1942). The distribution of the bicarbonate ion in mammalian muscle. J. Biol. Chem. 144:637–649.

Wallace, W. M., and Lowry, O. H. (1942). An *in vitro* study of carbon dioxide equilibria in mammalian muscle. J. Biol. Chem. 144:651–655.

Weber, F. (1924a). Methoden der Viscositätsbestimmung des lebenden Protoplasmas. Handbuch der biologischen Arbeitsmethoden (Abderhalden), Abt. 11, Teil 2; pp. 655–718.

Weber, F. (1924b). Plasmolyseform und Protoplasmaviskosität. Österreichische botanische Ztschr. 73:261–266.

Weisz, P. B. (1949). A cytochemical and cytological study of differentiation in normal and reorganizational stages of *Stentor coeruleus*. J. Morphol. 84:335–363.

Whitaker, D. M. (1940). The effect of shape on the developmental axis of the *Fucus* egg. Biol. Bull. 78:111–116.

Wiercinski, F. J. (1944). An experimental study of protoplasmic pH determination; I—*Amoebae* and *Arbacia punctulata*. Biol. Bull. 86:98–112.

Wilbrandt, W. (1946). Physiologie der Zell- und Kapillarpermeabilität. Helvet. med. acta 13:143–157.

Wille, N. (1897). Beiträge zur physiologischen Anatomie der Laminariaceen. Christiania.

Wilson, E. B. (1901a). Experimental studies in cytology; I—A cytological study of artificial parthenogenesis in sea urchin eggs. Arch. f. Entwcklungsmechn. d. Organ. 12:529–596.

Wilson, E. B. (1901b). Experimental studies in cytology; II—Some phenomena of fertilization and cell-division in etherized eggs. III—The effect on cleavage of artificial obliteration of first cleavage-furrow. Arch. f. Entwcklungsmechn. d. Organ. 13:353–395.

Wilson, K. (1955). The polarity of the cell-wall of *Valonia*. Ann. Botany, N. S. 19:289–292.

Wilson, W. L. (1951). The rigidity of the cell cortex during cell division. J. Cell. & Comp. Physiol. 38:409–415.

Wilson, W. L., and Heilbrunn, L. V. (1952). The protoplasmic cortex in relation to stimulation. Biol. Bull. 103:139–144.

Winer, B. J., and Moore, A. R. (1941). Reactions in the plasmodium *Physarum polycephalum* to physico-chemical changes in the environment. Biodynamica 3:323–345.

Wischnitzer, S. (1957). A study of the lateral loop chromosomes of amphibian oocytes by phase contrast microscopy. Am. J. Anat. 101:135–167.

Yamamoto, T. (1936). Studies on the rhythmical movements of the early embryo of *Oryzias latipes;* VII—Anaerobic movements and oxidation-reduction potential of the egg limiting the rhythmical movements. Journal of the Faculty of Science, Imperial University of Tokyo, Section IV, 4:233–247.

YATSU, N. (1908). Some experiments on cell division in the egg of *Cerebratulus lacteus.* Annotationes Zoologicae Japonenses 6:267–276.

YATSU, N. (1910). An experimental study on the cleavage of the ctenophore egg. Proceedings of the Seventh International Zoological Congress (1907). Cambridge Univ. Press, London.

ZIEGLER, H. E. (1898). Experimentelle Studien über die Zelltheilung; III—Furchungszellen von Beroë ovata. Arch. f. Entwcklungsmechn. d. Organ. 7:34–64.

ZIMMERMAN, A. M., LANDAU, J. V., AND MARSLAND, D. (1957). Cell division: A pressure-temperature analysis of the effects of sulfhydryl reagents on the cortical plasmagel structure and furrowing strength of dividing eggs (*Arbacia* and *Chaetopterus*). J. Cell. & Comp. Physiol. 49:395–435.

ZWEIFACH, B. W. (1954). The exchange of materials between blood vessels and lymph compartments. *In* Ragan, C. (ed.). Conference on Connective Tissues (Transactions of the Fifth Conference), pp. 39–77. Josiah Macy, Jr. Foundation, New York.

Index

Index